Nonlinear Dynamics and Chaos:
Where do we go from here?

Nonlinear Dynamics and Chaos: Where do we go from here?

Edited by

John Hogan, Alan Champneys,
Bernd Krauskopf, Mario di Bernardo,
Eddie Wilson, Hinke Osinga
and Martin Homer

Department of Engineering Mathematics,
University of Bristol, UK

CRC Press
Taylor & Francis Group
Boca Raton London New York

CRC Press is an imprint of the
Taylor & Francis Group, an **informa** business

CRC Press
Taylor & Francis Group
6000 Broken Sound Parkway NW, Suite 300
Boca Raton, FL 33487-2742

ISBN 13: 978-0-7503-0862-5 (hbk)

Visit the Taylor & Francis Web site at
http://www.taylorandfrancis.com

and the CRC Press Web site at
http://www.crcpress.com

Contents

Preface

This book was born out of the lingering suspicion that the theory and practice of dynamical systems had reached a plateau of development and that someone had to ask the question 'Where do we go from here?' The idea is not to look back warmly at 'chaos theory', give the main protagonists a slap on the back, and suggest how the mathematical state of the art can be tinkered with. Instead each invited contributor was asked to reflect on what has been learnt from the phenomenal rise in nonlinear dynamics and, more importantly, to address the question of what still needs to be learnt. Thanks to the efforts of our authors, we now have a plethora of suggestions, but the over-riding message is clear: if dynamical systems theory is to make a significant long-term impact, it needs to get smart, because most systems are ill-defined through either stochasticity, delay, spatial extent and inhomogeneity or the finite-time nature of real-world data.

How did we get here?

Since the pioneering works of Lorenz (1963)[1] and May (1976)[2], enormous mathematical strides have been taken[3], and almost every scientific field has been swamped by computations of nonlinear differential equations and maps. Most of the existing theory and its relation to applications lies in systems of low dimension. This is significant because the theoretical underpinning is often geometric in nature as many systems lend themselves to representation in three dimensions or less.

However, most systems of practical importance are of much higher dimension. Chemical systems can involve hundreds of coupled nonlinear ordinary differential equations, realistic neural systems are several orders of magnitude larger, and the partial differential equations involved in fluid mechanics are

[1] Lorenz E N 1963 Deterministic non-periodic flow *J. Atmos. Sci.* **20** 130–41

[2] May R M 1976 Simple mathematical models with very complicated dynamics *Nature* **261** 459–67

[3] For more introductory reading see, for example; Gleick J 1987 *Chaos—Making a New Science* (New York: Viking Penguin), Mullin T 1993 *The Nature of Chaos* (Oxford: Oxford University Press), Cvitanovic P (ed) 1989 *Universality in Chaos* (Bristol: Institute of Physics Publishing), Ruelle D 1991 *Chance & Chaos* (Princeton, NJ: Princeton University Press) and Stewart I 1990 *Does God Play Dice? The New Mathematics of Chaos* (Oxford: Blackwell).

infinite-dimensional; even the numerical representation of fluid mechanics by finite element or finite difference codes has dimensions running into many thousands. In these areas, theory advances understanding either by system reduction methods leading to low-dimensional systems[4] or by approaches involving lengthy numerical computations.

However, most large systems cannot be reduced simply, and numerical computation with many free parameters seldom leads to an understanding of the system under investigation. It is this general situation we felt must be addressed. However, it is not clear what the future problems should be, still less how they should be solved.

Hence we brought together theoreticians, experimentalists and the next generation of nonlinear scientists, so that they could together define an agenda for future research directions.

Thus our unique meeting 'Nonlinear dynamics and chaos: where do we go from here?' was held in Bristol in June 2001. Over three gloriously sunny and rain-free days, 70 scientists from over a dozen countries enjoyed the Victorian grandeur of Burwalls, a University of Bristol dedicated conference centre. Whether playing croquet, eating in the oak-panelled dining room, or looking out of the window at Brunel's majestic suspension bridge next door, the participants enjoyed a stimulating atmosphere that genuinely addressed the conference title's question. In addition to 14 invited speakers, the meeting was shaped by the contributions of a large number of graduate students and young researchers.

Where do we go from here?

This question is central to all 13 chapters of this book. Each chapter represents a write-up of an invited lecture, which has been edited (and in some cases transcribed) by us.

We recognized at the outset that the future directions and challenges for the theory of nonlinear dynamics are far too many and varied to be covered in one book. Therefore three broad inter-related themes have been selected:

(i) neural and biological systems,
(ii) spatially extended systems and
(iii) experimentation in the physical sciences.

It was a pleasant surprise to us that many of the resulting chapters of this book address two or more of these themes. As you will see, other synergies also emerge. Each author has offered suggestions as to what they see as the way forward, and each individual chapter gives an overview of its particular field and raises many specific and more general questions for future research.

[4] See, for example, Carr J 1981 *Applications of Centre Manifold Theory (Applied Math. Sci. 35)* (New York: Springer).

The chapters that follow are grouped together to approximately address (i)–(iii) thematically. But as already stated, there is a far greater degree of connectivity between chapters than would be suggested by any simple linear ordering. We shall not attempt to spell out that connectivity here, but encourage the reader to dip into the non-technical introduction section of each chapter. In fact each chapter can be read independently of the others and represents an essay in its own right, addressing the individual author's perspective on the question 'Where do we go from here?' Viewed in its entirety, this book presents a unique snapshot of current thinking on the challenges facing dynamical systems theory. We hope that you get from this book as much pleasure as we did in organizing and attending the meeting on which it is based.

Acknowledgments

This book is the result of a combined effort of a large number of people. We are very grateful to the authors for the enthusiasm and considerable effort that they invested in this project. We were very impressed how positively all invited speakers responded to the challenge of transforming their presentations into a book chapter, and how cheerfully they reacted to our various suggestions and queries.

We thank Andy Mackintosh for his professional technical help and Sue Payne for her peerless secretarial support. Both were invaluable for making the organization before and during our meeting seem effortless. We also thank our Bristol graduate students and research assistants for their help, and all participants for their contributions to the meeting.

We are very grateful to the Colston Research Society, hosted at the University of Bristol, for providing the initial support and acting as a catalyst for the symposium. The Society's mission is to bring people together to advance academic research, communicate new findings and enhance understanding in areas of common interest. The meeting received further funding from the Engineering and Physical Sciences Research Council and the London Mathematical Society.

John Hogan, Alan Champneys, Bernd Krauskopf, Mario di Bernardo, Eddie Wilson, Hinke Osinga, Martin Homer
Bristol, March 2002

Chapter 1

Bifurcation and degenerate decomposition in multiple time scale dynamical systems

John Guckenheimer
Cornell University

In keeping with the spirit of the Colston conference on Nonlinear Dynamics and Chaos, this chapter emphasizes ideas more than details, describing my vision of how the bifurcation theory of multiple time scale systems will unfold. Multiple time scale dynamical systems arise in modelling diverse phenomena. Examples include equations of chemical reactors, models of boundary layer phenomena in PDE, physical systems such as electrical circuits and lasers, and models within the life sciences. My work has been motivated especially by Hodgkin–Huxley-type models of neural systems [17]. In this setting, the time constants for gating different membrane channels vary by orders of magnitude. Individual neurons, as well as networks, display a rich variety of rhythms. Modelling the generation and modulation of these rhythmic behaviours is a substantial challenge that brings to the fore questions about the bifurcations of multiple time scale systems.

The analysis of dynamical systems with multiple time scales is a subject whose history interweaves three different viewpoints: non-standard analysis [6], classical asymptotics [27] and geometric singular perturbation theory [18]. These varied approaches reflect the complexity of the phenomena being studied. Slow–fast systems (defined later) give the simplest representation of multiple time scale systems. Nonetheless, the singular limits of $\varepsilon \to 0$ are awkward objects. The limits of families of solutions to the system as $\varepsilon \to 0$ do not converge to solutions of either form of the singular limit. Description of these limits requires a language that goes beyond the terminology used to describe solutions to systems of ordinary differential equations. Moreover, existence and uniqueness of solutions in the singular limit is problematic. Nonstandard analysis introduces a systematic way to describe all of this through the introduction of infinitesmal quantities. Computation of solutions is as problematic as their description. 'Exponentially small' quantities of magnitude $\exp(-c/\varepsilon)$ or smaller are prevalent, so the use of regular asymptotic series does not take one very far in solving slow–fast systems. This is also reflected in the development of numerical methods for solving initial value problems for 'stiff' equations. These are adapted to dealing with fast motion that is stable, giving transients that die rapidly and lead to motion on a lower dimensional slow manifold. In the phenomena we shall discuss, there is also unstable fast motion. Geometric methods enable the qualitative analysis of what types of solutions are robust to perturbation of initial conditions and perturbation of the system. Thus the different viewpoints emphasize different aspects of the subject. Nonetheless, it is difficult to sort one's way through the literature since few attempts seem to have been made to amalgamate or survey results coming from the different approaches. Moreover, I suggest that there is a paucity of examples and that the subject of multiple time scale dynamics has a richer plethora of phenomena than is readily found in its literature.

My perspective here is decidedly geometric. The success of dynamical systems theory in elucidating patterns of bifurcation in generic systems with a single time scale motivates the goal here, namely to extend this bifurcation theory to systems with two time scales. There are substantial obstacles to realizing this objective, both theoretical and computational. Consequently, the final shape that the theory will take is still fuzzy.

It may seem strange to talk about computational barriers to a mathematical theory, so I give some explanation for this. Much of the progress in dynamical systems theory throughout its history has been inspired by close examination of examples. This has been the case even before the widespread use of computers to study dynamical systems. As an early instance of an example involving multiple time scales, the concept of *relaxation oscillations* was introduced by Van der Pol [32] in the context of the equation that bears his name today. Although the analysis in his paper is restricted to this single system, there is a clear statement that the idea of relaxation oscillation applies to a broad class of multiple time scale systems. In recent decades, computer studies of models have been one of the foundations for the creation of new theory. Computer simulation and analysis enables extensive studies of model systems. Work on such systems as quadratic maps, the Hénon mapping [15], the standard map [22], the Lorenz system [25] and Duffing's equation [14] has been crucial in developing intuition about general properties of dynamical systems. General theory has been based upon what has been learned from these examples, in some cases incorporating computer proofs to establish facts that are not readily proved otherwise.

In the realm of multiple time scale systems, the strategy of working from computer studies has been less prevalent than in other areas of dynamical systems theory. This is due partly to the failure of 'standard' numerical integration algorithms to resolve important qualitative properties of multiple time scale systems. This failure is not only a consequence of the long times required to resolve the slow time scales in a system with two time scales, but also of the extreme instability and disparity of scales in the phase and parameter spaces that we encounter when studying these systems. Asymptotic analysis of local phenomena in multiple time scale systems demonstrates that the theory is subtle, frequently involving quantities that scale exponentially with the ratio of time scales. Consequently, our understanding of the phenomenology of multiple time scale systems is not yet at the level of our understanding of other important classes of dynamical systems. For example, much less is known about bifurcations of the forced Van der Pol equation than the bifurcations of the Lorenz system, even though the Van der Pol system has a much older heritage, including the monumental work of Cartwright and Littlewood [4, 5, 23, 24] that established the existence of chaotic dynamics in dissipative dynamical systems. Thus, I suggest that the development of better computational methods for studying multiple time scale systems is needed to help create more mathematical theory.

Bifurcation theory of dynamical systems classifies bifurcations by codimension and describes their unfoldings; see [1, 14] as general references. The primary objects of interest are families of C^r vector fields that depend upon k parameters. The theory examines dynamical phenomena that are *persistent*, meaning that C^s, $s \leq r$ perturbations of the family exhibit the same phenomenon. If a phenomenon occurs persistently in a k-parameter family but not in $(k-1)$-parameter families, then we say that it has *codimension* k. This definition generalizes the usage of codimension in singularity theory but is imprecise in this setting. Since formal definitions of the concept of codimension are complicated, I shall focus upon phenomenology here. We want to have a good understanding of a rich set of examples before attempting to formulate a comprehensive, rigorous theory. In the simplest cases, codimension-k bifurcations occur on a manifold of codimension k in the space of C^r vector fields. Equivalently, in these cases there will be a set of k *defining equations* for the bifurcations (if they are local bifurcations) [10].

An unfolding of a codimension-k bifurcation is a k-parameter family that contains the codimension-k bifurcation in a persistent way. The term unfolding also refers to a description of the dynamics within such a k-parameter family. The usefulness of bifurcation theory is due in part to the structural stability properties of unfoldings. However, those results come after our main task here of determining the unfoldings themselves. The local structure and existence of normal forms for multiple time scale systems has been studied both using asymptotic [11, 27] and geometric methods [1, 18]. Apart from the analysis of canards in the Van der Pol model [7], little attention has been given to bifurcation in this analysis. We emphasize here periodic orbits, especially *relaxation oscillations* in which the periodic motions include segments that alternately evolve on the slow and fast time scales. Bifurcations of relaxation oscillations add a new set of issues to those that have been addressed for multiple time scale systems. Theory formulated for systems with a single time scale is still valid, but there are enormous distortions in the unfoldings of systems with two time scales compared to those of single time scale systems. As we remarked earlier, this distortion can be sufficiently extreme that software designed for single time scale systems is unable to compute important qualitative aspects of these unfoldings. Instead of tackling the bifurcations of relaxation oscillations directly, we concentrate on the *slow–fast decomposition* of trajectories. Analysing this decomposition in terms of transversality properties enables us to gain insight into the bifurcations.

1.1 Definitions and background

We study *slow–fast* systems in this paper that are written in the form

$$\begin{aligned} \varepsilon\dot{x} &= f(x, y) \\ \dot{y} &= g(x, y) \end{aligned} \tag{1.1}$$

or

$$\begin{aligned} x' &= f(x, y) \\ y' &= \varepsilon g(x, y). \end{aligned} \tag{1.2}$$

Here $x \in R^m$ are the fast variables and $y \in R^n$ are the slow variables. In the limit $\varepsilon = 0$, system (1.1) becomes a system of *differential algebraic equations* (called the *reduced* system) on the slow time scale and system (1.2) becomes a parametrized family of differential equations (called the *layer equations*) on the fast time scale. For fixed y, the equations $\dot{x} = f(x, y)$ are a *fast subsystem*. We limit our attention in this chapter to systems in which the attractors of the fast subsystems are all equilibria. In this case, we say that the slow–fast system has no rapid oscillations. The *critical manifold* of the slow–fast system is the set of solutions of $f(x, y) = 0$ in R^{m+n}; see figure 1.1 for a one-dimensional example.

In slow–fast systems with no rapid oscillations, trajectories approach an ε-neighbourhood of the critical manifold on the fast time scale, and then move inside this neighbourhood on the slow time scale. This leads us to expect that as $\varepsilon \to 0$ the limit of a trajectory with a specified initial condition will be a finite union of trajectories of fast subsystems and curves that lie on the critical manifold. The curves on the critical manifold are themselves trajectories of a system of differential equations called the *slow flow* of the system. This image is made more explicit by several general theorems about the dynamics of slow–fast systems near their critical manifolds. We recall some of these results here, omitting details; see, e.g., [1, 18, 27, 28].

The first result is that the critical manifolds of generic slow–fast systems are indeed manifolds. This is a corollary of the Morse–Sard theorem, stating that the regular values of a a C^n mapping $f : R^{m+n} \to R^m$ form a residual set and are therefore dense in R^m [16]. For generic systems (1.1), zero is a regular value of f and its inverse image for f is a manifold of dimension n. However, note that generic one-parameter families of slow–fast systems may encounter parameters for which there are singularities of the critical manifold since the rank deficient linear maps $A : R^{m+n} \to R^m$ form a subset of codimension $n + 1$.

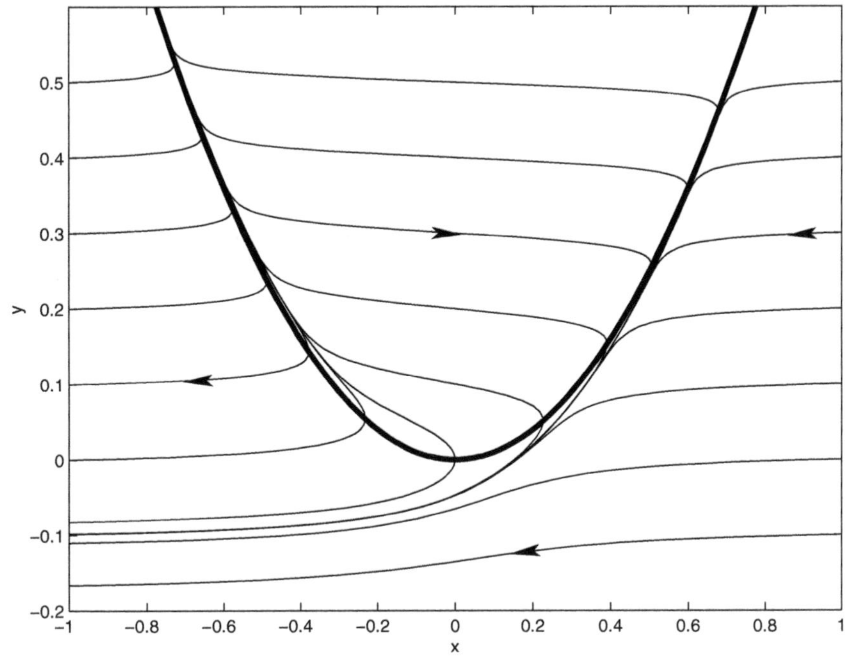

Figure 1.1. The flow in a neighbourhood of the origin for the system (1.3) with $\varepsilon = 0.01$. The critical manifold is the parabola drawn with a bold line.

Projection π_s of the critical manifold onto the space of slow variables plays a special role in the theory. At regular points of π_s, we can solve the equation $f(x, y) = 0$ implicitly to obtain a locally defined function $x(y)$ whose graph is an open region of the critical manifold. The equation $\dot{y} = g(x(y), y)$ then defines a vector field on the critical manifold that is called the *slow flow* of the slow–fast system. While the slow–fast system is not tangent to the critical manifold, the slow flow often yields an approximation to the flow on a nearby invariant manifold, called a *slow manifold* (or true slow manifold). Existence of the slow manifold and convergence of trajectories of the slow–fast system on the slow manifold to those of the slow flow on the critical manifold was proved by Tikhonov (see [27]) in the case of attracting slow manifolds and by Fenichel [8] in the hyperbolic case. The slow manifolds are not unique, but rather consist of a collection of slow manifolds that lie within a distance that is $O(\mathrm{e}^{-c/\varepsilon})$ from one another. The distance between the slow manifolds and the critical manifold is $O(\varepsilon)$.

At singular points of the projection π_s of the critical manifold, we cannot expect the slow flow to be defined. Consider the following system with one slow and one fast variable that is a standard example (see figure 1.1 and [1]) of a slow–

fast system with a fold in its critical manifold:

$$\begin{aligned} \varepsilon \dot{x} &= y - x^2 \\ \dot{y} &= -1. \end{aligned} \tag{1.3}$$

At the fold point $(x, y) = (0, 0)$, the equation $\dot{y} = -1$ is incompatible with the algebraic equation $y = x^2$ obtained by setting $\varepsilon = 0$. This is annoying, but does not seem like a big obstacle to understanding the reduced system. However, when the critical manifold of a slow–fast system is two-dimensional, new phenomena occur that prompt further analysis. This is illustrated by the system

$$\begin{aligned} \varepsilon \dot{x} &= y + x^2 \\ \dot{y} &= ax + bz \\ \dot{z} &= 1. \end{aligned} \tag{1.4}$$

Here the critical manifold is defined by $y = -x^2$ and the fold curve is defined by $x = y = 0$. At points where $bz < 0$, the slow flow is pointing toward the fold curve on the two sheets of the slow manifold. Where $bz > 0$, the slow flow is pointing away from the fold curve on the two sheets of the slow manifold. At the origin, there is a transition. To study this transition more carefully, we rescale the slow flow so that it has a continuous extension across the fold curve. We can use (x, z) as a regular system of coordinates near the fold. We differentiate the algebraic expression $y + x^2$ appearing in the reduced system of (1.4) to obtain $\dot{y} = -2x\dot{x}$ for motion of the reduced system on the critical manifold. We then replace $\dot{y}$ by $-2x\dot{x}$ in the slow flow equations. Rescaling the resulting equations by $-2x$ yields the system

$$\begin{aligned} \dot{x} &= ax + bz \\ \dot{z} &= -2x. \end{aligned} \tag{1.5}$$

This rescaled slow flow is clearly defined across the fold curve. It has the same trajectories as the slow flow except that their parametrizations are different and the direction of the trajectories is reversed in the half-plane $x > 0$. The origin is an equilibrium point of this system, called a *folded equilibrium*. Folded equilibria occur at the transition for motion in the slow flow toward a fold to motion away from the fold. Folded equilibria in systems with two-dimensional slow manifolds are classified by the type of equilibrium in the rescaled system, so that we speak of folded nodes, foci and saddles.

In the case of folds for the system (1.3) with one slow and one fast variable, the geometry of the slow–fast system for $\varepsilon > 0$ is evident: when trajectories reach the fold, they jump along a fast segment. With the scaling of variables

$$x = \varepsilon^{\frac{1}{3}} X \qquad y = \varepsilon^{\frac{2}{3}} Y \qquad t = \varepsilon^{\frac{2}{3}} T \tag{1.6}$$

ε disappears from the system (1.3). Thus, the geometry of the flows for different values of ε are all qualitatively the same, with features that scale as indicated by the coordinate transformation (1.6). Solutions of the equivalent *Riccati equation* $\dot{x} = t + x^2$ have been studied in terms of special functions [27]. There is a single solution that remains close to the parabola $y - x^2 = 0$ with $x > 0$ for all time $t < 0$. Figure 1.1 shows a phase portrait of the flow past the origin in system (1.3).

With two slow variables, the geometry of the solutions of (1.4) that pass near a folded singularity is much more complicated than that of system (1.3). In the case of a folded saddle, Benoît [3] proved that there are two algebraic solutions of (1.4) that remain close to the critical manifold for all time. These solutions have a hyperbolic structure and, hence, stable and unstable manifolds that divide the phase space into regions with similar asymptotic properties. It is noteworthy that one of the algebraic solutions passes from the stable sheet of the slow manifold to the unstable sheet. There are solutions in the unstable manifold of this trajectory that flow along the unstable sheet of the slow manifold and then leave the slow manifold, with a fast segment that either ends back at the stable sheet of the slow manifold or goes to infinity in finite time. Solutions that flow along the unstable sheet for a finite period on the slow time scale are called *canards*. (The name, the French word for duck, was chosen because of the shape of canard solutions in the Van der Pol equations [6].) Typically, the regions of initial conditions near the stable sheet of the slow manifold that give rise to canards are *exponentially thin*; i.e. of width $O(\mathrm{e}^{-c/\varepsilon})$ for a suitable positive constant c that is independent of ε. The reason for this behaviour is the fast instability of the slow manifold. This instability is also an impediment to numerical computation of canards. Deviations from the slow manifold are amplified exponentially on the fast time scale. Consequently, even tiny perturbations on the order of round-off errors are quickly magnified so much that numerically computed trajectories are very far from the canards.

Part of our message in this exposition is that canards play a central role in the bifurcations of relaxation oscillations. In the singular limit, changes in generic families of periodic orbits appear to be discontinuous. In many of these situations, canards form the 'connecting glue' that spans these discontinuities. There are several mechanisms that lead to the formation of periodic orbits in generic one-parameter families of slow–fast systems. In the next two sections, we discuss some of these, illustrating how the slow–fast decompositions of trajectories gives a starting point for the analysis of periodic orbits with canards. We end this section with comments about two open problems.

The characterization of canards that are associated with folded nodes in systems with two-dimensional critical manifolds has not been carried out fully. Benoît [3] proved the existence of canards near folded nodes, but he was unable to determine fully what the set of canard trajectories is in this case. Folded nodes present another problem for the general theory of relaxation oscillations in addition to the question of characterizing the canards. On the critical manifold, there is an entire open region of trajectories that flow to the folded node under the slow flow. Since the subsequent evolution of these trajectories from the folded node may involve canards beginning at the folded node, it is possible that there may be open regions of relaxation oscillations possessing canards in systems with two slow variables that do not collapse in the singular limit. I do not know of any examples that have been studied carefully to determine the properties of the trajectories that pass through regions with folded nodes. Relaxation oscillations passing through a folded node occur in the forced Van der Pol system (described in section 1.4) but the parameter regions and scale on which the relevant dynamics occur may make this a difficult example to analyse numerically.

I am unaware of any systematic studies of the slow–fast dynamics in systems with $n > 2$ slow variables. The critical manifolds of such systems are n-dimensional and the folds of their critical manifolds are $(n-1)$-dimensional, that is, at least of dimension two. Using singularity theory, we can introduce coordinates so that a regular fold is given by the equations $y_1 + x_1^2 = 0$; $x_i = 0$, $1 < i \leq m$. Following the argument described earlier, we use coordinates $(x_1, y_2, \ldots, y_n)$ near the fold and use the relation $\dot{y}_1 = -2x_1\dot{x}_1$ to obtain the slow flow in this coordinate system. Rescaling the system by the factor $2x_1$, we obtain a rescaled vector field on the critical manifold that has equilibria along the $(n-2)$-dimensional submanifold defined by $x_1 = g_1(0, y) = 0$, where g_1 is the right-hand side of the equation for y_1 in these coordinates. What happens near the folds of these generic slow–fast systems? I believe that little is known about such systems.

1.2 Slow–fast decompositions

Let γ_ε be a continuous family of trajectories in R^{m+n} for system (1.1) defined for $\varepsilon \geq 0$ small. For $\varepsilon = 0$, we understand that a trajectory is a union of trajectories of fast subsystems and the slow flow. A *slow–fast decomposition* of γ is defined by a partition of γ that depends continuously on ε with the property that γ_0 is partitioned into segments that lie in the critical manifold and segments that lie in its complement. To discuss the decomposition, we introduce some more terminology. First, we will continue to use the term trajectory to refer to a family of the type described earlier: a trajectory of a slow–fast system is a continuous family γ_ε that is (1) smooth for each $\varepsilon > 0$ and (2) solves the slow–fast equations when $\varepsilon > 0$. The *reduced trajectory* is the restriction of a trajectory to $\varepsilon = 0$. We partition a reduced trajectory into *slow segments* that lie on the critical manifold and *fast segments* on which the limiting differential-algebraic equation does not vanish; see figure 1.2 for examples. Without additional assumptions, there is little reason to believe that trajectories with finite slow–fast decompositions will exist. We approach this issue by seeking to characterize the settings in which we find stable relaxation oscillations whose slow–fast decomposition is stable under perturbation.

When a trajectory arrives at a fold that terminates a slow segment, the fold point is a non-hyperbolic equilibrium point of the fast subsystem. The fast

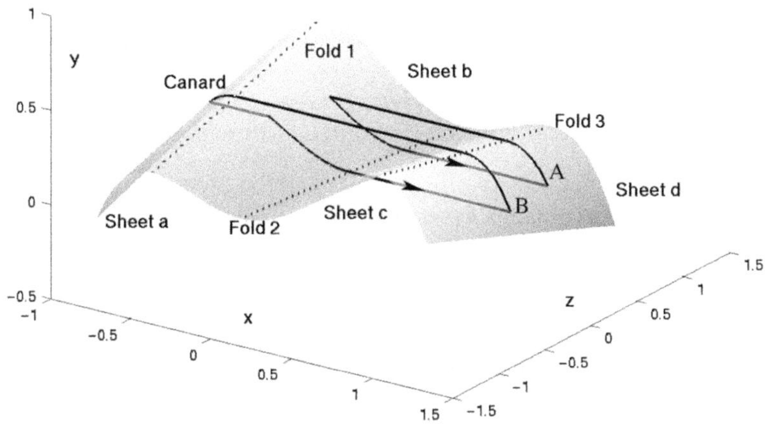

Figure 1.2. Schematic representation of a system having a periodic orbit with a degenerate slow–fast decomposition of the type in-fold → out-fold.

segment originating at this equilibrium will lie in its (weak) unstable manifold. The only circumstance in which we expect there to be a unique trajectory of the fast subsystem with this equilibrium point as its α-limit set is the case in which the fold point is a non-degenerate saddle-node with spectrum in the (closed) left half-plane. (I am unaware of any systematic study of examples of relaxation oscillations in which the fold points have higher dimensional unstable manifolds.) So we confine ourselves to this setting. We shall call a relaxation oscillation satisfying the following criteria ((i)–(iii)) a *simple* relaxation oscillation.

(i) The ends of the slow segments are fold points of the critical manifold with eigenvalues for the fast subsystem that are non-positive. These points are not folded singularities.
(ii) The ends of the fast segments are regular points of a stable sheet of the critical manifold.

To formulate the third property we need a new concept. Property (i) implies that the fold points near the trajectory are saddle-nodes of their fast subsystems with one-dimensional weakly unstable manifolds. From each of these fold points, there is a unique trajectory with the fold point as its α-limit set. We call the ω-limit sets of these points the *umbra* of the fold. The umbra is a codimension-one submanifold of the critical manifold since we assume that the ω-limit sets of all trajectories are equilibrium points of the fast subsystems.

(iii) The slow flow is transverse to the umbra of the fold.

If the slow flow is tangent to the fold, we shall say that there is an umbral tangency.

The following theorem is an immediate consequence of results of Levinson [20] analysing the properties of trajectories that jump at regular fold points.

Theorem 1.1. *Hyperbolic periodic orbits that are simple relaxation oscillations have slow–fast decompositions that vary continuously with perturbation.*

For trajectories that satisfy properties (i) and (ii), we can augment the slow flow of system (1.1) to obtain a *hybrid* system that represents the limit behaviour of trajectories as $\varepsilon = 0$. Hybrid dynamical systems [2] are ones in which there are

- a discrete set of bounded manifolds and flows on these manifolds; and
- mappings from the manifold boundaries into the manifolds.

Here the transition maps from the manifold boundaries will be the projection of a fold along the fast subsystem to its umbra. So the hybrid system that we obtain consists of the stable sheets of the critical manifold with its slow flow together with the maps of regular folds on the boundary to their umbra. The definition of this hybrid system will break down at folded equilibria, at trajectories whose fast segments do not end in the interior of a stable sheet of the critical manifold, and at boundary points more complicated than a fold. Return maps for this hybrid system need not be locally invertible in the neighbourhood of trajectories with umbral tangencies. In section 1.4 we illustrate these ideas with the forced Van der Pol system as a model example.

1.3 Degenerate decomposition and bifurcation

This section examines generic mechanisms by which a one-parameter family of stable, simple relaxation oscillations can reach a parameter value where its slow–fast decomposition becomes degenerate. We examine the slow–fast decomposition of the reduced periodic orbit and use transversality theory to determine the types of degeneracies that will persist with smooth perturbations of a one-parameter family. Persistence implies that the reduced orbits have at most one degeneracy in their slow–fast decomposition. Degeneracies that may occur include the following:

(A) A fast segment ends at a regular fold point. (There are two cases that differ as to whether the slow flow approaches or leaves the fold near this point.)
(B) A slow segment ends at a folded saddle.
(C) A fast segment encounters a saddle-point of the fast subsystem.
(D) There is a point of Hopf bifurcation at a fold.
(E) A slow segment ends at a cusp.
(F) The reduced system has a quadratic umbral tangency.

This list is incomplete, for example omitting cases with reduced trajectories that contain folded nodes. We leave as an open question the formulation of a classification of codimension-one degenerate decompositions that contains all of the degeneracies that occur in generic families of simple relaxation oscillations. For each of the cases on the list, we want to determine what the presence of a degenerate decomposition implies about nearby canards and bifurcations and to analyse the asymptotic properties of the families of periodic orbits related to the degenerate decomposition. There are several levels on which this analysis can be performed. The least precise level is to introduce specific models for portions of the flow and base the analysis upon these models. Elementary descriptions of homoclinic 'Shilnikov' bifurcation are a well-known example of this type of analysis [14]. Our ultimate goal is to give a rigorous analysis. In the case of Hopf bifurcation at a fold, Dumortier and Roussarie [7] have given a thorough analysis of one version of the Van der Pol system.

Kathleen Hoffman, Warren Weckesser and I have begun to formulate analyses for some of the cases on this list. I give here the barest sketch of the case in which there is a fast segment that begins at a fold and ends at a fold with slow trajectories flowing away from the fold. This type of degenerate decomposition occurs in a model of reciprocal inhibition of a pair of neurons studied by Guckenheimer *et al* [12].

Figure 1.2 is a three-dimensional plot of the slow manifold and two periodic orbits of the family we now describe; see (1.7). We work with a system that has two slow variables and one fast variable. We assume that the critical manifold, shown shaded in the figure, has three non-intersecting fold curves and is independent of the parameter in the family. The fold curves are labelled *Fold 1* to *Fold 3* and plotted as dotted lines. The fast variable is x; the axes are labelled in the figure. The projections of *Fold 1* and *Fold 3* onto the y–z plane intersect transversally. Cross-sections of the critical manifold orthogonal to the z-axis are quartic curves with two local maxima (on *Fold 1* and *Fold 3*) and a local minimum (on *Fold 2*). The fast flow is assumed to be in the negative x direction for points above the critical manifold and in the positive x direction for points below the critical manifold. *Fold 2* and *Fold 3* are in-folds with trajectories of the slow flow moving into the folds, while *Fold 1* is an out-fold with trajectories of the slow flow leaving the fold. The fold curves partition the critical manifold into four sheets, that are labelled *Sheet a* to *Sheet d* with increasing x. *Sheet b* and *Sheet d* of the critical manifold are stable, sheets *Sheet a* and *Sheet c* are unstable. The slow flow on *Sheet a* and *Sheet b* is downwards, the slow flow on *Sheet d* is upwards, and the slow flow on *Sheet c* changes direction from downwards along *Fold 2* to upwards along *Fold 3*.

Assume that at an initial value of the parameter, there is a stable periodic orbit with two slow segments on sheets *Sheet b* and *Sheet d* and two fast segments, the first from *Fold 3* to *Sheet b* and the second from *Fold 2* to *Sheet d*. In figure 1.2, this is the smaller closed curve A drawn with full lines. The orbit is drawn so that the slow flow is parallel to the (x, y)-plane. Assume further that, with the changing parameter, the orbits move in the direction of decreasing z. At a critical value of the parameter, the fast segment beginning on *Fold 3* hits *Fold 1* instead of the interior of *Sheet b*. This is the location of a degenerate decomposition. The continuation of the family from the degenerate decomposition will have periodic orbits with canards lying on *Sheet a*. The longer closed curve in figure 1.2 shows one of these. The canard B that is shown terminates at a fast segment that connects *Sheet a* with *Sheet b*. In the periodic orbit, the canard is followed by a slow segment along *Sheet b*, a fast segment from *Fold 2* to *Sheet d*, a slow segment on *Sheet d* and, finally, the fast segment from *Fold 3* to *Fold 1*.

Using models for different portions of the flow, we describe approximations of the return maps for the canard periodic orbits in this family. The return map is a composition of transition maps along slow and fast segments. Introduce uniformizing coordinates near *Fold 1* at the end of the fast segment from *Fold 3* to *Fold 1*:

$$\begin{aligned} \varepsilon \dot{x} &= y - x^2 \\ \dot{y} &= 1 \\ \dot{z} &= x. \end{aligned} \tag{1.7}$$

(Here, (x, y, z) are different coordinates from those depicted in the figure, but x is still the direction of fast motion.) In these coordinates, model the trajectories originating at *Fold 3* as the plane parametrized by (v, u, au). This set of trajectories has umbra on *Sheet b* along the curve $(x, y, z) = (\sqrt{u}, u, au)$, $u > 0$. In (x, z) coordinates on the critical manifold, $z = x^2/a$ on the curve of incoming trajectories from *Fold 3*. The slow trajectories of the model system (1.7) lie along curves $z = \frac{2}{3}x^3 + c$ for varying c. For points $(\sqrt{u}, au)$ we have $c(u) = au - \frac{2}{3}u^{3/2}$. On the critical manifold, these points reach $x = 1$ along the slow flow with $z = \frac{2}{3} + au - u^{3/2}$. The leading order term is regular, but the map is not C^2. The orbit with a degenerate slow–fast decomposition may still be stable.

The canards for this family occur along *Sheet a*, extending the slow flow trajectory with $z = \frac{2}{3}x^3$. If a canard jumps back to *Sheet b* with $x = -x_d > -1$, then the value of z along the jump is $z_d = -\frac{2}{3}x_d^3$ and the value of c is $c_d = -\frac{4}{3}x_d^3$. This arrival point on *Sheet b* reaches $x = 1$ with $z = \frac{2}{3} - \frac{2}{3}x_d^3$. Depending upon the sign of a, the intersections of the trajectories initiated by these canards overlaps the set of trajectories originating at $(x, y, z) = (\sqrt{u}, u, au)$, $u > 0$ or is disjoint from this set of trajectories.

Beyond the 'scaling region' of the model equations (1.7), trajectories that remain near the canard to reach $x = x_d$ come from an 'exponentially thin slab' along *Sheet a* of the critical manifold that flows across *Fold 3*. Let $y = y_c$ give the intersection of the infinite canards of the model system (1.7) with $x = 1$. (This is a set parallel to the z-axis since the normal form equations for $\dot{x}$ and $\dot{y}$ are independent of z.) Denoting $t_j = O(1)$ the time of a jump from the canard starting near *Fold 1*, the difference of its y coordinate from y_c is $y_p \approx \exp(-L/\epsilon)$ where L is given by an integral of the fast eigenvalue along the canard trajectory to the terminal time of the canard. The phase space coordinates of the end point of the canard are estimated by $y_j = y_c + y_p - \frac{\epsilon}{L} \ln y_p$, $x_j = \sqrt{y_j}$ and

$z_c = (-\frac{\epsilon}{L} \ln y_p)^{3/2}$. The derivative of (x_j, y_j, z_j) with respect to y_p has order $\epsilon / y_p = \epsilon \exp(-L/\epsilon)$ whose magnitude is large. Thus the formation of the canards leads to violent stretching of the return map for the cycles.

Assuming that the orbit with a degenerate decomposition is stable, there are now four cases to analyse. These are determined by the sign of a and orientation ($o+$ for orientable and $o-$ for non-orientable) of the reduced return map for the periodic orbit with degenerate decomposition. The case $(a+, o+)$ yields a monotone return map with a single saddle-node and the case $(a+, o-)$ yields a single period doubling. The case $(a-, o-)$ yields a period doubling cascade and chaotic invariant sets. The final case $(a-, o+)$ begins with a stable, orientation-reversing return map at the degenerate orbit but then changes orientation as the canards develop and there is a saddle-node bifurcation. In all of these cases, the bifurcations of the original periodic orbit occur as the canards begin to develop, not with the $O(\varepsilon)$ separation found in the case of Hopf bifurcation and the canard formation in the Van der Pol system [7].

This sketch of how relaxation oscillations bifurcate close to a degenerate decomposition with a fast segment joining two folds is far from a complete or rigorous analysis. Filling in the details of this analysis is a continuing objective for us. Some of the other cases on the list of degenerate decompositions have already been studied [28], but there is hardly a coherent theory of codimension-one bifurcation of stable relaxation oscillations analogous to what we know about bifurcations of systems with a single time scale. The cases of periodic orbits passing through a folded saddle and umbral tangencies occur in the forced Van der Pol system, described in the next section. Our discussion of this example gives the flavour of how the study of the reduced systems and degeneracies in the slow–fast decompositions of their trajectories gives new insight into the dynamics of multiple time scale dynamical systems.

1.4 The forced Van der Pol equation

The forced Van der Pol equation can be written in the form

$$\begin{aligned} \varepsilon \dot{x} &= y + x - \tfrac{1}{3}x^3 \\ \dot{y} &= -x + a \sin(2\pi\theta) \\ \dot{\theta} &= \omega. \end{aligned} \tag{1.8}$$

Here θ is a cyclic variable on the circle S^1 that is normalized to have length 1 and we restrict our attention to the regime in which $\varepsilon > 0$ is small. This system was introduced and studied by Van der Pol in the 1920s. He introduced the term relaxation oscillations to describe the solutions of the unforced system [32]. A generalization of the Van der Pol equation, the FitzHugh–Nagumo system [29], has been widely studied as a model of nerve impulses; see also chapter 11 by Bressloff and Cowan. The literature on the dynamics of the forced Van der

Pol system is dominated by the work of Cartwright and Littlewood during the period approximately 1940–55 [4, 5, 23, 24]. Levinson [21] and Levi [19] studied piecewise linear simplifications of system (1.8), establishing the existence of chaotic dynamics. Recently, Kathleen Hoffman, Warren Weckesser and I have begun to study this system further [13], using the conceptual framework described in the previous sections. In particular, we are interested in characterizing the bifurcations of periodic orbits and other invariant sets that take place in this system. Apart from a couple of numerical studies [9, 26] at values of ε that are large enough to make tracking canards feasible with initial value solvers and the work of Rand and Storti on a piecewise linear system [31], little has been done in this direction. Earlier work has primarily been based upon analysis of the return map to a section of constant θ. In contrast, we focus upon the slow–fast decomposition of trajectories. These are readily determined since the fold curves are given by $x = \pm 1$, $y = \mp 2/3$ and the fast subsystems flow parallel to the x-axis. Fast segments beginning at the fold with $x = \pm 1$ return to the critical manifold at $x = \mp 2$. In our view, the slow flow provides a scaffolding that will enable a comprehensive understanding of the dynamics for the system.

The rescaled slow flow on the critical manifold $y = x^3/3 - x$ has a global representation in (x, θ) coordinates as

$$\begin{aligned} x' &= -x + a\sin(2\pi\theta) \\ \theta' &= \omega(x^2 - 1). \end{aligned} \tag{1.9}$$

The equilibrium points of this system can be easily determined and its phase portraits produced numerically. Figure 1.3 shows a representative phase portrait for parameter values $a = 4$, $\omega = 1.55$. When $0 < a < 1$, the slow flow has no folded singularities. When $a > 1$, there are four folded singularities, two on each circle of the fold curve. The limits of the simplest relaxation oscillations in the system correspond to closed curves obtained from initial conditions that begin on the circle $x = 2$, flow to $x = 1$, connect to $x = -2$ along a segment $\theta = \text{const}$, flow from there to $x = -1$ and then connect to the initial point with another segment $\theta = \text{const}$. (There are also simple relaxation oscillations for some parameter values that have more than two slow segments and more than two fast segments.) All of these relaxation oscillations are stable in the fast directions of the flow; some are also stable along the slow manifold and some are unstable along the slow manifold.

We investigate the simple relaxation oscillations in more detail by constructing the return map to $x = 2$ for the reduced system. The analysis is made easier by exploiting the symmetry $S(x, y, \theta) = (-x, -y, \theta + 0.5)$ of the slow–fast system and its restriction to the critical manifold and rescaled slow flow. The half return map H is defined by following the slow flow from $x = 2$ to $x = 1$, jumping from $x = 1$ to $x = -2$ and applying the symmetry. If $(2, \theta_0)$ flows to $(1, \theta_1)$ under the slow flow, then $H(\theta_0) = \theta_1 + 0.5 \pmod 1$. When $a > 1$, the map H is discontinuous at points lying in the stable manifold of the folded saddle.

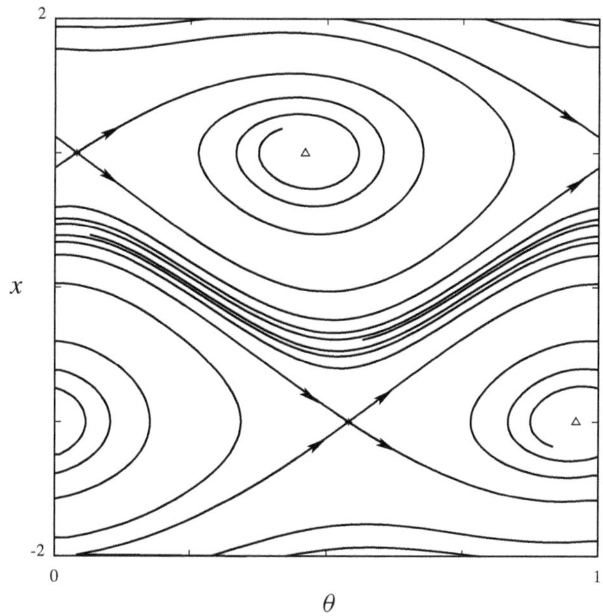

Figure 1.3. The phase portrait of the rescaled slow equation for the forced Van der Pol equation with $a = 4$ and $\omega = 1.55$. The stable and unstable manifolds of the folded saddles are drawn in the region $|x| < 2$. The folded foci are drawn as triangles.

The limit values of H at the two sides of the discontuity are $\theta_s + 0.5$ and $\theta_u + 0.5$ where the folded saddle has (x, θ) coordinates $(1, \theta_s)$, and $(1, \theta_u)$ is the (first) intersection of the unstable manifold of the folded saddle with $x = 1$. The map H has one local maximum and one local minimum when $a > 2$. These occur at the points of umbral tangency $(2, \frac{1}{2\pi}\sin^{-1}(2/a))$ and $(2, 0.5 - \frac{1}{2\pi}\sin^{-1}(2/a))$ where $\dot{x} = 0$ in the slow flow.

Fixed points of H correspond to trajectories with two fast and two slow segments, the fast and slow segments each forming a symmetric pair. Our investigations of these fixed points show two primary types of bifurcations: saddle-nodes and 'homoclinic' bifurcations where the fixed points approach a discontinuity of H. The saddle-node bifurcations do not involve degenerate decomposition of periodic orbits in a direct way. They correspond to saddle-nodes of periodic orbits in which a stable and unstable periodic orbit of the reduced system coalesce. The second type of bifurcation occurs when the periodic orbits in the slow flow approach heteroclinic orbits that connect the two folded saddles with two connecting segments that are related by the symmetry. There are two forms of heteroclinic orbit that differ as to whether the connecting orbits contain a segment that follows the unstable manifolds of the folded saddles or not. The periodic orbits that contain segments lying close to the unstable manifolds of the

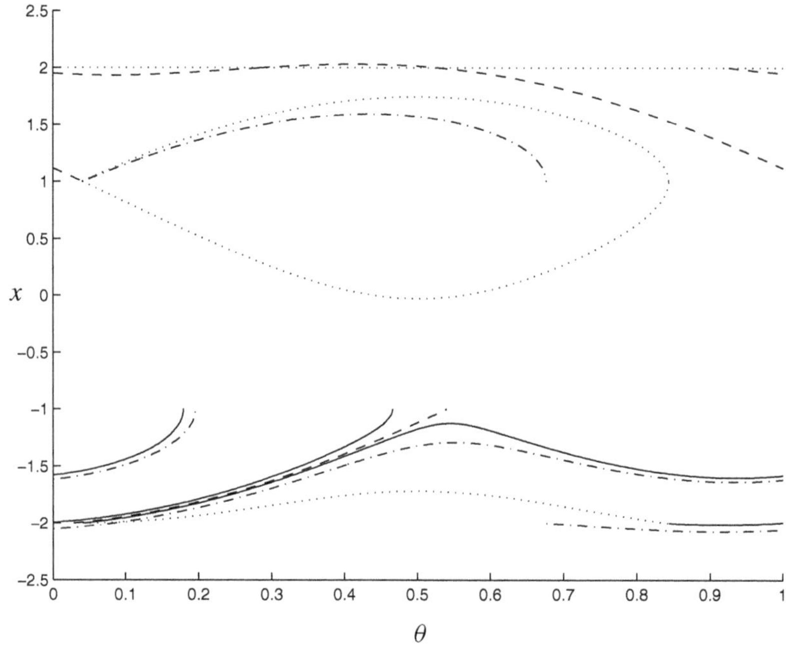

Figure 1.4. Information about canards in the forced Van der Pol equation with $a = 4$ and $\omega = 1.55$. Canard trajectories at the upper folded saddle follow the dotted stable manifold down to the right and jump vertically up or down to one of the two other dotted curves. The unstable manifold of the upper folded saddle and its prolongation after jumping from $x = 1$ to $x = -2$ is drawn as a dash-dotted curve. Branches of both folded saddles are drawn as dashed curves. The solid curves in the lower half of the figure begin at the ends of the canard umbra shown as the lower dotted curve.

folded saddle occur to the right of the stable manifolds of the folded saddle. There are additional distinctions that can be made that reflect the types of branches of the half return map that contain points from the homoclinic orbits.

Analysis of the half return map H does not address directly the issues of canards and bifurcations associated with degenerate decompositions that we discussed in previous sections. The discovery of chaotic dynamics in the forced Van der Pol system by Cartwright and Littlewood was a seminal event in the history of dynamical systems [4]. Their discovery was based upon results of Birkhoff in the case that there are two fixed points of H whose orbits in the reduced system have different periods. The subsequent analysis of Littlewood [23, 24] makes it evident that the chaotic trajectories involve canards.

We have begun to extend our analysis of the reduced system and its half return map to take account of the limiting behaviour of canard solutions in the forced Van der Pol system. Figure 1.4 shows a plot ($a = 4$, $\omega = 1.55$) that allows us to determine the limiting position of canards as $\varepsilon \to 0$ and the subsequent trajectories to their next return(s) with the circle $x = 2$. Trajectories that reach the folded saddle at $x = 1$ along its stable manifold (dashed line) can continue past the folded saddle along the stable manifold (dotted line) to smaller values of x. (Recall that our rescaling of the slow flow reversed time in the strip $|x| < 1$.) In the three-dimensional phase space, the canard trajectories follow the unstable sheet of the critical manifold, from which they can then jump to one of the two stable sheets of the critical manifold. The jumps are parallel to the x-axis, and their images on the two stable sheets are also drawn as dotted lines. These images act like the umbra of fold curves in that they are the initial points for slow segments of the trajectories that begin on the canards. The canards that jump up to $x > 1$ have trajectories that next reach the circle $x = 1$ in the interval between θ_u (at the end point of the dash-dotted unstable manifold of the folded saddle in figure 1.4) and the intersection point of the stable manifold with $x = 1$. From $x = 1$, these trajectories jump to $x = -2$. One of them lands in the stable manifold of the lower folded saddle (dashed line in the lower half of figure 1.4). The extension of the trajectory jumping from the unstable manifold of the (upper) folded saddle is shown as a dash-dotted line. The canards that jump down to $x < -1$ initiate slow segments that flow to $x = -1$ and then jump back to $x = 2$. One of these trajectories lands in the stable manifold of the upper folded saddle at its third intersection with the circle $x = 2$. The trajectories at the endpoints of the umbra from the canards that jump down are plotted as full curves.

Figure 1.5 plots the value of θ at returns of the canards to $x = 2$, as described in the previous paragraph. The dotted line is at an (integer translated) value of the third intersection of the stable manifold of the folded saddle with $x = 2$. Its intersection with the full curve comes from a canard that jumps down, flows to $x = -1$ and then jumps to a point in the stable manifold of the upper folded saddle. The vertical segment of the graph comes from the canard that jumps up, flows to $x = 1$ and then jumps to $x = -2$, landing in the stable manifold of the lower folded saddle. The points in this region will follow canards symmetric to the ones described earlier. In particular, there will be one canard that jumps down, flows to $x = -1$ and then jumps to $x = 2$, landing in a point of the stable manifold of the upper saddle. This description of the evolution of canards leads to the identification of the solenoid [30] (suspension of a horseshoe) discovered by Cartwright and Littlewood. A cross section to the stable manifold of the folded saddle will first flow along the slow manifold becoming compressed in the transverse direction to the slow manifold. A small portion will then follow canards and the evolution described earlier, the transverse direction to the vector field inside the slow manifold being stretched along the canards. Two portions of these canards will return to the upper sheet of the slow manifold, passing through the original cross section to the stable manifold of the folded saddle. The return map of this cross section appears to have a horseshoe.

We end with a conjecture and a question about this description of chaotic dynamics in the forced Van der Pol equation:

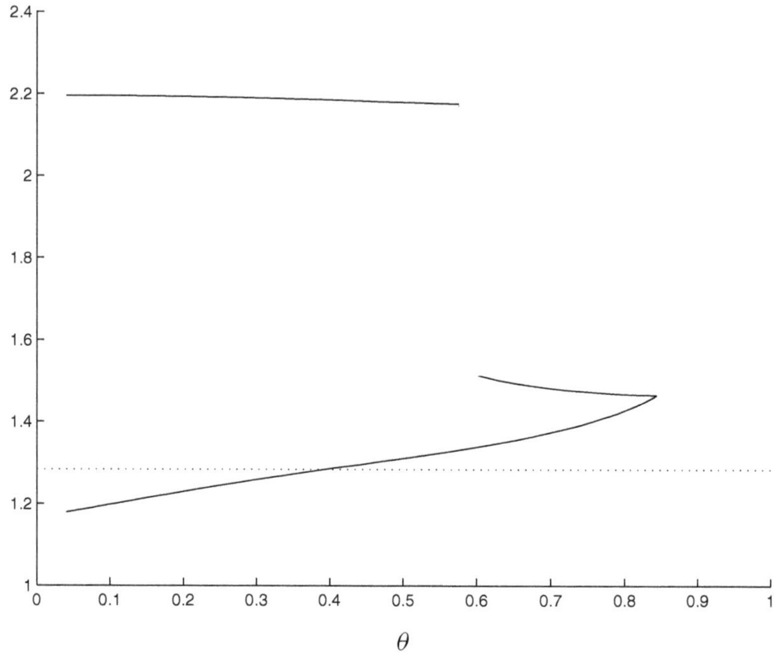

Figure 1.5. The θ coordinate of returns of canard trajectories to the circle $x = 2$ are plotted as a function of their initial θ coordinate. There is a discontinuity where the trajectories contain segments of the stable manifold of the lower folded saddle. The dotted line gives the θ coordinate for an intersection of the stable manifold of the upper folded saddle with $x = 2$. The extreme point at the right of the curve corresponds to the 'maximal' canard, the trajectory that follows the stable manifold of the folded saddle all the way to its intersection with $x = 1$ before jumping to $x = -2$.

Conjecture 1.2. *For the parameter values $a = 4$, $\omega = 1.55$ and $\varepsilon > 0$ sufficiently small, the non-wandering set of the forced Van der Pol equation (1.8) consists precisely of two stable periodic orbits and a hyperbolic solenoid that is the suspension of a full shift map on two symbols.*

Question 1.3. *What are the bifurcations that lead to the creation and destruction of the chaotic trajectories in the forced Van der Pol system?*

Acknowledgment

This research was partially supported by the Department of Energy and the National Science Foundation.

References

[1] Arnold V, Afrajmovich V, Ilyashenko Yu and Shil'nikov L 1994 *Bifurcation Theory: Dynamical Systems V (Encyclopaedia of Mathematical Sciences)* (New York: Springer)

[2] Guckenheimer B A and Myers J M 1993 A dynamical simulation facility for hybrid systems *Lect. Notes Comput. Sci.* **736** 255–67

[3] Benoît É 1990 Canards et enlacements *Publ. Math. IHES Publ. Math.* **72** 63–91

[4] Cartwright M and Littlewood J 1945 On nonlinear differential equations of the second order: I the equation $\ddot{y} - k(1 - y^2)\dot{y} + y = bk\cos(\lambda t + a)$, k large *J. London Math. Soc.* **20** 180–9

[5] Cartwright M and Littlewood J 1947 On nonlinear differential equations of the second order: II the equation $\ddot{y} - kf(y, \dot{y})\dot{y} + g(y, k) = p(t) = p_1(t) + kp_2(t)$, $k > 0, f(y) \geq 1$ *Ann. Math.* **48** 472–94 (addendum 1949 *Ann. Math.* **50** 504–5)

[6] Diener M 1984 The canard unchained or how fast/slow dynamical systems bifurcate *The Mathematical Intelligencer* **6** 38–48

[7] Dumortier F and Roussarie R 1996 Canard cycles and center manifolds *Mem. Am. Math. Soc.* **121**(577) x+100pp

[8] Fenichel N 1979 Geometric singular perturbation theory *J. Diff. Eqns* **31** 53–98

[9] Flaherty J and Hoppensteadt F 1978 Frequency entrainment of a forced Van der Pol oscillator *Stud. Appl. Math.* **58** 5–15

[10] Govaerts W 2000 *Numerical Methods for Bifurcation of Dynamical Equilibria* (Philadelphia, PA: SIAM)

[11] Grasman J 1987 *Asymptotic Methods for Relaxation Oscillations and Applications* (New York: Springer)

[12] Guckenheimer J, Hoffman K and Weckesser W 2000 Numerical computation of canards *Int. J. Bif. Chaos* **10** 2669–87

[13] Guckenheimer J, Hoffman K and Weckesser W 2001 Global analysis of periodic orbits in the forced Van der Pol equation in *Global Analysis of Dynamical Systems* ed H Broer, B Krauskopf and G Vegter (Bristol: IOP Publishing) pp 261–76

[14] Guckenheimer J and Holmes P 1983 *Nonlinear Oscillations, Dynamical Systems and Bifurcations of Vector Fields* (New York: Springer)

[15] Henon M 1976 A two-dimensional mapping with a strange attractor *Commun. Math. Phys* **50** 69–77

[16] Hirsch M 1994 *Differential Topology* (New York: Springer)

[17] Hodgkin A and Huxley A 1952 A quantitative description of membrane current and its applications to conduction and excitation in nerve *J. Physiol. (Lond.)* **116** 500–44

[18] Jones C 1994 Geometric singular perturbation theory *Lect. Notes Math.* **1609** 44–118

[19] Levi M 1981 *Qualitative Analysis of the Periodically Forced Relaxation Oscillations (Mem. Am. Math. Soc. 32(244))* (Providence, RI: American Mathematical Society)

[20] Levinson N 1949 A second order differential equation with singular solutions *Ann. Math.* **50** 127–53

[21] Levinson N 1950 Perturbations of discontinuous solutions of nonlinear systems of differential equations *Acta Math.* **82** 71–106

[22] Lichtenberg A and Lieberman M 1983 *Regular and Stochastic Motion* (New York: Springer)

[23] Littlewood J 1957 On nonlinear differential equations of the second order: III the equation $\ddot{y} - k(1 - y^2)\dot{y} + y = bk\cos(\lambda t + a)$ for large k and its generalizations *Acta Math.* **97** 267–308 (Errata at end of 1957, 2.)
[24] Littlewood J 1957 On nonlinear differential equations of the second order: III the equation $\ddot{y} - kf(y)\dot{y} + g(y) = bkp(\phi)$, $\phi = t + a$ for large k and its generalizations *Acta Math.* **98** 1–110
[25] Lorenz E 1963 Determinisitic non-periodic flow *J. Atmos. Sci.* **20** 130–41
[26] Mettin R, Parlitz U and Lauterborn W 1993 Bifurcation structure of the driven Van der Pol oscillator *Int. J. Bif. Chaos* **3** 1529–55
[27] Mischenko E and Rozov N 1980 *Differential Equations with Small Parameters and Relaxation Oscillations* (New York: Plenum)
[28] Mischenko E, Kolesov Yu, Kolesov A and Rozov N 1994 *Asymptotic Methods in Singularly Perturbed Systems* (New York: Consultants Bureau) (translated from the Russian by Irene Aleksanova)
[29] Rinzel J 1985 Excitation dynamics: insights from simplified membrane models *Theor. Trends Neurosci.* **44** 2944–6
[30] Smale S 1967 Differentiable Dynamical Systems *Bull. Am. Math. Soc.* **73** 747-817
[31] Storti D and Rand R 1988 Subharmonic entrainment of a forced relaxation oscillator *Int. J. Nonlin. Mech.* **23** 231–9
[32] Van der Pol B 1926 On relaxation oscillations *Phil. Mag.* **7** 978–92

Chapter 2

Many-body quantum mechanics

Robert S MacKay
University of Warwick

Nonlinear Dynamics and Chaos: Where do we go from here? is the title of this book. My answer for a proposed direction is *many-body quantum mechanics*. In keeping with the informal spirit of the presentation on which this chapter is based, I will pose several questions, though they vary in status from ill-defined to those for which I have a fairly complete answer.

I should start with some preliminaries about quantum mechanics. The experienced reader may find some of what I write to be naïve or even misconceived, so I would be grateful for feedback and I ask the inexperienced reader to take everything with a pinch of salt. In addition, I do not propose to give a scholarly account, so some of my references and credits will be a bit haphazard.

The traditional dogma is that quantum mechanics is linear, so there is no nonlinearity and *a fortiori* there is no chaos. Thus, it looks as if my proposal is a backwards step!

I believe that view of quantum mechanics is superficial, however. Any dynamical system, written in terms of its flow map

$$\phi : \mathbb{R} \times M \to M \qquad (t, x) \mapsto \phi_t(x)$$

on a state space M, can be viewed as linear via its action on functions $F : M \to \mathbb{R}$,

$$(\phi_t^* F)(x) = F(\phi_t x) \tag{2.1}$$

(evaluation along trajectories) or, equivalently, its inverse

$$(\phi_{*,t} F)(x) = F(\phi_{-t} x)$$

(advection of passive scalars). In the common case that M is a manifold and the dynamical system comes from a vector field

$$\dot{x} = v(x)$$

then (2.1) can alternatively be written as the linear partial differential equation (PDE)

$$\frac{\partial F}{\partial t} = DFv = v \cdot \nabla F \tag{2.2}$$

for how functions F evolve. For the special case of Hamiltonian vector fields v, those arising from frictionless classical mechanics, there is another way to write (2.2): its right-hand side is the *Poisson bracket* $\{H, F\}$,

$$\{H, F\} = \sum_{j=1}^{d} \frac{\partial H}{\partial q_j}\frac{\partial F}{\partial p_j} - \frac{\partial H}{\partial p_j}\frac{\partial F}{\partial q_j}$$

in canonical coordinates (p, q), where H is the Hamiltonian function and d is the number of degrees of freedom. So (2.2) can be written as

$$\frac{\partial F}{\partial t} = \{H, F\} \tag{2.3}$$

which is one of the standard formulations of Hamiltonian mechanics. In particular, from this point of view, Hamiltonian mechanics is linear. Nonetheless, all the nonlinear and chaotic phenomena are still present.

Quantum mechanics can be expressed in a similar form to (2.3), actually two alternative forms. The first is the Schrödinger form

$$\mathrm{i}\hbar\dot{\psi} = H\psi$$

for the evolution of a *wavefunction* ψ in some complex Hilbert space U, where H is a Hermitian operator and $\hbar$ is Planck's constant h divided by 2π. The second is the Heisenberg form (or maybe one should say Jordan–Born–Heisenberg form)

$$\dot{A} = -\frac{\mathrm{i}}{\hbar}[H, A]$$

for the evolution of any linear operator A on U, where $[H, A]$ denotes the commutator

$$[H, A] = HA - AH.$$

These are linear evolutions for ψ and A. The two are related: if every wavefunction evolves by Schrödinger's equation then for all ψ, $\mathrm{i}\hbar(A\psi)_t = H(A\psi)$, so $\mathrm{i}\hbar A_t\psi = HA\psi - \mathrm{i}\hbar A\psi_t = [H, A]\psi$. Conversely, if every operator evolves by Heisenberg's equation, then in particular the rank-1 projections $P = \psi\eta$, where $\psi \in U$ and $\eta \in U^*$ (the dual space, consisting of linear forms on U), evolve this way, so $\dot{\psi}\eta + \psi\dot{\eta} = -\frac{\mathrm{i}}{\hbar}(H\psi\eta - \psi\eta H)$ for all ψ, η, and the only solution is $\mathrm{i}\hbar\dot{\psi} = H\psi$, $-\mathrm{i}\hbar\dot{\eta} = \eta H$.

In particular, the Heisenberg form is closely analogous to (2.3), because $\{H, F\}$ and $[H, A]$ are 'derivations' on functions F and operators A, respectively, that is, they are linear and satisfy Leibniz' rule $[H, AB] = A[H, B] + [H, A]B$ (and similar for $\{H, FG\}$).

So we are not really justified in stating that quantum mechanics has no nonlinearity, because we know that in the case of (2.3) there is, in general, an underlying nonlinearity in the vector field. Who is to say that there is not some analogous underlying nonlinear vector field in the quantum case?

Indeed, the theory of non-commutative manifolds (e.g. [14]) provides a way of thinking about Heisenberg's equation as coming from an underlying vector field, but on a *non-commutative manifold*; see, e.g., [6]. Here is a rapid introduction to the subject (which is also known as *quantum geometry*). Consider the algebra $C(M)$ of (smooth) functions $F : M \to \mathbb{C}$ on a normal manifold M, with the usual operations of pointwise addition and multiplication and complex conjugation. For each point $x \in M$ we can define a linear functional ev_x on $C(M)$, namely the evaluation map $ev_x(F) = F(x)$. It has properties that algebraists call a *Hermitian character*, namely $ev_x(FG) = ev_x(F)ev_x(G)$ and $ev_x(F^*) = (ev_x(F))^*$, where * represents complex conjugation. Remarkably, every Hermitian character on $C(M)$ can be proved to be the evaluation map at some point $x \in M$. So one can think of M as being the set of Hermitian characters on $C(M)$. Now $C(M)$ is commutative, that is, $FG = GF$ for all F and $G \in C(M)$. It turns out that every commutative algebra $\mathcal{A}$ satisfying suitable properties (to get the smooth structure) is $C(M)$ for some M. The wonderful idea is to define a non-commutative manifold to be the set of Hermitian characters of a non-commutative algebra (with suitable smoothness properties). The basic example of a non-commutative algebra is the algebra of linear operators on a Hilbert space, with the operations of addition, multiplication and Hermitian conjugation. There are variations on the theme, using *prime ideals* of the algebra, or its *irreducible representations*. An unfortunate obstacle is that for even the simplest example of a quantum mechanical algebra, that generated by position x and momentum p operators for a one-dimensional (1D) oscillator, with $[p, x] = i\hbar$, there are no characters (nor finite-dimensional representations) and only one prime ideal. However, this does not seem to be regarded as serious by its proponents.

There is a growing literature on non-commutative manifolds, but I am not aware yet of anyone having tried to develop an analogue of the theory of dynamical systems for them, and this is in my opinion very important as it underlies quantum mechanics. So I am ready to pose my first question.

Question 2.1. *Can a theory of dynamics on non-commutative manifolds be developed?*

The first step is to define a vector field on a non-commutative manifold. This is easy. On an ordinary manifold M, one way of thinking of a vector field v is as a Hermitian derivation L_v on $C(M)$, defined by $L_v F = DFv$ (the right-hand side of (2.2)). Any Hermitian derivation L on $C(M)$ induces a vector field v_L on

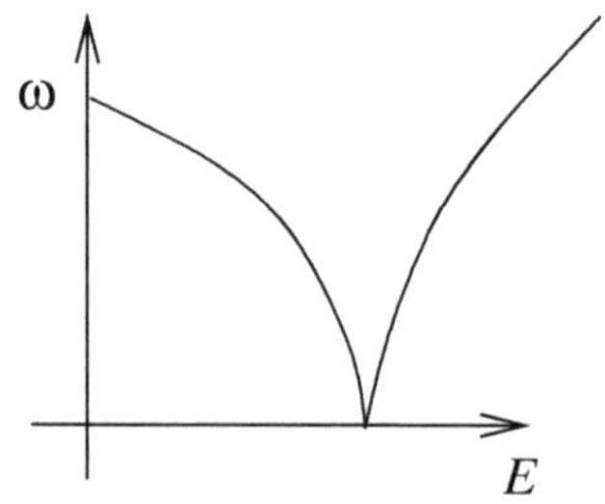

Figure 2.1. Frequency ω versus energy E for the ideal pendulum.

M, namely the unique v such that $LF = DFv$ for all F. So it is natural to define a vector field on a non-commutative manifold to be a Hermitian derivation on its algebra. The right-hand side of Heisenberg's equation is an example.

The next step is not so obvious. One might wish to analyse the initial value problem and derive a local flow, as for ordinary (Lipschitz) vector fields. But I do not know if this makes sense in the non-commutative case. An alternative could be to mimic the theory of uniformly hyperbolic systems, which can be formulated entirely in functional analytic terms, so it does not require solution of the initial value problem. Another promising direction would be to mimic the spectral theory of measure-preserving transformations; see, e.g., [3]. In any case, there is lots of room for work on this question.

2.1 Signs of nonlinearity in quantum mechanics

I am now going to move to a different approach, which is to look for signs of nonlinearity in quantum mechanics.

The basic nonlinearity in Hamiltonian mechanics is anharmonicity (which is better called non-isochronicity). An oscillator is said to be *anharmonic* if the frequency ω of oscillation varies non-trivially with the energy E. For example, a pendulum's frequency decreases until one reaches the energy when it can go over the top, as all bell-ringers know; see figure 2.1. Equivalently, when expressed in terms of the *action variable* I, the Hamiltonian $H(I)$ is not affine, since $\omega = \frac{\partial H}{\partial I}$. Note, however, that there are oscillators other than the harmonic oscillator for which the frequency is constant [11]!

We will say that a 1D quantum oscillator is *anharmonic* if its energy levels are not equally spaced. This is because neighbouring energy levels are thought of as corresponding to actions differing by $\hbar$, so the energy difference divided by $\hbar$ is a discretization of $\frac{\partial H}{\partial I}$. Most quantum oscillators (except for the harmonic one) have non-constant energy level spacing, so we see that in this sense nonlinearity is common in quantum mechanics.

To take this to more degrees of freedom, a classical Hamiltonian system is linear (in a coordinate system for which the Poisson brackets are constant) if and only if the Hamiltonian is a homogeneous quadratic in the coordinates (plus a constant). Similarly, we can say that a quantum system is linear (with respect to a generating set of operators whose commutators are constants) if the Hermitian operator H is a homogeneous quadratic (plus a constant) when expressed in this basis of operators. For example, if H is Hermitian and quadratic in operators a_j and their Hermitian conjugates $a_j^\dagger$, satisfying commutation relations $[a_j, a_k] = 0$, $[a_j, a_k^\dagger] = \delta_{jk}$, then there is a unitary change of generators to a set $\alpha_j, \alpha_j^\dagger$ satisfying the same commutation relations, such that H takes the form $\sum_j \lambda_j N_j$ for some real numbers λ_j, where $N_j = \alpha_j^\dagger \alpha_j$. Each N_j is called a *number operator* because its spectrum is precisely the non-negative integers $\mathbb{Z}_+$. They commute, so it follows that H commutes with them and hence the spectrum of H is $\{\sum_j \lambda_j n_j : n_j \in \mathbb{Z}_+\}$.

My next question is to make this more general.

Question 2.2. *How far can one generalize the previous example of a linear quantum system (e.g. fermionic operators are all right too), and when can a new generating set be chosen to make the system linear (cf Sternberg's linearization theorem near an equilibrium of a classical vector field [4])?*

We can go a bit further in our translations of properties from classical to quantum. A classical system is *integrable* if the Hamiltonian is a function only of a Poisson commuting set of integrals (conserved quantities), whose derivatives are linearly independent almost everywhere. It follows that its motion can be solved virtually explicitly and the dynamics fully comprehended (equivalent to constant velocity motion on tori). Similarly, a quantum system can be said to be *integrable* if the Hermitian operator H is a function only of some commuting and independent set of operators. This is the case in the previous linear example. Again, it allows one to solve the dynamics virtually explicitly. The only problem with this definition is that it is not yet agreed what counts as independent, but there is a large literature on quantum integrable systems; see, e.g., [30].

One can go still further. Many Hamiltonian systems, even if not exactly integrable, turn out to be close to integrable in certain parameter regimes or regions of the state space. For example, near a non-resonant equilibrium of a classical Hamiltonian system one can find canonical coordinate systems—here I have chosen complex coordinates z_j, z_j^*—to put H into the form

$$\begin{array}{ccc} \text{‘integrable’} & + & \text{‘remainder’} \\ H((|z_j|^2)_j) & + & O(\|z\|^{2N}) \end{array}$$

for arbitrary N, the so-called Birkhoff normal form. Similarly, in many quantum systems one can apply unitary transformations to put them into Birkhoff normal form, in the sense of a function only of a commuting set of operators, plus some ‘high-order’ remainder (Bogoliubov transformations, e.g. as rediscovered by [22]). Note that there are also quantum analogues of KAM theory (e.g. [15]) and Nekhoroshev theory (e.g. [10]).

In classical mechanics, Poincaré made the really important discovery that there are obstacles to integrability. Before his discovery the hope was that for every Hamiltonian system one could perform coordinate changes to bring them closer to integrable, and eventually push the remainder to zero, so every Hamiltonian system would be integrable. Poincaré realized that there is an obstacle, a very simple one: classical integrable systems of more than one degree of freedom generically have rational tori, a dense set of them in fact. He showed that they are fragile under general perturbation of the Hamiltonian. So the conclusion is that most Hamiltonian systems are not integrable. There is an analogue of this for quantum systems: if you have an integrable quantum system then degeneracy of energy levels is a codimension-one phenomenon—typically there are no degeneracies. However, in one-parameter families of integrable systems of more than one degree of freedom, energy levels can cross each other as the parameter varies, typically transversely, and so degeneracies occur at isolated parameter values and they are typically unremovable by small change in the family, preserving integrability. For general real symmetric quantum systems, however, Wigner pointed out that degeneracy is a codimension-two phenomenon, and it is of codimension three for general Hermitian ones: there is generically a repulsion of energy levels, leading to avoided crossings.

I regard this as a direct analogue of Poincaré's observation. It shows that most quantum systems are not integrable or, strictly speaking, most *one-parameter families* of quantum systems are not integrable. Actually, the latter comment highlights a subtle distinction between classical and quantum mechanics. I feel that in order to talk about nonlinearity in quantum mechanics it is probably usually necessary to think in terms of one-parameter families. A natural parameter might be Planck's constant or some equivalent scale of action, but it could be other parameters. There is a huge literature on the distribution of spacings of energy levels, in particular on the repulsion of energy levels as a sign of non-integrability, people at Bristol being key players; see, e.g., [8].

Poincaré went much further than just saying that most classical systems are not integrable—he showed what dynamics is substituted for the rational tori in a near integrable system, and he found the notion of homoclinic chaos that we are all familiar with. So my next question is:

Question 2.3. *Is there a quantum analogue of homoclinic chaos?*

Of course, one can answer that avoided crossings take the place of crossings, but I have a feeling that there should be some consequences of non-integrability of a more dynamical nature. For a snapshot of some views on this, see [7]. One avenue that I consider promising is adiabatic theory: on varying the parameter slowly a quantum system can switch energy levels near an avoided crossing, best described probabilistically (e.g. [28]), which recalls probabilistic interpretations of transitions across a separatrix in classical mechanics; see, e.g., [41]. In both cases the transition rate is very small for small perturbation from integrable (exponentially so for analytic systems).

OK, that is all by way of introduction. For an interesting source of other directions in the dynamics of simple quantum systems, see [42]. Now I want to concentrate on the many-body case of quantum mechanics. I am going to discuss in detail one particular phenomenon in many-body quantum mechanics—*quantum discrete breathers*—but before I begin on that I should say that there are many well-known approximations in many-body quantum mechanics that produce obviously nonlinear equations. Hartree–Fock theory is a basic example, where a many-electron problem is replaced by some effective one-electron problem which is determined by a nonlinear self-consistency condition (e.g. [47]). It is the basis for understanding the periodic table of the elements, for example, though its validity is questionable. Another example is Bose–Einstein condensates, where the Gross–Pitaevskii equation (the nonlinear Schrödinger equation in an external potential) is derived for the wavefunction that is supposed to describe the condensate [33]. Thus, many-body problems, if you make certain approximations, can boil down to nonlinear few-body problems.

Another direction one could pursue in many-body quantum mechanics is another type of answer to Question 2.3. Arbitrarily small perturbation of some quantum many-body problems can produce qualitatively new effects, an example being superconductivity when electron–phonon interactions are added. The effects even have the property of being exponentially small in the perturbation parameter.

But these are not the directions in which I wish to go. I want to pursue the anharmonicity idea, as we did not have to do anything fancy to see that there is anharmonicity in quantum mechanics, but its implications for many-body systems are less obvious.

2.2 Discrete breathers

First I should tell you what discrete breathers are classically; for one survey see [35]. A classical *discrete breather* (DB) is a time-periodic spatially localized vibration in a Hamiltonian network of oscillators. There are dissipative analogues, but I am going to concentrate entirely on the Hamiltonian case. Imagine that we have some network—it could be sites in a crystal lattice or it could be some amorphous thing like a globular protein, and at each site we have a classical oscillator with Hamiltonian

$$H(x_s, p_s) = \sum_{s \in S} \left(\frac{1}{2} p_s^2 + V(x_s) \right) + \frac{\varepsilon}{2} \sum_{r,s \in S} \Delta_{rs} (x_r - x_s)^2 \tag{2.4}$$

for some network S, local potential V, coupling strength ε and coupling matrix Δ. We will suppose that the potential V is not harmonic, as is generically the

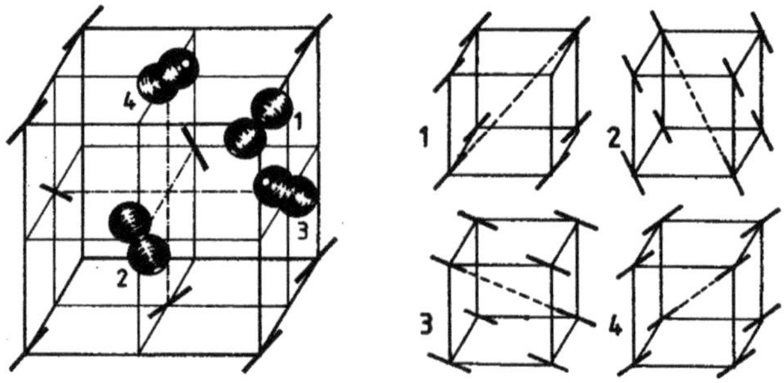

Figure 2.2. The crystal structure of orthotropic H_2 at less than 3 K. Reproduced from I F Silvera, The solid molecular hydrogens in the condensed phase: fundamentals and static properties, 1980 *Rev. Mod. Phys.* **52** 393–452 ©1980 by the American Physical Society.

case, and that the coupling is relatively weak and satisfies some decay condition like

$$|\Delta_{rs}| \leq C e^{-\lambda d(r,s)}$$

or even some summable power law. Many other forms of coupling can also be considered. To avoid confusion let us just think of the case of linear nearest-neighbour coupling.

Such models are proposed for vibrations of molecular crystals and I want to mention a few examples here. First, figure 2.2 shows a molecular hydrogen crystal, solid H_2, where the molecules arrange themselves in a lattice and orient themselves along the four main diagonals. Now, what is the degree of freedom that I wish to consider here? I am thinking of the stretch of the molecule. The stretches in nearby molecules are coupled by Van der Waals and quadruple–quadruple interactions. Of course, stretch of one molecule also makes forces which shift the centre of mass and change the orientation of nearby molecules, but we will ignore that for present purposes.

Another example, sketched in figure 2.3, is potassium bicarbonate $KHCO_3$. The bicarbonates pair up by hydrogen bonding and one gets a degree of freedom where the protons can switch sides simultaneously between configurations (a) and (b) in figure 2.3 [26] (though there is now evidence in similar materials for *four*-well dynamics [21]).

A third example is 4-methylpyridine; see figure 2.4. This is a crystal of a molecule that has pyridine rings, like benzene but with nitrogen at the bottom, and a methyl group attached to the top. The methyl groups are relatively free to rotate, so it is that rotation that is the degree of freedom here [17]. But each methyl group has another one facing it and they interact quite strongly (along the dotted lines in figure 2.4(b)), so really we have to consider two-degree-of-freedom quantum units consisting of such pairs of methyl groups. Methyl groups in neighbouring

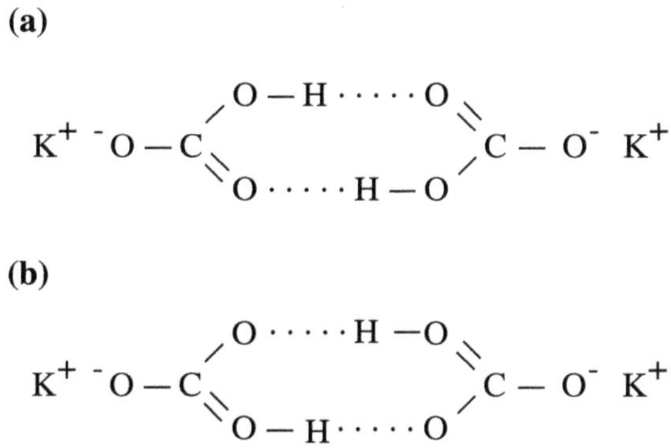

Figure 2.3. The two configurations (a) and (b) of the potassium bicarbonate dimers.

pairs interact with each other (e.g. via the dashed lines in the figure, but also along similar lines in the b direction), linking the pairs in a 3D lattice.

The final example is a platinum chloride material, $Pt(en)_2Cl_2\,Pt(en)_2\,(ClO_4)_4$ where en stands for ethylenediamine, which contains long chains of alternate platinum and chloride ions. The chains dimerize (meaning that instead of the Pt and Cl being equally spaced along these chains, the equilibrium position of the nth Cl is displaced by $(-)^n\delta$, and δ is relatively large for this material). This is because the electrons reduce their energy by doing so, more than the energy required to move the Cl ions (Peierls' instability). Essentially one ends up with alternate 'molecules' of $PtCl_2(en)_2$ and $Pt(en)_2$; see figure 2.5 and [46] for more details. The degree of freedom that is relevant here is the symmetric stretch of the Cl_2 in the $PtCl_2(en)_2$ molecules, though the interaction with neighbours forces one to include at least the antisymmetric stretch as well. So again the units have two degrees of freedom.

Now if we suppose that the potential for the local degree of freedom in solid H_2 or the PtCl material, say, is harmonic, meaning a pure parabola,

$$V(x) = \tfrac{1}{2}\omega_0^2 x^2$$

and we suppose the system has crystal periodicity, so the coupling is translation invariant, then there are no DBs for any positive coupling because, as is well known, any localized initial condition will disperse. For example, if you look at the dispersion relation for a 1D nearest-neighbour-coupled chain

$$\omega^2 = \omega_0^2 + 4\varepsilon \sin^2 \frac{k}{2}$$

it is non-trivial (meaning $\partial\omega/\partial k$ is not constant) and so any localized initial condition will spread out. Therefore, one never gets any time-periodic spatially localized solutions.

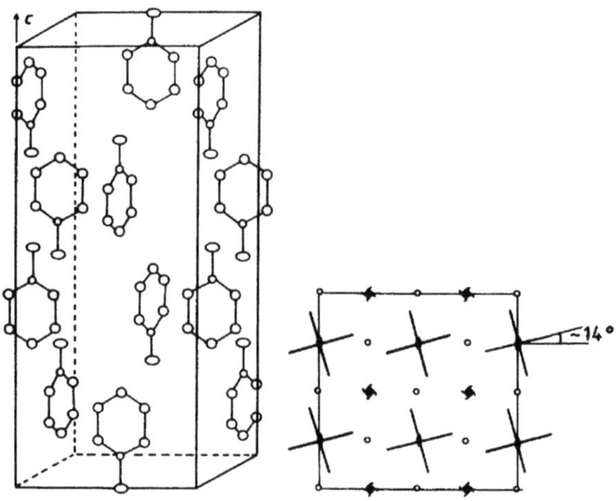

(*a*)

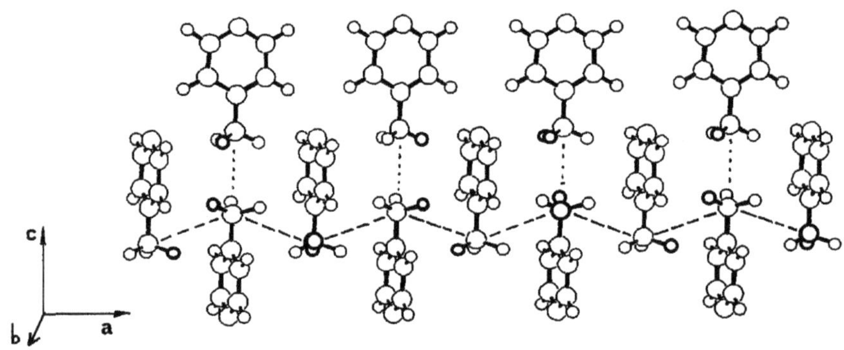

(*b*)

Figure 2.4. Two views of the crystal structure of 4-methylpyridine: the 3D structure (*a*) (reprinted with permission from N Le Calvé, B Pasquier, G Braathen, L Soulard and F Fillaux 1986 Molecular mechanism determining phase transitions in the 4-methylpyridine crystal *J. Phys. C* **19** 6695–715 ©1986 American Chemical Society), and a section (*b*) (reproduced from F Fillaux, C J Carlile and G J Kearley 1991 Inelastic neutron scattering study at low temperature of the quantum sine-Gordon breather in 4-methylpyridine with partially deuterated methyl groups, *Phys. Rev.* B **44** 12 280–93 ©1991 by the American Physical Society).

Real molecules are anharmonic; frequency *does* vary with amplitude. Morse found that the so-called Morse potential

$$V(x) = \tfrac{1}{2}(1 - e^{-x})^2$$

gives a good fit to a lot of spectroscopic data. See, for example, figure 2.6 for

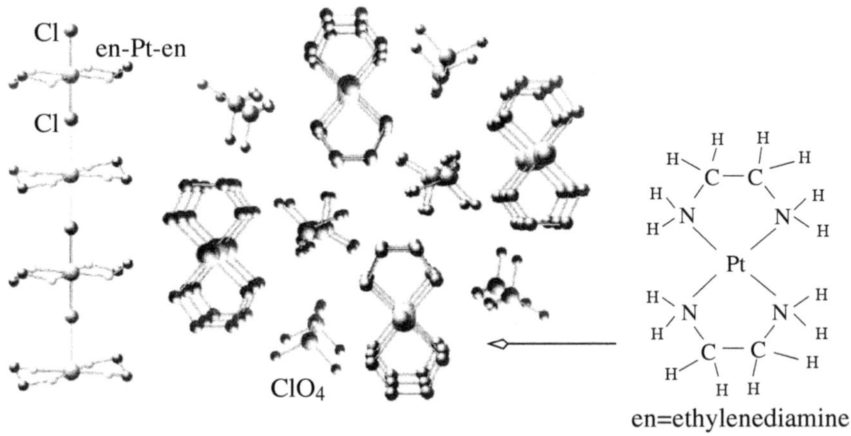

Figure 2.5. The 'molecules' involved in the platinum chloride example. Reproduced from B I Swanson, J A Brozik, S P Love, G F Strouse, A P Shreve, A R Bishop, W-Z Wang and M I Salkola 1999 Observation of intrinsically localized modes in a discrete low-dimensional material *Phys. Rev. Lett.* **82** 3288–91 ©1999 by the American Physical Society.

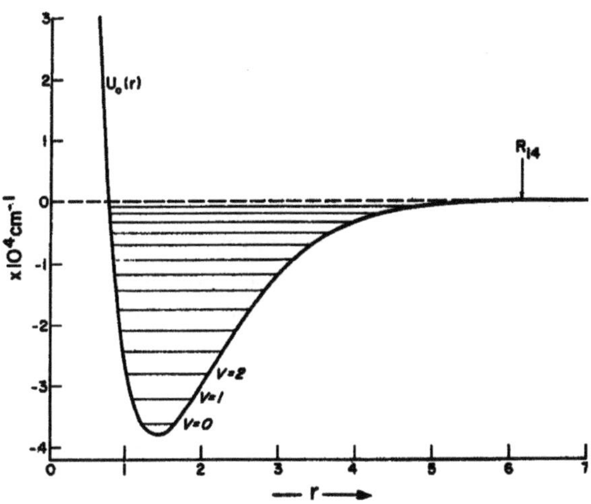

Figure 2.6. The effective potential and the 15 vibration levels of stretch of an H_2 molecule. The arrow indicates the classical turning point in level 14 which is not shown. Reproduced from J van Kranendonk 1983 *Solid Hydrogen* Plenum ©1983 by Plenum Publishing Corporation.

Table 2.1. Anharmonicities of some diatomic molecules in gas phase.

Molecule	ω (cm^{-1})	Anharmonicity x (%)
H_2	4401.21	2.757
HD	3813.1	2.404
D_2	3115.5	1.984
$H\,^{35}Cl$	2990.946	1.766
O_2	1580.19	0.758
CO	2169.81	0.612
N_2	2358.03	0.599
$^{35}Cl_2$	599.71	0.482
Br_2	323.07	0.361
I_2	214.50	0.268

the case of solid hydrogen. Table 2.1 indicates the amount of anharmonicity, of certain gas-phase diatomic molecules (from [25,45]). These figures are obtained by fitting spectroscopic data to a formula of the form

$$E = (n + \tfrac{1}{2})\hbar\omega - x(n + \tfrac{1}{2})^2\hbar\omega + \cdots .$$

From my results with Aubry [38], if V in (2.4) is anharmonic and the coupling is weak enough compared with the anharmonicity, i.e. $\varepsilon < \varepsilon_0$ for some $\varepsilon_0 > 0$, then we have proved the existence of DBs in the classical system (extended to multi-degree-of-freedom oscillators in [43]). If the coupling is nearest neighbour or exponentially decaying then the amplitude of the DBs decays exponentially in space [43]. (Similarly, if the coupling is a (summable) power law, we obtain power-law localization of the DB [9].) The existence proof is very constructive—we just continue from the uncoupled case using the implicit function theorem, and that means that the proof can essentially be used numerically. Furthermore, the DBs are stable in a sense: if ε is small enough, $\varepsilon < \varepsilon_1$ (where maybe ε_1 is smaller than ε_0), then the DBs are l_2-linearly stable, i.e. stable under the linearized evolution of finite energy perturbations [39]. This is the best form of stability one could hope for. So it suggests they could be physically observable.

But real molecules are *quantum mechanical*. Just to give an illustration, look again at figure 2.6 showing the effective potential for the stretch of a hydrogen molecule. Classically, any energy of vibration between the minimum and the dissociation energy is possible. But spectroscopists observe only 15 energy levels, labelled from 0...14. (Actually they observe transitions between them and infer the levels.) So real molecules are highly quantum mechanical and we cannot ignore quantum mechanics in trying to understand vibrations in molecular crystals. Note also the anharmonicity in figure 2.6: the spacing between levels is

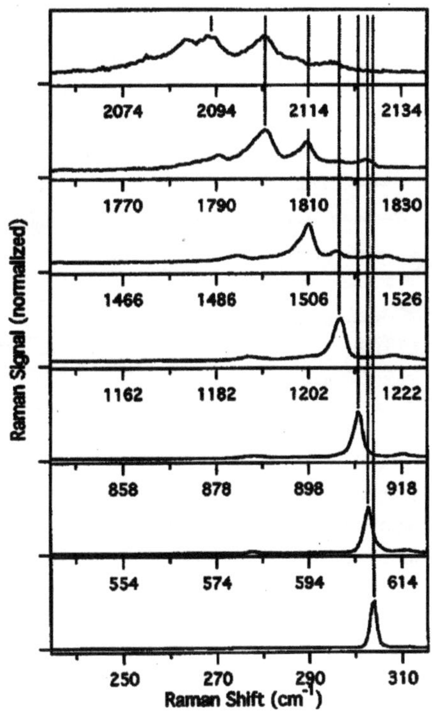

Figure 2.7. Raman scattering results on Pt^{35} Cl. Reproduced from B I Swanson, J A Brozik, S P Love, G F Strouse, A P Shreve, A R Bishop, W-Z Wang and M I Salkola 1999 Observation of intrinsically localized modes in a discrete low-dimensional material, *Phys. Rev. Lett.* **82** 3288–91 ©1999 by the American Physical Society.

relatively large at the bottom and shrinks as we go up. The question that poses itself, and the key question to be addressed in this chapter is:

Question 2.4. *Is there a quantum analogue of discrete breathers?*

2.3 Experimental evidence for quantum discrete breathers

Experiments suggest that the answer is yes. Figure 2.7 shows the experimental results of Swanson *et al* [46] on the platinum-chloride material that was represented in figure 2.5.

What they do is fire in photons from a laser and have a look for any photons that come out with less energy; the idea is that if they have lost energy they have created a vibration in the crystal. This is called Raman scattering. Spectroscopists measure energy in cm^{-1}! Sorry about that but it is just units. We see that they

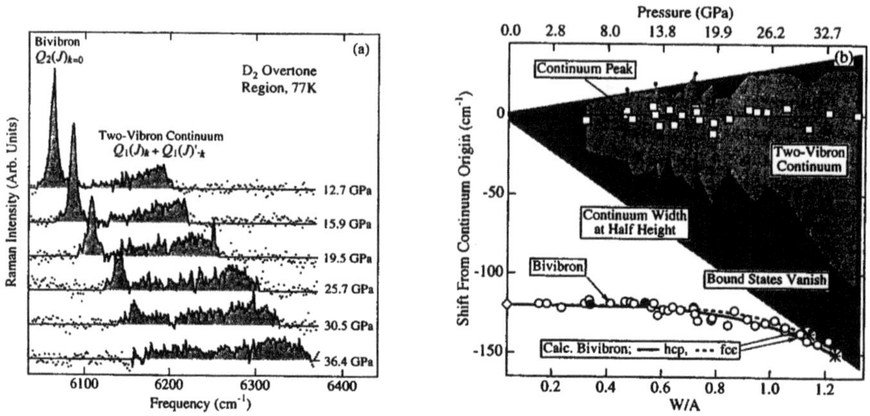

Figure 2.8. Pressure-induced bi-vibron bound–unbound transition of D_2; shown are Raman spectra of D_2 in the overtone region as a function of pressure (*a*), and a comparison between experimental results and theoretical calculations (*b*). Reproduced from J H Eggert, H-K Mao and R J Hemley 1993 Observation of a two-vibron bound-to-unbound transition in solid deuterium at high pressure *Phys. Rev. Lett.* **70** 2301–4 ©1993 by the American Physical Society.

can create an excitation at 312 cm^{-1}, and they can also create one at nearly twice that. To see the difference, they cut up the energy axis and align the pieces above, shifted by multiples of 312 cm^{-1}. The second panel from the bottom shows a peak at slightly less than twice the original, the third panel up shows a peak at yet less than three times, the next even less than four times, and so on. So they say ‘Ah-ha! we are creating some anharmonic vibration here, because we are seeing successive levels of some anharmonic vibration’. They assume that it is a localized $PtCl_2$ symmetric stretch. The $PtCl_2$ symmetric stretch assumption is fine because for the chosen material that is the only Raman active mode in that range of energies. But the localization inference is not justified in my opinion. They do not observe spatial localization at all; they just say that their results are consistent with one molecule that is anharmonic and is picking up various amounts of energy.

I think it would be really interesting to do an experiment to observe the spatial structure of the created excitations. In principle, one could do this by scattering neutrons, instead of photons, because their energy can be chosen to make their de Broglie wavelength comparable to or shorter than the lattice spacing. So I tried it [37], but it turned out that the neutrons excited far too many other modes in the same energy range to even see the $PtCl_2$ stretch.

Another experiment that I like is the one by Eggert, Mao and Hemley [16] on solid deuterium D_2, the results of which are depicted in figure 2.8. The great thing about solid deuterium (and solid hydrogen) is that one can compress it relatively easily, and thereby change the coupling strength between the molecules. This is really fun. What they do is change the pressure, from 12.7 GPa (gigapascals) to 36.4 GPa (for reference, atmospheric pressure is about 0.1 MPa). At 12.7 GPa, they can create an excitation that they call a bi-vibron, and a continuum of excitations with a range of slightly higher energies. They interpret the bi-vibron as a second excited state of one deuterium molecule and the continuum as the formation of a pair of first excited molecules but where the excitations propagate in opposite directions. As they increase the pressure, the bi-vibron peak merges into the continuum; see figure 2.8(*a*). In figure 2.8(*b*) the continuum is depicted above the bi-vibron, but as the pressure increases one can see the bi-vibron energy decreases slightly and then it is absorbed into the continuum. This is nice because this shows that just as in the classical case some localized excitation exists, provided that coupling strength is smaller than or the same order of magnitude as a measure of the anharmonicity. In fact, in the appropriate dimensionless units, which are plotted on the lower horizontal scale of figure 2.8(*b*), the threshold is at about 1.2. Note again, however, that the experiment did not measure the spatial structure of the excitation. This material would be much better adapted than PtCl to the inference of spatial structure by neutron scattering and I would very much like to find someone willing to try it. The only problem is that solid deuterium is much harder to handle than PtCl because one needs to keep it cold and under pressure.

These experiments (and others, for example bi-vibrons were reported in solid hydrogen in 1957 [23]!) suggest that there is a quantum analogue of the classical discrete breather concept. But we need a mathematical formulation.

2.4 Towards a mathematical theory of quantum discrete breathers

A quantum system is specified by a Hermitian operator H on a Hilbert space U. A large part of the game, though not all, is to find the eigenvalues and eigenvectors of this operator. For example, if we had a single degree of freedom x, like stretch of a hydrogen molecule, then our Hilbert space would be L^2 functions of one variable and our operator would be the standard Schrödinger operator

$$H\psi = -\frac{\hbar^2}{2m}\frac{\mathrm{d}^2\psi}{\mathrm{d}x^2} + V(x)\psi \qquad \text{for } \psi \in U = L_2(\mathbb{R}, \mathbb{C}).$$

A typical spectrum would consist of a ground-state energy (which, without loss of generality, I will choose to be the origin of energy), first excited energy etc, as depicted in figure 2.9. In practice, there would be a continuum corresponding to dissociation for higher energies, but I am not really interested in the high excitations and, to avoid technicalities to do with unbounded operators (or even with infinite-dimensional bounded ones), I will simplify by supposing that my

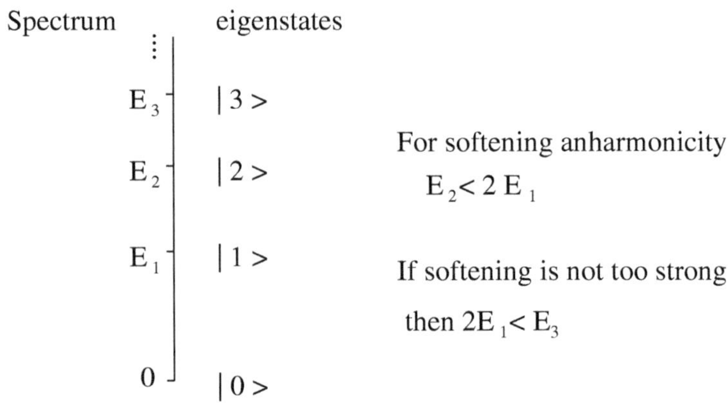

Figure 2.9. A typical spectrum of a one-degree-of-freedom quantum system.

Hilbert spaces are finite-dimensional. The important thing is that typically these spectra have softening anharmonicity as we saw earlier. So the gaps between the eigenvalues decrease as you go up the spectrum.

Now, how do we describe quantum mechanically the whole molecular crystal? Mathematically, the Hilbert space for N molecules is the *tensor* product of the individual spaces,

$$U_1 \otimes U_2 \otimes \cdots \otimes U_N.$$

It is defined to be the set of multilinear maps ψ from $U_1^* \times U_2^* \times \cdots U_N^*$ to $\mathbb{C}$ (where U^* is the dual of U), but the easiest way to think about a tensor product is to describe a basis for it. A natural basis is given by states of the form $|n_1, \ldots, n_N\rangle$, which represents that molecule 1 is in its n_1 state, molecule 2 in its n_2 state etc, and is defined by

$$|n_1, \ldots, n_N\rangle \quad \text{on} \quad \langle m_1|\langle m_2| \ldots \langle m_N| \quad \text{gives} \quad \prod_s \delta_{m_s n_s}.$$

(This ignores the degrees of freedom describing the relative positions and orientations of the molecules; including these is a project which goes beyond my present thinking, though approximate theory features in many books and it has been treated rigorously in the classical case [2, 36].)

The mathematical difficulties in many-body quantum mechanics come about because the dimension of a tensor product is the product of the dimensions rather than the sum, that is

$$\dim(\times_s U_s) = \sum_s \dim U_s$$

but

$$\dim \otimes_s U_s = \prod_s \dim U_s.$$

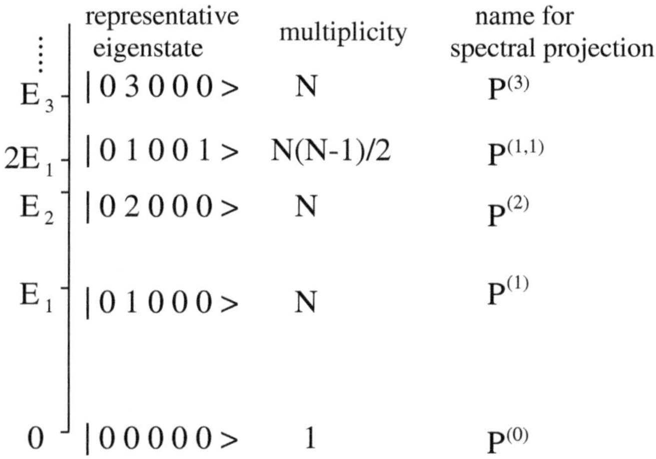

Figure 2.10. Multiplicity of the spectrum of an uncoupled N-unit network.

Actually, related difficulties occur in probability theory of large networks, so I am developing a parallel theory there. We will see where this difficulty starts to play a role in a moment, but if we think just of an *uncoupled* network then the Hamiltonian is just the sum of the individual Hamiltonian operators

$$H^0 = \sum_s \text{`}H_s\text{'},$$

where a superscript '0' corresponds to 'uncoupled' and 'H_s' is defined on $\otimes U_s$ by

$$\text{`}H_s\text{'}\psi(\psi_1^*, \dots) = \psi(\psi_1^*, \dots, H_s\psi_s^*, \dots).$$

Hence the eigenvalues of H^0 are just sums of eigenvalues of H_s, taking one for each unit.

So what do we get for the spectrum of the uncoupled network? We get a ground state where each molecule is in its ground state. We get a first excited subspace where one molecule is first excited and the rest are in their ground state. But, of course, there are N ways of choosing which molecule is first excited so this has dimension N. There is a second excited subspace where one molecule is second excited and the rest are in their ground state, and again that has dimension N. Then there is a subspace where two molecules are first excited and the rest are in their ground state, and that has dimension $N(N-1)/2$; see figure 2.10. You can go up as far as you want but this is going to be enough for our present purposes.

Let me give names to the spectral projections corresponding to these subspaces; $P^{(0)}$, $P^{(1)}$, $P^{(2)}$, $P^{(1,1)}$ and so on, as indicated in figure 2.10. Just to remind you, a *spectral projection* P (for a finite-dimensional Hermitian operator)

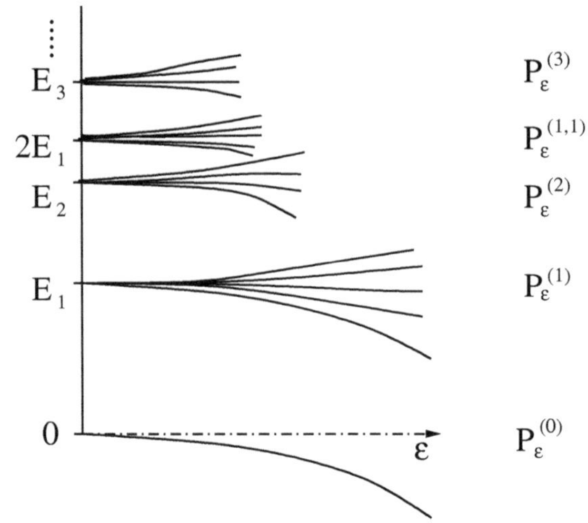

Figure 2.11. The spectrum of the N-unit quantum system for weak coupling ε.

is the orthogonal projection onto the span of eigenvectors corresponding to an isolated part of the spectrum. The question that I want to ask is:

Question 2.5. *What happens to the spectrum of H^0 when one adds some coupling operator, i.e. $H^\varepsilon = H^0 + \varepsilon\Delta$?*

We will suppose the coupling operator takes the form

$$\Delta = \sum_{r,s\in S} \Delta_{rs}$$

together with some decay property, as in the classical case.

Now if the size $\varepsilon\|\Delta\|$ of the coupling (measured using the Hilbert norm, i.e. the largest eigenvalue in absolute value) is small, then standard perturbation theory tells one that the spectral projections move smoothly and the spectral gaps cannot change dramatically in size; in particular they can shrink by at most $2\varepsilon\|\Delta\|$ if at all. So as we turn on coupling we should get a situation such as those sketched in figure 2.11.

The ground-state energy may move but will do so smoothly and it will remain non-degenerate. The first excited subspace, in general, splits into a variety of energy levels. However, the spectral projection corresponding to the subspace still moves smoothly, and we keep a nice gap of almost the same size between these N states and the ground state. Similarly, there is a gap between this family and the family of states emerging from the energy of the second excited state, etc.

Here is a nice proof of the smooth persistence of spectral projections, which so far I have not seen in the literature anywhere. (If someone has then please let me know.) It is this strategy of proof that will be useful to me in what follows. The idea is that P is a spectral projection for H if and only if it is a non-degenerate equilibrium for the vector field

$$F(P) = \mathrm{i}[H, P]$$

on the manifold M of orthogonal projections ($P = P^2 = P^\dagger$). (It is a manifold, called Grassmannian, though it has connected components of many different dimensions.) It is simple to check that $F(P)$ is indeed tangent to M at P and that the amount of non-degeneracy of an equilibrium is inversely proportional to the spectral gap g, that is, the shortest distance to the rest of the spectrum. In fact, using the operator norm (induced by the Hilbert norm) on DF, which takes an infinitesimal change in P on M to a change in the tangent vector to M, we have precisely that

$$\|DF^{-1}\| = 1/g.$$

What we are faced with is just the question of the persistence of a non-degenerate equilibrium P_0. If you are a dynamical systems person, then you know that that is easy to answer. By the implicit function theorem, the non-degenerate equilibrium persists under small changes of vector field F, in particular, if we change the Hamiltonian so that

$$F_\varepsilon(P) = \mathrm{i}[H^\varepsilon, P].$$

So we get a continued spectral projection, P_ε for H^ε and, in fact, we can continue it as long as the equilibrium remains non-degenerate, that is, $DF_\varepsilon = \partial F/\partial P$ remains invertible. By the chain rule

$$\frac{\mathrm{d}}{\mathrm{d}\varepsilon} DF_\varepsilon^{-1} = -DF^{-1} \frac{\mathrm{d}}{\mathrm{d}\varepsilon} DF DF^{-1}.$$

Hence,

$$\frac{\mathrm{d}}{\mathrm{d}\varepsilon} \|DF^{-1}\|^{-1} \le 2\|\Delta\|$$

as long as it is defined, and so

$$g_\varepsilon \ge g_0 - 2\varepsilon\|\Delta\|$$

as long as this remains positive, i.e. for $\varepsilon < g_0/2\|\Delta\|$. This ends the proof.

2.5 Obstructions to the theory

You might say 'well that is great, persistence of spectral projections explains those Eggert *et al* results in figure 2.11'. The bi-vibron would be the zero momentum part (because photons contribute negligible momentum compared with their energy) of the second excited spectral projection, and the continuum band would be that for the $P^{(1,1)}$ spectral projection. We see that they continue for some range of ε, namely up to 1.2. So you might say 'that is OK, it just means that we started with a gap larger than 2.4 times the size of the coupling'. But the big problem is that if one works out the Hilbert norm of the coupling operator, it grows like the system size: it is of the order of NE, where E is the energy change to a single unit due to coupling to the rest. This is because the quantum mechanics of a product system is a tensor product not a direct product. So the standard persistence result of the last section gives a result only for ε up to $g_0/(2NE)$, which is useless if we have $N \approx 10^{23}$ molecules in our crystal.

Question 2.6. *Can this persistence of spectral projections result be extended up to* $\varepsilon = O(1)$*?*

Well actually, there are some genuine obstructions to extending the result.

(i) The first is that a typical coupling operator Δ changes all energies by order $N\epsilon$. The easy way to see this is to take the trivial coupling operator

$$\Delta_{rs} = \begin{cases} I & r = s \\ 0 & r \neq s \end{cases}$$

that just adds the identity at each site. That shifts all the energy levels of each unit. Then we add up our sums of energy levels, one per unit, and the whole energies are going to shift by $N\varepsilon$. If all the energies are moving roughly by $N\varepsilon$ then it looks very dubious that we would keep any gaps. It is certainly not going to be an easy argument to show that we keep any gaps for ε greater than $O(1/N)$ because some might move slightly faster with ε than others.

(ii) Another obstruction is that typical coupling turns the ground state through $O(45°)$ as soon as $\varepsilon \sim 1/\sqrt{N}$, and nearly 90° as soon as it becomes significantly bigger than $1/\sqrt{N}$. So this means we cannot expect spectral projections to move uniformly smoothly in the standard Hilbert norm, that is, uniformly in system size, because we go from 0–45° in $O(1/\sqrt{N})$, which is non-uniform in system size. An easy example of this problem is a set of spin-$\frac{1}{2}$ particles in a magnetic field at angle ε, which have a ground state ϕ_ε that is always aligned with the field. So we can work out that the overlap with the unperturbed ground state is

$$\langle \phi_0 | \phi_\varepsilon \rangle = (\cos \varepsilon)^N$$

which tends to zero as soon as ε is significantly bigger than $1/\sqrt{N}$.

(iii) Furthermore, there are examples, admittedly with a degenerate ground state (unlike the molecular crystals I am talking about), for which spectral gaps are indeed lost for $\varepsilon \approx 1/\sqrt{N}$. For example, if you take a half-filled 1D Hubbard model (e.g. [40]), you lose spectral gaps between the ground-state band and the first excited-state band when the coupling ε reaches the Hubbard constant U divided by $\sqrt{N}$; see figure 2.12.

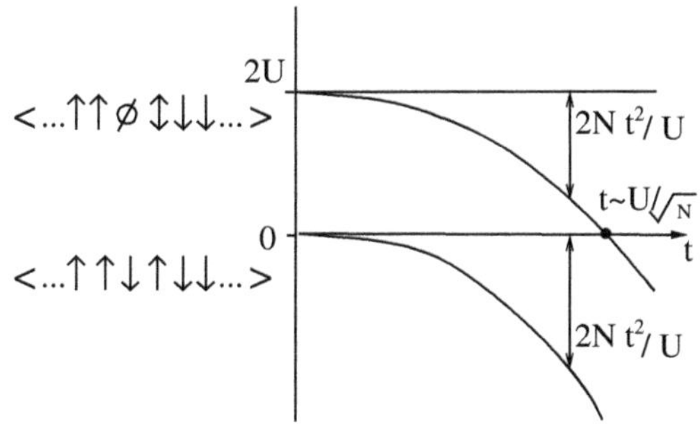

Figure 2.12. Spectrum for the half-filled Hubbard model, indicating typical states in the $t = 0$ limit.

So there are real obstacles to extending the result to one that is uniform in system size. However, there is one result which partially extends it: Kennedy and Tasaki [29] proved persistence of a non-degenerate ground state, with its spectral gap, uniformly in system size. But they did not claim any smoothness for how their ground state moves (fortunately, because it is probably false!). Nor does their method (cluster expansions of $e^{-\beta H}$) look easy to extend to higher spectral projections. If you like, in figure 2.12 the Kennedy and Tasaki result tells you that the ground state persists as a non-degenerate ground state and that the gap between it and the rest of the spectrum remains roughly of the same order. But to understand the Eggert *et al* results and the PtCl work I am interested in higher spectral gaps.

2.6 A proposed solution

My proposed solution is to introduce new norms on tangent vectors to the manifold M of projections on $\otimes_{s\in S}U_s$, instead of using the standard Hilbert norm. We will find we also need to introduce something I have called a *domination condition* on the coupling in order to rule out systems like the half-filled Hubbard model. So here is the idea. I have to call it an 'idea' because I have to still fill in a couple of details in the proof. I am hoping they are just technical details but I never get around to them so I apologise to those who have seen this before and you must think it is about time I finished it.

The idea goes as follows.

- We start by taking the manifold M of orthogonal projections on our tensor product

$$P : \otimes_{s\in S}U_s \to \otimes_{s\in S}U_s$$

and we might as well restrict attention to the component of this manifold composed of projections which have the same rank as the unperturbed case P_0.

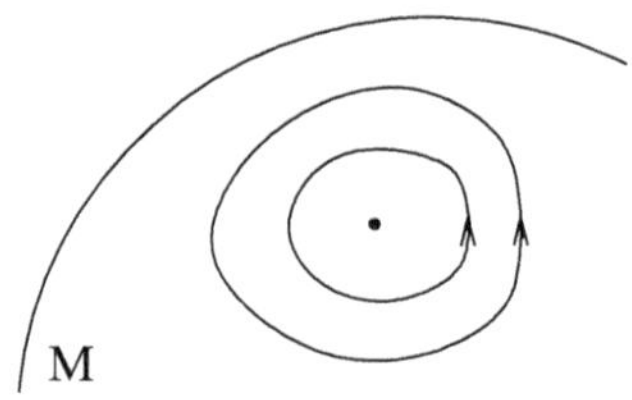

Figure 2.13. The typical dynamics on M.

- As I said earlier, we have a vector field F_H on M
$$\dot{P} = F_H(P) = \mathrm{i}[H, P] \in T_P M.$$
- P is a spectral projection if and only if it is a non-degenerate zero of F_H. I should say that the typical dynamics is that a spectral projection is an equilibrium and nearby projections rotate around it as depicted in figure 2.13. They rotate at angular rates equal to the differences between eigenvalues corresponding to the spectral projection and its complement, as is easy to see by going into an eigenbasis of H.
- We want to get results that are uniform in system size. We are going to need to quantify things so we are going to need find norms (and a condition on Δ) such that this vector field is C^1 in our parameter and state space. That is, we require
$$\left.\begin{array}{l} F \text{ is } C^1 \text{ in } (\varepsilon, P) \\ DF_{H_0, P_0} \text{ is invertible} \end{array}\right\} \text{uniformly in } N.$$
Note that I use the notation DF for the derivative of F with respect to P, and I will use $\partial F/\partial \varepsilon$ for the derivative with respect to ε.
- Then we just invoke the implicit function theorem, with whatever these norms might be, to obtain that P_0 persists uniformly in N with respect to ε.

Now, to quantify F being C^1 we have to come up with

(a) the norms and
(b) a coordinate chart around P_0 in M.

2.6.1 Norms

Choosing new norms is the key idea. This means assigning a length to each tangent vector π to the manifold M of projections on $\otimes_{s\in S} U_s$. For tangent vectors π at $P \in M$, we write $\pi \in T_P M$. First we define, for every subset $\Lambda \subset S$ of our molecular crystal, an operator π^Λ on $\otimes_{s\in\Lambda} U_s$, which I call the *density matrix* of π on Λ. This is a standard thing in physics—one just takes the trace over the complementary sites:
$$\pi^\Lambda = \mathrm{Tr}_{S\setminus\Lambda}\, \pi : \otimes_{s\in\Lambda} U_s \to \otimes_{s\in\Lambda} U_s.$$

The only difference is that usually density matrices are defined only for operators π which are non-negative and have $\operatorname{Tr}\pi = 1$. It turns out that I need to define two norms.

- I define the *0-norm* of a tangent vector π to be the supremum over non-empty choices of the subset Λ of the standard Hilbert norm of the density matrix, divided by the size of the subset

$$|\pi|_0 := \sup_{\phi \neq \Lambda \subset S} \frac{\|\pi^\Lambda\|}{|\Lambda|}.$$

  This is sensible. For example, take the spin-$\frac{1}{2}$ particles in the magnetic field at angle ε, and take π to be the rate of change of the projection onto the ground state with ε, which corresponds to simultaneously rotating the state space for each spin. If one works out the Hilbert norm of the density matrix of π for a subset Λ of spins, it turns out to be precisely equal to the size of Λ. So if we take the ratio, we get a nice supremum of 1. This means that the 0-norm in this example—which is of the size of the rate at which the ground state actually moves—is uniform in the system size. This is very promising.
- The *1-norm* is just the 0-norm of the commutator with the unperturbed Hamiltonian

$$|\pi|_1 := |\mathrm{i}[H_0, \pi]|_0.$$

  One can check that it is a norm on $T_P M$, because it is at least as big as the gap multiplied by the 0-norm. Why do I need the 1-norm? It is because F takes a P and gives us $\mathrm{i}[H, P]$, but the commutator with H is like taking a derivative (it is a derivation, as we saw near the beginning of this chapter). Also, if we think about the proof that Aubry and I did for the classical case [38], then what we had to do was to take a loop, that is a candidate for a periodic orbit, and evaluate the equations of motion on it, which should give zero if it is a periodic orbit. But evaluating equations of motion involves looking at $\dot{x}$, so it means taking a derivative. If our loops are in C^1 then we have to take our tangent loops in C^0, so there is a loss of one derivative. It is exactly the same here, you have to think of a commutator with H like taking a derivative, and so regard F as taking M with the 1-norm to TM with the 0-norm.

2.6.2 Chart

The next thing is that, unfortunately, M is a nonlinear manifold, so if we are going to quantify F being C^1 then we are going to have to do something like use a coordinate chart. There is an alternative, namely to embed M into a linear space by adding the equation $P^2 = P$, but the chart route is the one I will take here. Figure 2.14 shows a chart around P_0 in M, which is constructed as follows.

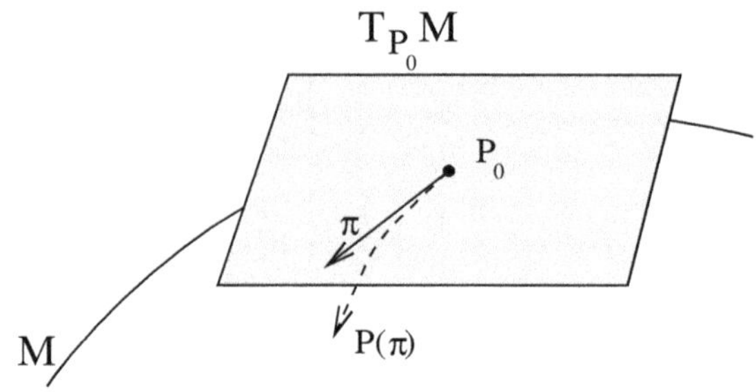

Figure 2.14. The chart.

- First define J to be the difference between P_0 and its complementary projection $Q_0 = I - P_0$,
$$J := Q_0 - P_0.$$
- We note that if π is in $T_{P_0}M$, then $\mathrm{e}^{J\pi}$ is a unitary operator.
- Now we define $\mathcal{B}$ to be a not too large ball with respect to the 1-norm:
$$\mathcal{B} := \{\pi \in T_{P_0}M : |\pi|_1 \leq \delta\}.$$
- Then we introduce coordinates on P within this ball by writing
$$\begin{aligned} P : \mathcal{B} &\to M \\ \pi &\mapsto e^{J\pi} P_0 e^{-J\pi}. \end{aligned}$$

2.6.3 The formulation

Having defined appropriate norms and a coordinate chart, we can now return to a formulation of the continuation problem. If you express the continuation problem in the chart it turns into this: find zeros of

$$G_H : \mathcal{B}_{(1)} \times \mathbb{R} \mapsto T_{P_0}M_{(0)} \tag{2.5}$$
$$(\pi, \varepsilon) \mapsto \mathrm{i}[A_\pi(H^\varepsilon), P_0] \tag{2.6}$$

where

$$A_\pi(H) := \mathrm{e}^{-J\pi} H \mathrm{e}^{J\pi}. \tag{2.7}$$

Here the subscripts (0) and (1) are used to stress that we have to consider the left-hand side as being in the space of the 1-norm and the right-hand side as being in the space of the 0-norm. Now, if you take a tangent vector at P_0 and you work out the right-hand side of (2.6), then to get the uniform persistence of the spectral projection P it will be enough to show that

$$\left.\begin{array}{l} G \text{ is } C^1 \text{ in } (\varepsilon, \pi) \\ DG \text{ is invertible at } (0, 0) \end{array}\right\} \text{uniformly in } N. \tag{2.8}$$

There are several steps to showing this, I checked most of the things—well the easy parts—which can be summarized as follows:

2.6.4 Part of a proof of (2.8)

Here is a formula for DG:

$$DG(\sigma) = \mathrm{i}[[A_\pi(H), B_\pi(\sigma)], P_0]$$

where

$$B_\pi(\sigma) = \int_0^1 A_{t\pi}(J\sigma)\,\mathrm{d}t.$$

Let us see what we need.

- First of all, DG is invertible, and both it and its inverse are bounded, that is, at $\varepsilon = \pi = 0$,

$$DG(\sigma) = \mathrm{i}[[H_0, J\sigma], P_0] = \mathrm{i}[H_0, \sigma]$$

  so $|DG|_{1\to 0} = |DG^{-1}|_{0\to 1} = 1$ where again the subscripts refer to the appropriate norms.
- If we assume that the coupling is uniformly summable:

$$\sup_{r\in S} \sum_{s\in S} \|\Delta_{rs}\| \le K$$

  then the rate of change of G with the coupling parameter is bounded uniformly in system size, at least at the unperturbed projection

$$\frac{\partial G(\pi)}{\partial \varepsilon} = \mathrm{i}[A_\pi(\Delta), P_0], \qquad \text{so} \left|\frac{\partial G(0)}{\partial \varepsilon}\right|_0 \le K \qquad \text{at } \pi = 0.$$

- The next thing is that we have to show that the derivative is continuous, because for the implicit function theorem it is not enough just to show that F is differentiable, we have to show it is C^1. Furthermore, to obtain uniform results we have to show this uniformly in system size, which means showing a uniform module of continuity for the derivative. The easiest way to do that is to bound the second derivative. First consider the mixed second derivative

$$\frac{\partial}{\partial \varepsilon} DG(\sigma) = \mathrm{i}[[A_\pi(\Delta), B_\pi(\sigma)], P_0].$$

  If we assume a *'domination' condition*:

$$|\mathrm{i}[[\Delta, \sigma], P_0]|_0 \le \frac{K}{g} |\mathrm{i}[H_0, \sigma]|_0 \tag{2.9}$$

  where K is as before and g is the unperturbed spectral gap, which I convinced myself once is satisfied for low-lying spectral projections of molecular crystal models with units having non-degenerate groundstate (though I cannot reconstruct the proof now!), then we obtain

$$\left|\frac{\partial}{\partial \varepsilon} DG\right| \le \frac{K}{g} \qquad \text{at } \pi = 0.$$

Interestingly, the domination condition is not satisfied for the ground state of the half-filled Hubbard model, for which we know the result cannot apply, so it is nice to see where it gets excluded. But really I need a bound like this in a uniform neighbourhood, that is an estimate of the form:

$$\left|\frac{\partial DG}{\partial \varepsilon}\right|_0 \leq f(|\pi|_1)|\sigma|_1 \qquad \text{for some function } f. \tag{2.10}$$

I have not yet obtained such an estimate, but I think it is very likely to be possible.

- Similarly, we had better check the second derivative D^2G:

$$\begin{aligned} D^2G(\sigma,\tau) &= \mathrm{i}[[[A_\pi(H), B_\pi(\tau)], B_\pi(\sigma)], P_0] \\ &\quad + \left[\left[A_\pi(H), \int_0^1 [A_{t\pi}(J_\sigma), B_{t\pi}(\tau)]\,\mathrm{d}t\right], P_0\right]. \end{aligned}$$

It is zero at the unperturbed case, that is, at $\pi = \varepsilon = 0$,

$$D^2G(\sigma,\tau) = \mathrm{i}[[LH_0, J_\sigma], J_\tau], P] = 0.$$

But we have to show that it remains uniformly small in some neighbourhood. That is, we would like to obtain

$$|D^2G(\sigma,\tau)|_0 \leq h(|\pi|_1, \varepsilon)|\sigma|_1|\tau|_1 \tag{2.11}$$

for some function h. Again, I have not yet done this.

So if we were to obtain the estimates (2.10) and (2.11) (and check that the domination condition really holds) then the implicit function theorem would give the persistence of a spectral projection from P_0 uniformly in the system size N.

Question 2.7. *Can one prove estimates of the forms (2.10) and (2.11)?*

If so, I could explain why we should expect the Eggert *et al* result to happen up to 1.2 rather than just 1.2×10^{-23}, and similarly why we should expect the Swanson *et al* results for PtCl, as follows.

2.6.5 Application to quantum discrete breathers

Here is the application to quantum discrete breathers (QDBs). I will define a first excited QDB to be any element of the range of the continued first spectral projection, and so on thus:

$$\begin{aligned} \text{first excited QDB} &= \text{elements of Range } P_\varepsilon^{(1)} \\ \text{second excited QDB} &= \text{elements of Range } P_\varepsilon^{(2)} \\ (1,1)\text{ multi QDB} &= \text{elements of Range } P_\varepsilon^{(1,1)} \\ &\text{etc.} \end{aligned}$$

Many might say 'Well OK, that is great, I have got some analogue of the continued periodic orbits that we had in the classical case. But in the classical case we also got that the amplitude decays exponentially in space away from some site, which leads to ask whether we get a localization result for quantum discrete breathers.'

There is a standard dogma in quantum mechanics that says that in a periodic crystal nothing is localized, therefore it is ridiculous even trying to prove a general localization result because it would have to include the periodic case. Well again, that is based on a total misconception. It is true that the eigenfunctions in a periodic crystal are, without loss of generality, Bloch waves, so they are not localized. However, there are very strong localization results, not of wavefunctions about a particular molecule, but a spatial decorrelation for the spectral projections; see, e.g., [24, 27].

Here is a weak localization result for QDBs. One can look at the density matrix for a pair of states $\{r, s\}$

$$P_{\varepsilon}^{(1)\{r,s\}} = \begin{bmatrix} N-2 & 0 & 0 & \mathbf{0} \\ 0 & 1 & 0 & \mathbf{0} \\ 0 & 0 & 1 & \mathbf{0} \\ \mathbf{0} & \mathbf{0} & \mathbf{0} & \mathbf{0} \end{bmatrix} + O(2\varepsilon) \qquad \text{in basis} \quad \begin{matrix} |0_r 0_s\rangle \\ |1_r 0_s\rangle \\ |0_r 1_s\rangle \\ \textbf{the rest}. \end{matrix}$$

This means that the probability to see $\{r, s\}$ in state $|0_r 0_s\rangle$ is approximately $1 - (2/N)$, in state $|1_r 0_s\rangle$ is $\sim 1/N$, in state $|0_r 1_s\rangle$ is $\sim 1/N$, and in state $|1_r 1_s\rangle$ is $\sim \varepsilon/N$.

Hence, the spectral projections indeed have some localization properties. One should also note that the total probability is 1, so the spatial correlation function is summable with respect to distance between sites, which implies a weak form of decay. But one could hope for a stronger result if the coupling decays suitably in space, a feature that we have not used so far.

Question 2.8. *Can we prove that the spatial correlation decays exponentially with the distance between r and s if the coupling Δ_{rs} does?*

An analogue of the method in [5] could perhaps be developed.

2.7 A tentative application to 4-methylpyridine

Before I wrap up I would like to say some final words about 4-methylpyridine.

Question 2.9. *Could quantum discrete breathers explain experimental results on 4-methylpyridine?*

I think so, if for the quantum units one takes coupled pairs of rotors. Independent coupled pairs of rotors have successfully explained experiments on lithium acetate [13], but this was dismissed as a possible explanation for 4-methylpyridine in [17], mainly because the effects in isotopic mixtures (with a

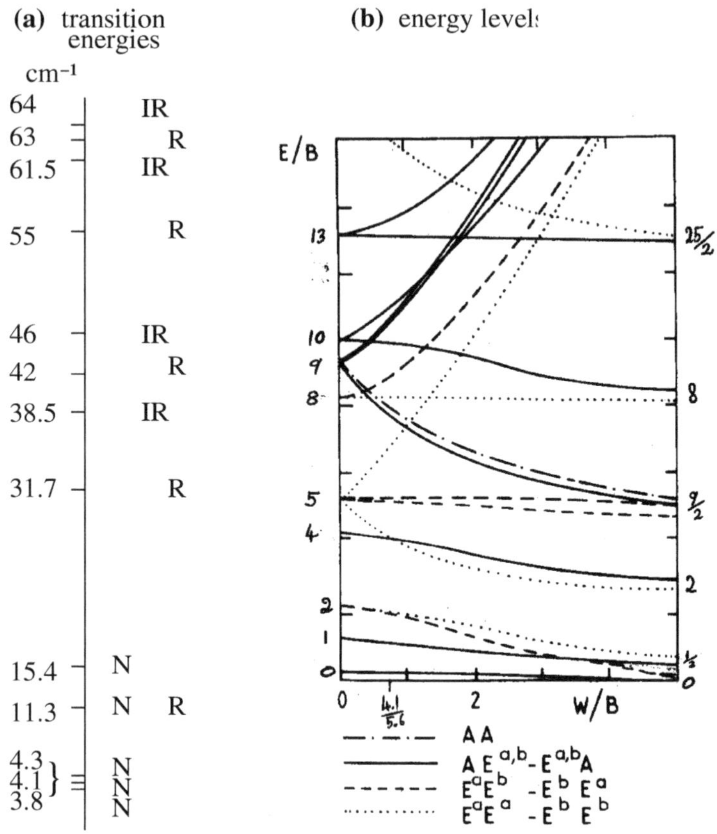

Figure 2.15. Observed transition energies (measured in cm^{-1}) for 4-methylpyridine, mostly at 5 K (*a*), where N stands for neutron scattering (from [1, 17]), R for Raman scattering and IR for infrared absorption (from [32]). Approximate energy levels for a pair of CH_3 rotors coupled by a potential $\frac{1}{2}W(1+\cos 3\delta)$, scaled to the rotation constant B for a single methyl group (*b*) (reproduced from S Clough, A Heidemann, A H Horsewill and M N J Paley 1984 Coupled tunnelling motion of a pair of methyl groups in Lithium acetate studied by inelastic neutron scattering, *Z. Phys.* B **55** 1–6 ©1984 by Springer Verlag). The different line styles correspond to different symmetry types of excitation.

proportion p of CH_3 groups replaced by CD_3) do not fit. However, I think that it is possible that the continuation of the excitations of a single coupled pair of rotors to QDBs, when coupling between neighbouring pairs of rotors is taken into account, could lead to a good explanation. On the other hand, Fillaux recently showed me some very interesting further experimental results on this material that cast doubt on my interpretation, so this section should be taken with a large dose of salt (or better, with a molecular crystal).

Figure 2.15 shows a summary of various observations of transition energies for 4-methylpyridine, by three different methods, mainly at 5 K. Next to this, I show an approximate spectrum for a pair of methyl rotors coupled by a potential $\frac{1}{2}W(1 + \cos 3\delta)$, where δ is the angle between a chosen arm in each rotor (from [13]). One could obtain a much more accurate spectrum by separating the Schrödinger equation into mean and relative coordinates (θ, δ), which reduces to a free rotor in θ and a standard Mathieu problem in δ though one must be careful with the boundary conditions. However, the picture from [13] will do for present purposes. The energy scale B is the rotation constant $\hbar^2/2I \approx 5.6\ \text{cm}^{-1}$, for a single methyl group of moment of inertia I. Calculations of [17] using an empirical proton-proton potential suggest that $W \approx 4.1\ \text{cm}^{-1}$, thus the appropriate value of W/B to look at is 0.73. It looks to me as if the spectrum is in the right ballpark to interpret all the observed transition energies as excitation energies from the ground state. (One could also allow some transitions from the first excited state, however, as in [1, 17], since 5 K corresponds to a thermal energy of 3.3 cm^{-1}, a 4.1 cm^{-1} excited state would have a 30% thermal population. Furthermore, one should pay attention to the selection rules governing which types of transitions between symmetry classes are possible via the three spectroscopic methods.)

To make a more precise comparison with experiments, there are several further steps that would be required. First, one should probably add to the pair of rotors a potential $V(\theta, \delta) = \alpha \sin 6\theta \sin 3\delta + \gamma \cos 12\theta \cos 6\delta$, representing the effects of the pyridine rings. This form is dictated by symmetry considerations, which I think make all lower harmonics cancel. But because we are left with only relatively high harmonics, one can expect the amplitudes to be of relative order e^{-12C} for some order one constant C, by analyticity arguments, and thus they will probably not be very important.

Much more importantly, one should include the effect of coupling between different pairs. This is estimated *ab initio* to have an energy barrier of 6.6 cm^{-1} in [17], so it is highly significant. But I hope that it is not too large compared with anharmonicity (for which a reasonable measure would be $B \approx 5.6\ \text{cm}^{-1}$) so that my continuation theory still applies. (Remember that the theory appears to apply to solid deuterium up to a dimensionless coupling of 1.2.) Thus, one would obtain QDBs which continue each of the low-lying excitations of a single pair of methyl rotors. These would have a significant width in space, however, so their energies would be affected significantly by the choices of isotopes in the four neighbouring methyl pairs. In particular, I suggest that the 'smooth' shift of a transition energy from 516 μeV $= 4.1\ \text{cm}^{-1}$ to 350 μeV observed in neutron scattering in [17,20] as the deuterated proportion p increases from 0 towards 1, is an effect created by populating a spectrum of discrete energies corresponding to $(CH_3)_2$ and CH_3CD_3 rotors in the full range of environments (e.g. the four nearest neighbours are all $(CH_3)_2$ or one or more is a CH_3CD_3 or a $(CD_3)_2$ etc.), weighted by the appropriate binomial coefficients, which of course vary significantly with p.

Finally, the splitting of the 4.1 cm^{-1} transition into three distinct energies (around 3.8, 4.1 and 4.3 cm^{-1}) can be explained by the crystal point group. There are four pairs of rotors per unit cell, and from an examination of the crystal structure given in [12] (which differs little from that of figure 1.4, though in [32] it is mentioned that there might be a phase transition at around 100 K between the two cases), I believe it has D_4 as its point group symmetry, generated by a 1/4 screw rotation and inversion in a point not on the screw rotation axis (not the C_{4h} symmetry claimed in [17]). The three distinct transitions could correspond to the three different symmetry types contained in the relevant representation of D_4, so-called A_g, B_u and E, though a calculation would be required to assign them and to explain the splitting ratio. Similar splitting should presumably occur for every excitation, but perhaps the splitting is less for higher excitations or some of the symmetry types are less visible.

But the whole of my interpretation may be wrong. The preferred interpretation of [17] is in terms of moving breathers in the quantum sine-Gordon model

$$H = \int \mathrm{d}x \, \frac{p^2}{2IL} + \frac{n^2 V_c L}{4} \theta_x^2 - \frac{V_0}{2L} \cos n\theta$$

with commutation relation $[p(x), \theta(x')] = \mathrm{i}\hbar L \delta(x - x')$ (one can scale p by L to obtain the standard commutation relation if one replaces the first term in H by $p^2 L/2I$). This is a continuum model (for $n = 1$, think of a heavy elastic band with one edge fixed along a horizontal wire, though free to rotate around it) for the angles of a 1D chain of rotors of moment of inertia I with $n = 3$ arms in a cosine potential with barrier energy V_0, and coupled to nearest neighbours with lattice spacing L by a cosine force with barrier energy V_c. Classically, it has breather (and kink) solutions, and the model is Lorentz invariant (with respect to the relevant sound speed c, $c^2 = V_c L^2 n^2/2I$) so they can move too. For the quantum version, it is convenient to introduce the rotational constant $B = \hbar^2/2I$, the dimensionless quantum parameter $\beta = \sqrt{2n}(B/V_c)^{1/4}$ and associated quantity $Z = 8\pi/\beta^2 - 1$, and the energy $\lambda = \sqrt{nV_0/2}(V_c B)^{1/4}$. If $Z > 0$ it is proved [34] that the quantum model has elementary excitations analogous to the classical breathers and kinks, but the quantum breathers are restricted to having rest energies $E_k = (2\beta\lambda Z/\pi)\sin(\pi k/2Z)$, for $k = 1, \ldots, N_b$, where N_b is the integer part of Z (and the kink rest energy is $\beta\lambda Z/\pi$). A breather moving with linear momentum P has energy $\sqrt{E_k^2 + c^2 P^2}$. Fillaux and Carlile proposed in [17] that the principal effect of spatial discreteness is to quantize the momentum P to the values j/L, $j \in \mathbb{Z}$ (actually, I think they intended hj/L), and interpreted the transition energy of 4.1 cm^{-1} as corresponding to exciting a pre-existing $k = 1$ breather from $j = 0$ to $j = 1$ (though they interpret the transition energies at 3.8 and 4.3 cm^{-1} as the extremes of the tunnelling band).

I think their interpretation has fundamental problems. First, where does the quantisation rule come from (e.g. electrons in metals are not restricted to such values of momentum)? Second, they make a parameter fit which is a factor seven out from their prior estimates of the interaction energies of methyl groups. Third, they compute breather energies based on an assumption that the quantum parameter $\beta = 3$, but I think this is due to a misunderstanding of the theoretical papers, most of which use scaled versions of the sine-Gordon model like $H = \int \mathrm{d}X\, \frac{1}{2}(P^2 + U_X^2) - \lambda^2 \cos \beta U$ with $[P(X), U(X')] = \mathrm{i}\delta(X - X')$ and which admittedly do not explain how to get to this from physical variables; the fitted parameter values of [17] would give a value of $\beta = 1.46$; the difference is huge because N_b for the former is 1, whereas for the latter it is 10. Fourth, using their parameters, the rest energy of a $k = 1$ breather would be 45.7 cm^{-1}, so its thermal population at 5 K would be only 10^{-6}, so it is surprising that transitions between its momentum states would be observable (in their favour, Fillaux *et al* do observe that the strength of the transition decays on a timescale of about 70 hr after cooling [18]). Fifth, the chains of methyl groups in the a-direction are strongly coupled to those in the b-direction, and thus all together (including in the c-direction), so it is doubtful that a 1D model would suffice, a point acknowledged in [17]. Indeed they would have liked to use a theory that allowed for a 3D network of interactions but none was available at the time: hopefully mine will provide what is required, though I would be the first to admit that it is not yet sufficiently mature to provide quantitative predictions.

Nonetheless, further experimental results that Fillaux showed me just after I presented this material at the workshop in June 2001 lend very strong support to the idea that the 4.1 cm^{-1} transition energy does correspond to setting something in motion along the a- or b-directions, so the problem is still wide open! Perhaps I have to generalize my theory to moving QDB.

Question 2.10. *Can one generalize QDB to mobile QDB?*

This may not be too hard, as the range of the spectral projection $P^{(2)}$, for example, already contains combinations (Bloch waves) that have any desired (pseudo-) momentum, and I guess one gets true momentum by usual group velocity analysis of the dispersion relation.

2.8 Conclusion

Now, I *will* conclude. Modulo checking a couple of technicalities, I think I have a proof of existence (which provides a definition as well) of QDBs in models of molecular crystals, provided the coupling strength is sufficiently less than the anharmonicity. I think this will make a firm foundation for the interpretation of many experiments like the two mentioned in section 2.3, and maybe the methylpyridine that I have just attempted to explain and $KHCO_3$ which I have not started to analyse (Fillaux tells me the interaction between different dimers is very weak, but this would make my theory all the more applicable). Furthermore, I think the mathematical approach will have many other applications to quantum many-body problems. There is a huge number of quantum many-body problems, e.g. spins, bipolarons, excitons, etc, where what theoretical physicists do currently is work out some power series and then chop things off, despite the fact that the terms grow like powers of N. I think the new norms I have introduced will sort out what is going on there and justify the procedures ... or maybe not always justify them? It would be interesting if discrepancies arise!

Acknowledgments

I would like to thank the organizers of the Colston workshop for giving me the opportunity to present my half-baked ideas. I am particularly grateful to Alan Champneys for typing up a first draft of this chapter from the audio-transcript and preparing the figures. I am grateful to Aidan Schofield (Bristol) for filling me in on current thinking in non-commutative geometry, to François Fillaux (Thiais) for many discussions about his experiments, and to Nuno Catarino (Warwick) for calculating the formulae for the quantum sine-Gordon parameters β and λ.

References

[1] Alefeld B, Kollmar A and Dasannacharya B A 1975 The one-dimensional CH_3-quantum rotator in solid 4-methylpyridine studied by inelastic neutron scattering *J. Chem. Phys.* **63** 4415–17

[2] Aubry S 1998 Discrete breathers in anharmonic models with acoustic phonons *Ann. Inst. H. Poincaré Phys. Theor.* **68** 381–420

[3] Arnol'd V I and Avez A 1968 *Ergodic Problems of Classical Mechanics* (Reading, MA: Addison-Wesley)

[4] Arnol'd V I and Ilyashenko Yu S 1988 Ordinary differential equations, chapter 3 *Dynamical Systems* vol I, ed D Anosov and V I Arnol'd (New York: Springer)

[5] Baesens C and MacKay R S 1997 Exponential localization of linear response in networks with exponentially decaying coupling *Nonlinearity* **10** 931–40

[6] Bellissard J 1994 Non-commutative methods in semiclassical analysis *Transitions to Chaos in Classical and Quantum Mechanics (Lecture Notes in Mathematics 1589)* (New York: Springer)

[7] Bellissard J, Bohigas O, Casati G and Shepelyanski D L (ed) 1999 Classical chaos and its quantum manifestations *Physica* D **131** (special issue)

[8] Berry M V 1987 Quantum chaology *Proc. R. Soc.* A **413** 183–98

[9] Baesens C and MacKay R S 1999 Algebraic localization of linear response in a network with algebraically decaying interaction, and application to breathers in dipole–dipole networks *Helv. Phys. Acta* **72** 23–32

[10] Bambusi D 1995 Uniform Nekhoroshev estimates on quantum normal forms *Nonlinearity* **8** 93–105

[11] Bolotin S and MacKay R S 1999 Isochronous potentials, unpublished

[12] Carlile C J, Ibberson R M, Fillaux F and Willis B T M 1990 The crystal structure of 4-methylpyridine at 4.5 K *Z. Kristallog.* **193** 243–50

[13] Clough S, Heidemann A, Horsewill A H and Paley M N J 1984 Coupled tunnelling motion of a pair of methyl groups in Lithium acetate studied by inelastic neutron scattering *Z. Phys.* B **55** 1–6

[14] Connes A 1994 *Non-Commutative Geometry* (London: Academic Press)

[15] Duclos P, Stovicek P and Vittot M 1999 Perturbation of an eigenvalue from a dense point spectrum: a general Floquet Hamiltonian *Ann. Inst. H. Poincaré* **71** 241–301

[16] Eggert J H, Mao H-K and Hemley R J 1993 Observation of a two-vibron bound-to-unbound transition in solid deuterium at high pressure *Phys. Rev. Lett.* **70** 2301–4

[17] Fillaux F and Carlile C J 1990 Inelastic neutron scattering study of methyl tunnelling and the quantum sine-Gordon breather in isotopic mixtures of 4-methylpyridine at low temperature *Phys. Rev.* B **42** 5990–6006

[18] Fillaux F, Carlile C J, Cook J C, Heidemann A, Kearley G J, Ikeda S and Inaba A 1995 Relaxation kinetics of the sine-Gordon breather mode in 4-methylpyridine crystal at low temperature *Physica* B **213&214** 646–8

[19] Fillaux F, Carlile C J and Kearley G J 1991 Inelastic neutron scattering study at low temperature of the quantum sine-Gordon breather in 4-methylpyridine with partially deuterated methyl groups *Phys. Rev.* B **44** 12 280–93

[20] Fillaux F, Carlile C J and Kearley G J 1998 Inelastic-neutron-scattering study of the sine-Gordon breather interactions in isotopic mixtures of 4-methylpyridine *Phys. Rev.* B **58** 11 416–19

[21] Fillaux F, Limage M H and Romain F 2002 Quantum proton transfer and interconversion in the benzoic acid crystal: vibrational spectra, mechanism and theory *Chem. Phys.* **276** 181–210

[22] Gramespacher T and Weigert S 1996 Small denominators, frequency operators, and Lie transformations for nearly integrable quantum spin systems *Phys. Rev.* A **53** 2971–82

[23] Gush H P, Hare W F J, Allin E J and Welsh H L 1957 Double transitions in the infrared spectrum of solid hydrogen *Phys. Rev.* **106** 1101–2

[24] He L and Vanderbilt D 2001 Exponential decay properties of Wannier functions and related quantities *Phys. Rev. Lett.* **86** 5341–4

[25] Huber K P and Herzberg G 1979 *Constants of Diatomic Molecules* (New York: van Nostrand)

[26] Ikeda S and Fillaux F 1999 Incoherent elastic-neutron scattering study of the vibrational dynamics and spin-related symmetry of protons in the $KHCO_3$ crystal *Phys. Rev.* B **59** 4134–45

[27] Ismail-Beigi S and Arias T A 1999 Locality of the density matrix in metals, semiconductors and insulators *Phys. Rev. Lett.* **82** 2127–30

[28] Joye A and Pfister C-E 1999 Exponential estimates in adiabatic quantum evolution *XIIth Int. Congr. Math. Phys. ICMP'97* ed D De Wit *et al* (Somerville, MA: International) pp 309–15

[29] Kennedy T and Tasaki H 1992 Hidden symmetry breaking and the Haldane phase in $S = 1$ quantum spin chains *Commun. Math. Phys.* **147** 431–84

[30] Korepin V E, Bogoliubov N M and Izergin A G 1993 *Quantum Inverse Scattering Method and Correlation Functions* (Cambridge: Cambridge University Press)

[31] van Kranendonk J 1983 *Solid Hydrogen* (New York: Plenum)

[32] Le Calvé N, Pasquier B, Braathen G, Soulard L and Fillaux F 1986 Molecular mechanism determining phase transitions in the 4-methylpyridine crystal *J. Phys.* C **19** 6695–715

[33] Lieb E H, Seiringer R and Yngvason J 2000 Bosons in a trap: a rigorous derivation of the Gross–Pitaevskii energy functional *Phys. Rev.* A **61** 043602

[34] Luther A 1976 Eigenvalue spectrum of interacting massive fermions in one dimension *Phys. Rev.* B **14** 2153–9

[35] MacKay R S 2000 Discrete breathers: classical and quantum *Physica* A **288** 174–98

[36] MacKay R S 2000 Optic discrete breathers in Euclidean invariant systems *Int. J. Nonlin. Sci. Num. Sim.* **1** 99–103

[37] MacKay R S 2000 Quantum breathers *ISIS2000 CLRC Rutherford Appleton Laboratory Experimental Report* RB10545

[38] MacKay R S and Aubry S 1994 Proof of existence of breathers for time-reversible or Hamiltonian networks of weakly coupled oscillators *Nonlinearity* **7** 1623–43

[39] MacKay R S and Sepulchre J-A 1998 Stability of discrete breathers *Physica* D **119** 148–62

[40] Montorsi A 1992 *The Hubbard Model: A Reprint Collection* (Singapore: World Scientific)

[41] Neishtadt A I 1975 Passage through a separatrix in a resonance problem with slowly-varying parameter *J. Appl. Math. Mech.* **39** 594–605

[42] Oppo G-L, Barnett S M, Riis E and Wilkinson M (ed) 1996 *Quantum Dynamics of Simple Systems* (Bristol: IOP Publishing)

[43] Sepulchre J-A and MacKay R S 1997 Localized oscillations in conservative or dissipative networks of weakly coupled autonomous oscillators *Nonlinearity* **10** 679–713

[44] Silvera I F 1980 The solid molecular hydrogens in the condensed phase: fundamentals and static properties *Rev. Mod. Phys.* **52** 393–452

[45] Suchard S N and Melzer J E 1976 *Spectroscopic Data* vol 2 (New York: Plenum)

[46] Swanson B I, Brozik J A, Love S P, Strouse G F, Shreve A P, Bishop A R, Wang W-Z and Salkola M I 1999 Observation of intrinsically localized modes in a discrete low-dimensional material *Phys. Rev. Lett.* **82** 3288–91

[47] Szabo A and Ostlund N S 1983 *Modern Quantum Chemistry: Introduction to Advanced Electronic Structure Theory* (London: Macmillan)

Chapter 3

Unfolding complexity: hereditary dynamical systems—new bifurcation schemes and high dimensional chaos

Uwe an der Heiden
University of Witten/Herdecke

> *The world of difference equations, which has been almost hidden up to now, begins to open in all its richness.*
>
> A N Sharkovsky

In this chapter we discuss some dynamical systems that are a paradigm for complexity. More specifically, we are concerned with systems where the future is not only determined by their present state, but by part of their history. Such systems can be formally described by seemingly simple difference-differential equations. They not only play an important role in applications, e.g those featuring nonlinear delayed feedback, but are also very suitable objects for a numerical and substantial analytical discussion giving insight into complex dynamics. This includes new types of bifurcation patterns, multi-stability of highly structured periodic orbits, and high dimensional strange attractors. The aim of the present paper is two-fold: first, to briefly give the state of the art in the field and problems that have remained unanswered until now and, second, to open the way to the discovery of new kinds of complex dynamics.

3.1 Hereditary systems

Recent decades have shown the discovery of complexity in low-dimensional dynamical systems. Complexity can be seen as phenomena like multiple steady states, limit cycles and extensive bifurcation patterns, including period-doubling bifurcations and deterministic chaos (strange attractors). Well-known examples are the Lorenz attractor, chaos in the discrete logistic equation, the Hénon attractor, the motions of coupled pendula and the Mandelbrot set; see [3, 33, 36] and further references therein.

In spite of the ongoing difficulties in analysing them [36], these systems are 'simple' in the sense that in the real world there are only very few processes that can adequately be described by two, three or four ordinary differential equations, or by one- or two-dimensional maps. For example, the phenomenon of turbulence is generally not yet understood via low-dimensional deterministic chaos and still forms a major challenge to theoretical physicists and mathematicians; see also chapter 7 by Mullin.

It is often believed that the now well-known bifurcation schemes or scenarios, such as pitchfork, saddle-node and Hopf bifurcation, period-doubling bifurcation or bifurcations to chaos from quasi-periodic motion, are generic in the sense that essentially no other bifurcation patterns usually occur. However, we will indicate that this is far from true when one considers *hereditary systems* that depend not only on their present state but also on their history. This opens the way to more possibilities of explaining complex phenomena.

Furthermore, delays appear naturally in many models arising in applications. Examples are lasers subject to optical feedback, as discussed in the chapter by Roy, and many biological systems; see [10] and the chapters by Ermentrout and Bressloff and Cowan. An example from economics is given in [24]; indeed an open-minded person will find delayed interactions in essentially all disciplines and areas.

Systems with delays are very suitable for making steps toward understanding complexity in higher-dimensional systems since, in a certain sense, they lie between low-dimensional ordinary differential equations and systems that must be described by partial differential equations. This may be illustrated by the concrete example of the hyperbolic system

$$i_s + Cv_t = 0 \qquad v_s + Li_s = 0$$

with boundary conditions

$$\begin{aligned} v(0,t) = 0 \qquad & v(s,0) = v_0(s) \\ i(l,t) = g(v(l,t) + E) \qquad & i(s,0) = i_0(s) \end{aligned}$$

modelling an electric medium with a tunnel diode which has a nonlinear voltage–current characteristic g; for details see [33]. This system can be transformed into the seemingly simple single equation

$$x(t) = f(x(t-\tau)) \qquad t \geq 0, \tag{3.1}$$

i.e. a so called *difference equation with continuous argument* or a *functional difference equation* [33]. The positive constant τ denotes the time needed by a signal to travel from one end of the medium to the other.

Equation (3.1) appears to be nearly the same as the *difference equation*

$$x_n = f(x_{n-1}) \qquad n \in \mathbb{N} \tag{3.2}$$

where $f : I \to I$ is a map defined on some interval $I \subset \mathbb{R}$. However, solutions to (3.2) and (3.1) are drastically different. Equation (3.2) may have a strange attractor, the fractal dimension of which, however, is bounded from above by one. The reason is that the state space, i.e. the space of initial conditions, is some subset of $\mathbb{R}$ and, thus, at most one-dimensional. A concrete example is the famous *logistic equation*

$$f(\xi) = \lambda\xi(1 - \xi) \tag{3.3}$$

for a constant parameter $\lambda \in [0, 4]$ with the interval [0,1] as its state space. For $\lambda = 4$ there is a strange attractor with a fractal dimension exactly equal to one; this attractor is dense in the whole state space [0,1].

In contrast, for (3.1) an initial condition is an arbitrary function $\varphi : [-\tau, 0) \to I$. Therefore, the state space is $C^{-1}([-\tau, 0), I)$, i.e. the space of all functions with domain $[-\tau, 0)$ and range I. Taking in (3.1) again the quadratic function (3.3) with $\lambda = 4$ there exists a strange strange attractor, which in fact has dimension infinity!

Before going into details with respect to the aforementioned equations, we address another 'closely related' type of equation that has complicated behaviour: equations (3.1) and (3.2) have been considered as singular perturbation problems in the context of delay differential equations of the type

$$\varepsilon \frac{\mathrm{d}x}{\mathrm{d}t}(t) + x(t) = f(x(t-1)) \tag{3.4}$$

in the formal limit $\varepsilon \to 0$.

Equation (3.4), the so-called *Mackey–Glass equation*, has found many applications in physics, biology and economics; see, e.g., [5, 28]. The reason for the importance of this equation in applications is that it falls under the general scheme where the rate of change $\mathrm{d}x/\mathrm{d}t$ of some time-dependent quantity $x(t)$ is the net effect of two factors: a productive factor p and a destructive factor q [16]:

$$\frac{\mathrm{d}x}{\mathrm{d}t}(t) = p - q.$$

When there is feedback, both p and q may depend on the quantity x itself. Often, the production needs considerable time, e.g., in commodity markets (in particular agricultural ones [24]), in population growth, or in hormonal systems. In such situations the production p may be a functional $p(t) = P(x_t)$ of the history $x_t : (a, 0) \to \mathbb{R}$, $x_t(s) := x(t+s)$ of the variable x, where the constant a is either a negative number or $-\infty$. Similarly one obtains a functional $q = q_t$.

Thus, the most general approach would be to write

$$x(t) = F(x_t).$$

For most applications the dependence on the past may be made explicit by an integral equation of the type

$$x(t) = \int_{-\infty}^{t} f(x(t' - \tau))g(t - t')\,\mathrm{d}t' \tag{3.5}$$

where $t \geq 0$, $f : I \to I$, τ is a constant delay, and $g : [0, \infty) \to [0, \infty)$ denotes a weight function (or more generally a distribution) [13, 16].

Taking for g the δ-distribution, (3.5) becomes (3.1) which thus appears as an extreme case of (3.5).

In applications and for analytical reasons it is useful to consider weight functions $g = g_k$ of the type

$$g_k(t) := \alpha t^{k-1}\mathrm{e}^{-\alpha t}/(k-1)!$$

for $k \in \mathbb{N}$. With such a weight function (3.5) can be transformed into a system of differential equations. Define

$$\begin{aligned} x_i(t) &:= \int_{-\infty}^{t} f(x(t' - \tau))g_i(t - t')\,\mathrm{d}t' \qquad \text{for } i = 1, 2, \ldots, k, \\ x_0(t) &:= \alpha^{-1} f(x(t - \tau)). \end{aligned} \tag{3.6}$$

Then, because of

$$\mathrm{d}x_i(t)/\mathrm{d}t = \alpha x_{i-1}(t) - \alpha x_i(t) \qquad \text{for } i = 1, 2, \ldots, k$$

we arrive at the system

$$\begin{aligned} \mathrm{d}x_1(t)/\mathrm{d}t &= f(x_k(t - \tau)) - \alpha x_1(t) \\ \mathrm{d}x_i(t)/\mathrm{d}t &= \alpha x_{i-1}(t) - \alpha x_i(t) \qquad \text{for } i = 2, 3, \ldots \end{aligned} \tag{3.7}$$

Since $g = g_k$ (3.5) together with (3.6) implies $x_k = x$ and, thus, x obeys system (3.7). Vice versa, (3.5) can be retrieved from (3.7) by successively integrating the ith equation, $i = k, k-1, \ldots, 1$.

We note that system (3.7) represents a feedback model for the regulation of protein synthesis introduced by [6]. In this context, system (3.7) has been studied intensively in recent years; for a review in the case $\tau = 0$ see [30]. In particular, it has been proved that in the case of negative feedback, i.e. if the feedback function f is monotone decreasing, system (3.7) has non-constant periodic solutions provided that f is bounded and differentiable and the equilibrium is unstable. This result can be generalized to arbitrary $\tau \geq 0$; for the case $k = 1$ see [7], for the case $k = 2$ see [9], and for the case of arbitrary $k \in \mathbb{N}$ see [8, 25].

By forming the kth derivative of $x_k(t)$ and using (3.6), one can show that system (3.7), and thus also (3.5) with $g = g_k$, is equivalent to the kth order delay differential equation

$$\sum_{i=0}^{k} \binom{k}{i} \alpha^i x^{(k-i)}(t) = \alpha^k f(x(t - \tau)) \tag{3.8}$$

where $x^{(i)}$ denotes the ith derivative of x.

Equation (3.8) is a special case of the very interesting class of kth order delay-differential equations [12] of the form

$$\sum_{i=0}^{k} a_i d^i x(t)/\mathrm{d}t^i = f(x(t-\tau)) \qquad a_i \in \mathbb{R}. \tag{3.9}$$

Note that (3.1) is obtained from (3.9) by choosing $k = 0$, and (3.4) by choosing $k = 1$. Formally (3.9) is in the 'vicinity' of (3.1), and thus also indirectly connected to (3.2), if $a_i = \varepsilon_i$ are small numbers for $i = 1, 2, \ldots, k$; see [8,13]. In particular, the kth-order equation

$$\varepsilon\, \mathrm{d}^k x(t)/\mathrm{d}t^k + x(t) = f(x(t-\tau)) \tag{3.10}$$

formally approaches (3.1) as $\varepsilon \to 0$.

There is a striking contrast between equations (3.2) and (3.1) on the one hand, and equations (3.4) and (3.10) on the other, concerning what we know about the complex behaviour of solutions and bifurcation patterns. To cite a statement of A N Sharkovsky from 1986: 'In spite of the apparent simplicity of (3.4), the investigation of it is not an easy task'. For any sufficiently small ε, this equation can no longer possess solutions of the turbulent type, because

$$\varepsilon|\mathrm{d}^k x(t)/\mathrm{d}t^k| \leq \frac{1}{\varepsilon}|-x(t) + f(x(t-\tau))|.$$

Thus, we arrive here at the principal question that still has no answer: What happens with these solutions when $\varepsilon > 0$ and $t \to \infty$? The remark by Sharkovsky [33] that 'unfortunately our understanding of this process leaves much to be desired' still holds even in view of the recent great progress in the analysis of equations like (3.4); compare [4, 19, 20, 38, 39]. The detailed knowledge of difference equations cannot simply be extended by continuity arguments to differential equations, in spite of the formal limit transition $\varepsilon \to 0$ between these two classes of equations. This impossibility has become evident from the work of Mallet-Paret, Nussbaum, and the Russian group around Sharkovsky, who showed that there is a 'bifurcation gap' between the two classes and that they differ drastically in the asymptotic behaviour of solutions; see [18, 27, 29]. An important reason for the mathematical difficulties is the tremendous difference in the dynamics between (3.1) and (3.2), as we have already mentioned in the beginning and will describe in more detail later on.

In the following sections the reader is invited to travel along a pathway of increasing rank and complexity of equations, like wandering through a landscape of different levels with different perspectives and different scenarios. By pointing to areas where until now there has been little insight, we would like to stimulate the reader's interest in the field of hereditary systems.

3.2 Difference equations with continuous argument: idealized turbulence

In this section we briefly recall recent results by Sharkovsky and co-workers concerning equations of type (3.1): the so-called *difference equations with continuous argument* [33, 35]. The character of their solutions is very different from that of the difference equation (3.2), in spite of the strict relationship

$$x(t + n\tau) = f^n(x(t)) \qquad \text{for all } n \in \mathbb{N}$$

which means that for each fixed $t \in [-\tau, \infty)$ the sequence $(x(t + n\tau))_{n\in\mathbb{N}}$ is a solution of (3.2).

A striking difference can already be seen with the simple example of the sigmoid function $f : \mathbb{R} \to \mathbb{R}$, $f(\xi) = \arctan(\alpha\xi)$ with a constant $\alpha > 1$. With this nonlinearity, (3.2) has two attracting fixed points (say $\bar{a}$ and $\underline{a}$) and no periodic solutions, whereas (3.1) has infinitely many periodic solutions (though it also has two attracting constant solutions). Namely, e.g., for each $m \in \mathbb{N}$, m even, the initial condition $\varphi_m : [-1, 0] \to \mathbb{R}$ (we assume $\tau = 1$ here!) gives rise to a periodic solution x_{φ_m} with minimal period $2/m$ if φ_m is given as follows:

For $0 \le i \le m - 1$ and $t \in [-1 + \frac{i}{m}, -1 + \frac{i+1}{m})$ define

$$\varphi(t) = \begin{cases} \bar{a} & \text{if } i \text{ is even} \\ \underline{a} & \text{if } i \text{ is odd.} \end{cases}$$

Of course, these periodic solutions are discontinuous which seems to be a consequence of the fact that equation-type (3.1) is still, in some sense, near to equation-type (3.2).

In general, and following the classification in Sharkovsky *et al* [33], (3.1) has two main types of solution:

(a) *relaxation type*: smooth, bounded solutions, converging, as $t \to \infty$, to discontinuous periodic solutions with finitely many discontinuity points per period.

(b) *turbulent type*: smooth, bounded solutions, converging, as $t \to \infty$, to 'limiting generalized solutions' having infinitely many discontinuities per unit time interval. The frequency of the oscillations on the time interval $[n\tau, (n + 1)\tau]$ increases towards infinity as $n \to \infty$. (One may observe something of this sort at a smoothly shaped ocean beach when a smooth wave is breaking.)

Which type of solution occurs, with given nonlinearity $f : I \to J$, depends on the so-called *separator set* $\mathrm{D}(f) := \{\xi \in I : f^i(\xi), i = 1, 2, \ldots,$ is an unstable trajectory of (3.2)$\}$. Note that the closure of $\mathrm{D}(f)$ is the *Julia set* of f. A solution of (3.1) corresponding to a continuous initial condition $\varphi : [-\tau, 0) \to I$ is of relaxation type if $T := \varphi^{-1}(\mathrm{D}(f))$ is finite, and it is of turbulent type if T is infinite.

Following [33], one can distinguish between three subclasses of (b):

(b1) the *pre-turbulent type* where T is countable,
(b2) the *turbulent type* where T is uncountable but nowhere dense in I (that is, T is a Cantor set) and
(b3) the *strong turbulent type* where T contains cyclic intervals with absolutely continuous invariant measures with respect to the map f.

In the pre-turbulent case (b1) the number of oscillations (alternating increase and decrease of $x(t)$) on the time interval $[n\tau, (n+1)\tau]$ increases according to a power law as $n \to \infty$. However in the turbulent cases (b2) and (b3) it increases exponentially. Moreover, on the shift sets $nT = \{nt : t \in T\}$ the slopes of the solutions tend to infinity (in modulus).

Thus, the classification into (a), (b1), (b2), and (b3) is an indication of the enormous richness of the solution structures inherent in the deceptively simple equation (3.1); for more details see [34,35]. Of course, such an equation is still an extreme caricature of real processes. However, the fact that its attractors may be of arbitrary high dimension makes them more suitable for discussing the inherent properties of turbulence than the well-known low-dimensional strange attractors. It remains a challenge to study these phenomena in detail [35].

There are two intimately related unrealistic features in (3.1). First, as $t \to \infty$, the solutions become arbitrarily steep on arbitrarily small intervals, such that, asymptotically, the solutions become discontinuous. Second, there is no dissipative or friction term in (3.1). Both of these deficits are overcome (at least theoretically) by introducing the friction term $\varepsilon \frac{dx}{dt}(t)$, thus arriving at (3.4).

3.3 First-order difference-differential equations: a singular perturbation problem with bifurcation gaps

There are some results which support the belief that for small ε the solutions of (3.4) should be very close to the solutions of (3.1). For example, if $\bar{x}$ is an attracting or repelling fixed point of f, then $x(t) = \bar{x}$ is an attracting or repelling constant solution of both (3.1) and (3.4), respectively. The following result states that solutions to (3.1) and (3.4) can stay arbitrarily close together for arbitrarily long, but finite times, provided ε is small enough.

Proposition 3.1. (Continuous dependence of solutions on the parameter ε for finite time intervals; [18]) *Let $f : I \to I$ be continuous on the closed interval I and let $\varphi : [-\tau, 0] \to \mathbb{R}$ be continuous. Then for each $T > 0$ and for each $\delta > 0$ there is a positive number $\varepsilon^* = \varepsilon^*(T, \varphi, \delta)$ such that the solution x_φ^0 of (3.1) and the solution x_φ^ε of (3.4) corresponding to the initial condition φ obey*

$$\|x_\varphi^\varepsilon(t) - x_\varphi^0(t)\| < \delta \qquad \textit{for all } t < T \textit{ whenever } \varepsilon < \varepsilon^*.$$

We note that this proposition can be generalized to piecewise continuous f if the number of jumps is finite. However, proposition 3.1 cannot be generalized to hold for $T = \infty$. Although solutions to (3.4) and (3.1) with the same initial condition may stay close to each other for a very long but finite time, asymptotically they can differ substantially. For example, one may converge to a constant, while the other converges to a non-constant periodic solution with an amplitude independent of ε.

Solutions to (3.10) can be both simpler and more complicated compared with those of (3.1) and also to those of (3.2) as will turn out in the following discussion.

Example 3.2. Let the function $f : \mathbb{R} \to \mathbb{R}$ be defined by

$$f(\xi) = \begin{cases} a & \text{if } \xi < \Theta \\ b & \text{if } \xi > \Theta \end{cases} \tag{3.11}$$

where $a, b, \Theta \in \mathbb{R}$ are constant parameters and $a \neq b$.

Note that non-smooth nonlinearities like those in this example occur in dynamical systems where one of the variables can only attain a finite number of discrete values. Examples are heating or cooling machines which are either 'on' or 'off', electric circuits with relays which are 'open' or 'closed' or neurons which are either 'fire' or are 'silent'. In fact, in some types of *artificial neural networks* equation (3.4) with f given by (3.11) is used to model neurons [31]. Another reason for studying systems with non-smooth nonlinearities is that they often allow for much more and much easier mathematical analysis and provable results than is the case for smooth nonlinearities. Often solutions can be explicitly and exactly calculated as compositions of piecewise solutions of linear systems.

Proposition 3.3. (Infinitely many unstable high-frequency periodic solutions; [1, 18]) *Consider (3.4) with $\varepsilon > 0$ and f given by (3.11) with $a \neq b$. Without loss of generality assume $\Theta = 0$. Then (3.4) has infinitely many periodic solutions with pairwise different minimal periods. The countable set of periods can be written as a sequence converging to zero. In the case $a < 0 < b$ of positive feedback all of these periodic solutions are unstable. In the case $b < 0 < a$ of negative feedback the periodic orbit with the largest minimal period is asymptotically orbitally stable, while all other periodic orbits are unstable.*

A proof of this proposition 3.3 was given in [1]; see also [4, 18]. In [1] it was not only shown that most of the periodic solutions are unstable but also that the asymptotic behaviour of almost all solutions is very simple:

Proposition 3.4. [1, 18] *Let the assumptions of proposition 3.3 be satisfied.*

(i) Let $a < 0 < b$. Then almost all solutions of (3.4) satisfy $\lim_{t\to\infty} x(t) = a$ or $\lim_{t\to\infty} x(t) = b$.

(ii) Let $b < 0 < a$. Then (3.4) has an asymptotically stable periodic solution (with period $> 2\tau$) and almost all solutions converge towards this periodic solution (in the sense of orbital convergence).

Here 'almost all' means that the corresponding set of initial conditions is open and dense in the state space $C([-1, 0], \mathbb{R})$. A result similar to proposition 3.4 has been obtained in [38] for the equation

$$\mathrm{d}x(t)/\mathrm{d}t = f(x(t-1)) \tag{3.12}$$

for a rather general class of *continuous* functions f satisfying the condition $xf(x) < 0$ if $x \neq 0$.

In contrast to proposition 3.4, which states that (3.4) possesses at most two stable orbits, (3.1) with the nonlinearity f satisfying (3.11) has infinitely many asymptotically stable periodic solutions with pairwise different minimal periods. Note, however, that the corresponding stable orbits lie in $\mathrm{PC}([-\tau, 0), I)$, the extended state space of piecewise constant functions $\varphi : [-\tau, 0) \to I$.

3.3.1 Mixed feedback and chaos

In the case of monotone feedback functions f the phenomenon of deterministic chaos seems to be excluded. This is apparent for difference equations like (3.2) and has been shown for first-order difference-differential equations in [26]. There is a long history of investigations concerning the chaotic behaviour of solutions of difference equations (3.2) if f is non-monotone, e.g., the discrete logistic equation (3.3). If one tries to prove the existence of chaos for systems of ordinary differential equation or for difference-differential equations like (3.4), then results are not easily obtained. One of the earliest successes was presented in [16, 17, 37] under the simplifying assumption that f is piecewise constant, or smooth but close to a piecewise constant function. Recently the existence of chaos was also proved for smooth nonlinearities [19, 20], for equations of type (3.12). Here we only summarize some results with respect to non-smooth functions f defined as follows:

$$f(\xi) = \begin{cases} 0 & \text{if } \xi < 1 \\ c & \text{if } 1 < \xi < \Theta \\ d & \text{if } \xi > \Theta \end{cases} \tag{3.13}$$

where the constants are assumed to obey

$$c > 0 \qquad \Theta > 1 \qquad d < c. \tag{3.14}$$

We introduce the following notion of chaos which is adapted from the definition in [22] for difference equations.

Definition 3.5. A difference-differential equation of type (3.9) is called *chaotic in the sense of Li and Yorke* [22] if

(i) there are countably many periodic solutions with pairwise different minimal periods; and
(ii) there is an uncountable set S of aperiodic solutions such that
(a) if x is a periodic solution and $\tilde{x} \in S$ then $\limsup_{t\to\infty} \|x_t - \tilde{x}_t\| > 0$
(b) if $x, \tilde{x} \in S$ and $x \neq \tilde{x}$ then $\liminf_{t\to\infty} \|x_t - \tilde{x}_t\| = 0$ and $\limsup_{t\to\infty} \|x_t - \tilde{x}_t\| > 0$.

Here, as usual, $x_t : [-\tau, 0] \to \mathbb{R}$ is the shift function defined by $x_t(s) = x(t+s)$, and $\|x_t\| := \sup\{|x_t(s)| : s \in [-\tau, 0]\}$.

The following theorem has been proved in [16] after the earlier work in [17] where a more complicated f (with three discontinuities) had been used.

Theorem 3.6. [16] *Let the function f be defined by (3.13) with the parameters c, Θ and d obeying (3.14). Assume, moreover, that ε and c satisfy*

$$c/(c-1)^2 + z < 1$$

where z is the positive root of the quadratic

$$z^2 - (c - r - c^2)z - cr = 0$$

with $r := (c-1)\exp(-1/\varepsilon)$.

Then there are positive numbers $\mu = \mu(c, \varepsilon)$ and $d^* = d^*(c, \varepsilon)$ such that (3.4) is chaotic in the sense of Li and Yorke whenever Θ and d satisfy

$$(c-z)/(c-1) < \Theta < (c-z)/(c+1) + \mu \qquad \text{and} \qquad d \leq d^*.$$

It is worth noting that with f given by (3.13), neither (3.2), nor (3.1) (which is the limiting case of (3.4) for $\varepsilon = 0$), exhibits this kind of chaos. In contrast, in (3.4) this chaos may occur for arbitrarily small $\varepsilon > 0$. For further results about chaos in first-order differential delay equations see [11, 18–21].

3.4 Prime number dynamics of a retarded difference equation

If one tries to solve the difference equation with continuous argument (3.1) with the help of a digital computer, the simplest way would be to discretize time t into discrete values $t_n = nh$, $n \in \mathbb{N}$ with some fixed positive step size h. If one takes $h = \tau/k$ then the resulting discretization of (3.1) is (see [13]):

$$x_n := x(t_n) = f(x(t_n - \tau)) = f(x(nh - kh) = f(x((n-k)h)) = f(x_{n-k})$$

leading to the difference equation

$$x_n = f(x_{n-k}). \tag{3.15}$$

Note that the discrete approximation (3.15) is independent of the value of τ, corresponding to the fact that without loss of generality in (3.1) one can assume $\tau = 1$.

When $k > 1$ (3.15) is called a *retarded difference equation*. Of course, with $k = 1$ we arrive again at (3.2) which, viewed this way, is a crude discretization of (3.1).

While (3.15) lies somehow in between (3.2) and (3.1), it is of interest to consider the relation of (3.15) to equations of the form of (3.1), (3.2) and (3.4). Before we come back to this question we show first that the structure of the solutions of (3.15) may be considerably more complex in the case $k > 1$ than in the case $k = 1$.

The following result by A N Sharkovsky, which includes the famous period-doubling bifurcation and the statement 'period three implies chaos', is one of the deepest concerning the case $k = 1$.

Theorem 3.7. [32] *Let $f : I \to \mathbb{R}$ be a continuous function defined on some interval $I \subseteq \mathbb{R}$. Let the set $\mathbb{N}$ of natural numbers be ordered in the following way*

$$3 \rhd 5 \rhd \cdots \rhd 2 \cdot 3 \rhd 2 \cdot 5 \rhd \cdots \rhd 2^2 \cdot 3 \rhd 2^2 \cdot 5 \rhd \cdots \rhd 2^2 \rhd 2 \rhd 1. \qquad (3.16)$$

If the equation $x_n = f(x_{n-1})$ has a periodic orbit of period p then it also has periodic orbits with period p' for all p' positioned to the right of p in the ordering (3.16).

In the ordering (3.16), generally referred to as the *Sharkovsky ordering*, each natural number appears exactly once. Since the number three is the left-most number, theorem 3.7 implies that if there is a periodic solution of period 3, then there is a periodic solution of period p for every $p \in \mathbb{N}$; cf [22].

We generalized theorem 3.7 to cover (3.15) for all $k \in \mathbb{N}$ [15], and this will be discussed now. For arbitrary $k \in \mathbb{N}$ the state space of (3.15) is the k-dimensional cube I^k (recall that $f : I \to I$) and an initial condition is a vector $x^{(0)} = (x_{-k}, x_{-k+1}, \ldots, x_{-1}) \in I^k$. A *solution* of (3.15) corresponding to the initial condition $x^{(0)}$ is a sequence (x_n) with $n \in \mathbb{N} \cup \{-k+1, \ldots, 0\}$, satisfying (3.15) for all $n \in \mathbb{N}$ and obeying $(x_{-k+1}, \ldots, x_0) = x^0$. A solution (x_n) of (3.15) is called *periodic* with period $p \in \mathbb{N}$ if

$$x_{n+p} = x_n \qquad \text{for all } n \in \mathbb{N}.$$

Theorem 3.8. [15] *Let $f : I \to \mathbb{R}$ be a continuous function defined on some interval $I \subset \mathbb{R}$. Let $k \in \mathbb{N}$. If the difference equation*

$$x_n = f(x_{n-1})$$

has a periodic orbit of minimal period p then the difference equation

$$x_n = f(x_{n-k}) \qquad (3.17)$$

has periodic orbits with minimal period p' for all numbers $p' \in S_k(m)$ whenever $S_k(m)$ is either equal to $S_k(p)$ or to the right of $S_k(p)$ in the Sharkovsky ordering of order k *defined by*

$$S_k(3) \rhd S_k(5) \rhd \cdots \rhd S_k(2 \cdot 3) \rhd S_k(2 \cdot 5) \rhd \cdots \rhd S_k(2^2 \cdot 3) \rhd S_k(2^2 \cdot 5) \rhd \\ \rhd \cdots \rhd S_k(2^2) \rhd S_k(2) \rhd S_k(1).$$

Here $S_k(p)$ denotes the set given by

$$S_k(p) := \begin{cases} \{1\} & \text{for } p = 1 \\ \{l \cdot p \mid l \in \mathbb{N},\ l \text{ divides } k \text{ and } (\frac{k}{l}, p) \text{ coprime}\} & \text{for } p \in \mathbb{N} \backslash \{1\}. \end{cases}$$

If f has more than one fixed point then (3.17) also has periodic orbits with minimal period p' for all $p' \in S_k(*) := \{l \mid 2 \leq l \leq k, l \text{ divides } k\}$.

Recall that a pair (m, n) of natural numbers is called coprime if 1 is the only common divisor of m and n. For illustration of this theorem we give the following example.

Example 3.9. Let $k \in \mathbb{N}$ be a prime number. Then l divides k if and only if $l = 1$ or $l = k$; moreover $(\frac{k}{l}, p)$ is coprime if and only if p is not a multiple of k. Thus for $p \in \mathbb{N}$, $p > 1$ we have

$$S_k(p) = \begin{cases} \{kp\} & \text{if } p \text{ is a multiple of } k \\ \{p, kp\} & \text{if } p \text{ is not a multiple of } k. \end{cases}$$

As an example, the Sharkovsky sequence of order $k = 11$ is

$$\{3, 33\} \rhd \{5, 55\} \rhd \cdots \rhd \{9, 99\} \rhd \{111\} \rhd \{13, 143\} \rhd \cdots \rhd \{2 \cdot 3, 2 \cdot 3 \cdot 11\} \\ \rhd \cdots \rhd \{4, 44\} \rhd \{2, 22\} \rhd \{1, 11\}.$$

It follows from this ordering that if $x_n = f(x_{n-1})$ has, say, a 4-periodic orbit, then $x_n = f(x_{n-11})$ has periodic orbits with minimal periods 1, 11, 2, 22, 4, and 44.

The following theorem tells us something about the number of periodic orbits of (3.17).

Theorem 3.10. **[15]** *Let $M, N \subset \mathbb{R}$, $f : M \to N$, and let $S = \{s_1, \ldots, s_p\}$ be a periodic orbit of minimal period p of the difference equation $x_n = f(x_{n-1})$. Let $k \in \mathbb{N} \backslash \{1\}$. Then the number $\mathcal{N}(p, k)$ of pairwise different periodic orbits (x_n) of the difference equation*

$$x_n = f(x_{n-k})$$

which obey $x_n \in S$ for all $n \in \mathbb{N}$ is given by

$$\mathcal{N}(p,k) = \frac{1}{p} \sum_{i \in A_k^{(p)}} \frac{p^i}{i} \Upsilon(\frac{k}{i})$$

where $A_k^{(p)} := \{i \in \mathbb{N} \mid i \text{ divides } k, \text{ and } (k/i, p) \text{ coprime}\}$ *and*

$$\Upsilon(m) := \begin{cases} 1 & \textit{if } m = 1 \\ \prod_{\iota=1}^{\kappa} \frac{m_\iota - 1}{m_\iota} & \textit{if } m \in \mathbb{N}\backslash\{1\}, \textit{ and } \{m_1, \ldots, m_\kappa\} \textit{ is the set of} \\ & \qquad \textit{pairwise different prime factors of } m. \end{cases}$$

Example 3.11. Let $k = 11$ and $x_n = f(x_{n-1})$ have a 4-periodic orbit $\{s_1, s_2, s_3, s_4\}$. Then $A_k^{(p)} = A_{11}^{(4)} = \{i \in \mathbb{N} : i$ divides 11, and $(\frac{11}{i}, 4)$ coprime$\} = \{1, 11\}$. Thus

$$\begin{aligned} \mathcal{N}(p,k) = \mathcal{N}(4,11) &= \frac{1}{4}(\Upsilon(11) + \frac{4^{11}}{11}\Upsilon(1)) = \frac{10}{11} + \frac{4^{10}}{11} \\ &= 95\,326 = (\text{number of periodic orbits of } x_n = f(x_{n-11}) \\ &\qquad \text{with } x_n \in \{s_1, s_2, s_3, s_4\} \text{ for all } n \in \mathbb{N}). \end{aligned}$$

Some further interesting applications to the theory of prime numbers can be found in [23].

Proposition 3.12. *Let the equation $x_n = f(x_{n-1})$ have a strange attractor with Hausdorff dimension H. Then for each $k \in \mathbb{N}$ the equation $x_n = f(x_{n-k})$ has a strange attractor with Hausdorff dimension $k \cdot H$.*

Proof. Let $A \subset \mathbb{R}$ be a strange attractor of $x_n = f(x_{n-1})$. Then $A^k \subset \mathbb{R}^k$ is a strange attractor of $x_n = f(x_{n-k})$. □

3.5 Second-order non-smooth difference-differential equations

We now turn to the case $k = 2$ of (3.10):

$$\frac{\mathrm{d}^2 x(t)}{\mathrm{d}t^2} = f(x(t-\tau)) - \alpha x(t) \tag{3.18}$$

where $\alpha > 0$.

Without loss of generality we assume $\alpha = 1$. Then (3.18) can be transformed into the system

$$\begin{aligned}\frac{\mathrm{d}x}{\mathrm{d}t}(t) &= y(t)\\ \frac{\mathrm{d}y}{\mathrm{d}t}(t) &= f(x(t-\tau)) - x(t).\end{aligned} \tag{3.19}$$

Let us first consider the situation of non-smooth feedback with f defined by (3.11). Without loss of generality we take $a = \frac{1}{2}$ and $b = -\frac{1}{2}$, thus, there remain just the two parameters τ and Θ. An initial condition of (3.19) is a pair $(\varphi, y_0) \in C^1([-\tau, 0], \mathbb{R}) \times \mathbb{R}$ such that $\varphi'(0) = y_0$. Solutions $(x(t)), y(t)), t \geq 0$, of (3.19) can be represented as continuous trajectories $t \to (x(t), y(t))$ in the (x, y)-plane, that is, in $\mathbb{R}^2$. If f is of type (3.11) and if, moreover, the set $\{t : \varphi(t) = \Theta\}$ is finite then these trajectories are composed piecewise of arcs (or sectors) of circles having their centres at either $(a, 0)$ or $(b, 0)$. For example, the centre is $(a, 0)$ for all $t \in [t_1, t_2]$ if for all $t \in (t_1 - \tau, t_2 - \tau)$ the inequality $x(t) < \Theta$ is satisfied. Note that the angular length of the arc associated with $[t_1, t_2]$ is just $t_2 - t_1$, because the angular velocity of the trajectory point $(x(t), y(t))$ is always one, independent of the varying radii of the arcs which make up the trajectory. For the situation of negative feedback the following theorem, whose proof can be found in [2], shows that for $k = 2$ in (3.10) there are substantially more periodic solutions than in the case $k = 1$ in (3.10).

Theorem 3.13. **[2]** *Let f be given by (3.11) and let $a = \frac{1}{2}$ and $b = -\frac{1}{2}$ (which can be assumed without loss of generality whenever $b < a$). Let $\Theta \in [0, \frac{1}{2}]$. Then*

(i) for each $n \in \mathbb{N}$ and for each $\tau \in (0, 2n\pi)$ system (3.19) (and thus also (3.18)) has a periodic solution with minimal period τ/n; and
(ii) for each $n \in \mathbb{N}$, n odd, and for each $\tau \in (n\pi, 2n\pi)$ system (3.19) (and thus also (3.18)) has periodic solutions with minimal period $2\tau/n$.

Figure 3.1 illustrates the bifurcation diagram for these periodic solutions [2]. By distinguishing between stable and unstable orbits, figure 3.1 also illustrates that, contrary to the first-order case ($k = 1$) in (3.10), there may coexist more than one asymptotically orbitally stable periodic orbit for fixed values of the parameters τ and Θ.

3.5.1 Chaos

We still consider (3.19), however now with a mixed feedback nonlinearity given by

$$f(\xi) = \begin{cases} a & \text{if } \xi < \Theta_1 \\ b & \text{if } \Theta_1 < \xi < \Theta_2 \\ c & \text{if } \xi > \Theta_2 \end{cases} \tag{3.20}$$

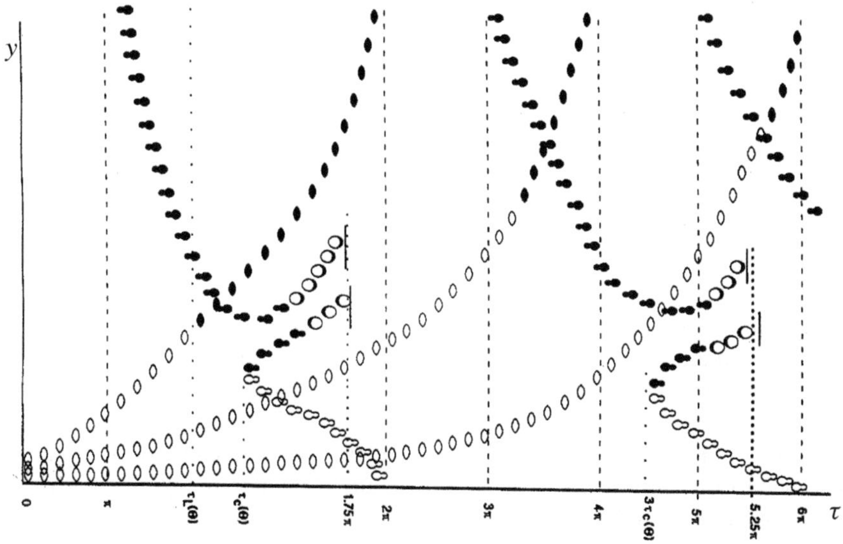

Figure 3.1. Bifurcation diagram for the second-order differential equation system (3.19), with assumptions on f and parameters as in theorem 3.13. The bifurcation parameter is τ (plotted horizontally), while y measures the diameter of the periodic orbits which are identified by their shape in the (x, y)-plane. Filled symbols denote stable orbits and open symbols denote unstable orbits. For this bifurcation diagram the parameter Θ is positive and not too large; other bifurcation diagrams can be obtained for different values of Θ. Reproduced from W Bayer and U an der Heiden 1998 Oscillation types and bifurcations of nonlinear second order differential-difference equations *J. Dynam. Diff. Eqs* **10** 303–26 ©1998 by Plenum Publishing Corporation.

with constants $\Theta_1 < \Theta_2$, $a < b$, $c < b$. Without loss of generality we assume that

$$\alpha = 1 \qquad \Theta_1 = 0 \qquad a = -\tfrac{1}{2} \qquad b = \tfrac{1}{2} \qquad c < \tfrac{1}{2}. \tag{3.21}$$

Hence, the only free parameters are τ, Θ_2, and c. A proof of the following theorem can be found in [14].

Theorem 3.14. *Let the function f be given by (3.20). Without loss of generality let the parameters α, Θ_1, a, b and c satisfy (3.21). Moreover, let the parameters τ and Θ_2 obey the conditions*

$$0 < \tau < \pi/2$$
$$(1 + \cos \tfrac{1}{2}\tau)/2 < \Theta_2 < 1$$
$$1 + (2\Theta_2 - 1)\cos\tau > 2\sqrt{\Theta_2(1-\Theta_2)}\sqrt{\sin^2\tau + \frac{1-\cos\tau}{1+\cos\tau}}.$$

Then there are numbers $c_1 = c_1(\tau, \Theta_2)$ *and* $c_2 = c_2(\tau, \Theta_2)$, $c_1 < c_2$, *such that (3.18) is chaotic in the sense of Li and Yorke whenever the parameter* c *satisfies* $c_1 < c \leq c_2$.

In conclusion, we have learned that equation (3.4) displays different types of behaviour for $k = 0$, 1, and 2, no matter how large ε is. For $k > 2$ almost nothing is known.

3.6 Outlook

We have not covered all of the literature about the types of equations considered here. Nevertheless, we hope that the reader has got the impression that the pathway we followed opens the perspective to many further and very interesting investigations. The promise is that, in the spirit of Sharkovsky's quote at the beginning of this chapter, there is still an even richer realm of fascinating phenomena to be discovered. In particular, these include many new forms of bifurcation patterns and high-dimensional strange attractors.

Acknowledgment

I thank M-L Liang for helping me very much with the preparation of this manuscript.

References

[1] Aliev S Y, Ivanov A F, Maistrenko Y L and Sharkovski A N 1984 Singular perturbations of difference equations with continuous time *Preprint* no 84.33. Inst. of Math. of the Ukrainian Acad. Sciences (Russian)

[2] Bayer W and an der Heiden U 1998 Oscillation types and bifurcations of nonlinear second order differential-difference equations *J. Dynam. Diff. Eqs* **10** 303–26

[3] Collet P and Eckmann J-P 1980 *Iterated Maps on the Intervals as Dynamical Systems* (Boston, MA: Birkhäuser)

[4] Fridman L M, Fridman E M and Shustin E I 1993 Steady modes in an autonomous system with break and delay *J. Diff. Eqs* **29** 1161–6

[5] Glass L and Mackey M C 1988 *From Clocks to Chaos—The Rhythms of Life* (Princeton, NJ: Princeton University Press)

[6] Goodwin B C 1965 Oscillatory behavior of enzymatic control processes *Advances in Enzyme Regulation* vol 3, ed G Weber (Oxford: Pergamon) pp 425–39

[7] Hadeler K P and Tomiuk J 1977 Periodic solutions of difference-differential equations *Arch. Rational Mech. Anal.* **65** 87–95

[8] Hale J K and Ivanov A F 1993 On higher order differential delay equations *Math. Anal. Appl.* **173** 505–14

[9] an der Heiden U 1979 Periodic solutions of a nonlinear second order differential equation with delay *J. Math. Anal. Appl.* **70** 599–609

[10] an der Heiden U 1979 Delays in physiological systems *J. Math. Biol.* **8** 345–64

[11] an der Heiden U 1983 Periodic, aperiodic, and stochastic behaviour of differential-difference equations modelling biological and economical processes *Differential-Difference Equations* ed L Collatz, G Meinardus and W Wetterling (Boston, MA: Birkhäuser) pp 91–108

[12] an der Heiden U, Longtin A, Mackey M C, Milton J and Scholl R 1990 Oscillatory modes in a nonlinear second order differential equation with delay *J. Dynam. Diff. Eqs* **2** 423–49

[13] an der Heiden U 2001 Nonlinear feedback systems with memory: From 0-th to higher order, discrete and continuous *Fields Inst. Commun.* **20** 223–34

[14] an der Heiden U and Bayer W 2001 Chaos proved for a second order difference–differential equation *Nonlin. Anal.—Theory, Meth. Appl.* **48** 464–74

[15] an der Heiden U and Liang M-L Sharkovsky-sequences of higher order difference equations *Preprint*

[16] an der Heiden U and Mackey M C 1982 The dynamics of production and destruction: Analytic insight into complex behaviour *J. Math. Biol.* **16** 75–101

[17] an der Heiden U and Walther H O 1983 Existence of chaos in control systems with delayed feedback *J. Diff. Eqs* **47** 273–95

[18] Ivanov A F and Sharkovsky A N 1992 Oscillations in singularly perturbed delay equations In *Dynamics Reported—Expositions in Dynamical Systems* vol 1, ed C K R T Jones, U Kirchgraber and H O Walther (Berlin: Springer-Verlag) pp 164–224

[19] Lani-Wayda B and Walther H O 1995 Chaotic motion generated by delayed negative feedback. Part I: A transversality criterion *Diff. Int. Eqs* **8** 1407–52

[20] Lani-Wayda B and Walther H O 1996 Chaotic motion generated by delayed negative feedback. Part II. Construction of nonlinearities *Math. Nachrichten* **180** 141–211

[21] Lasota A and Mackey M C 1994 *Chaos, Fractals and Noise: Stochastic Aspects of Dynamics* (Berlin: Springer)

[22] Li T Y and Yorke J A 1975 Period three implies chaos *Am. Math. Monthly* **82** 985–92

[23] Liang M-L and an der Heiden U 2002 Multistability of the difference equation $x_n = f(x_{n-k})$ with an application to prime number factorization of large natural numbers *Preprint*

[24] Mackey M C 1989 Commodity fluctuations: price dependent delays and nonlinearities as explanatory factors *J. Econ. Theory* **48** 497–509

[25] Mahaffy J M 1980 Periodic solutions for certain protein synthesis models *J. Math. Anal. Appl.* **74** 72–105

[26] Mallet-Paret J and Smith H 1990 The Poincaré-Bendixson theorem for monotone cyclic feedback systems *J. Dyn. Diff. Eqs* **2** 367–421

[27] Mallet-Paret J and Nussbaum R D 1986 A bifurcation gap for a singularly perturbed delay equation *Chaotic Dynamics and Fractals* ed F D Barnsley and S G Demko (New York: Academic) pp 263–86

[28] Milton J 1996 *Dynamics of Small Neural Populations* (Providence, RI: American Mathematical Sociecty)

[29] Nussbaum R D and Mallet-Paret J 1986 Global continuation and asymptotic behaviour of periodic solutions of a differential delay equation *Ann. Mater. Pura Appl.* **145** 33–128

[30] Othmer H G 1976 The qualitative dynamics of a class of biochemical control circuits *J. Math. Biol.* **3** 53–78

[31] Ritter H, Martinetz T and Schulten K 1991 *Neuronale Netze* (New York: Addison-Wesley)

[32] Sharkovsky A N 1964 Coexistence of cycles of a continuous transformation of a line into itself *Ukrain. Mat. Zhurn.* **16** 61–71 (Russian)

[33] Sharkovsky A N, Maistrenko Y L and Romanenko E Y 1986 *Difference Equations and their Applications* (Kiev: Nauka Dumka) (Russian) (Engl. transl. 1993 (Amsterdam: Kluwer))

[34] Sharkovsky A N, Kolyada S F, Sivek A G and Fedorenko V V 1997 *Dynamics of One-Dimensional Maps* (Kiev: Nauka Dumka) (Engl. transl. 1997 (Amsterdam: Kluwer))

[35] Sharkovsky A N, Romanenko E Y and Vereikina M B 1999 *Ideal Turbulence—Structures, Self-Similarity and Random Evolution Birth in Deterministic Systems* (London: World Scientific)

[36] Viana M 2000 What's new on Lorenz strange attractors? *The Math. Intelligencer* **22** 6–19

[37] Walther H O 1981 Homoclinic solution and chaos of $\dot{x}(t) = f(x(t-1))$ *Nonlin. Anal., Theory, Meth. Appl.* **5** 775–88

[38] Walther H O 1981 Density of slowly oscillating solutions of $\dot{x}(t) = -f(x(t-1))$ *J. Math. Anal. Appl.* **79** 127–40

[39] Walther H O 1995 *The 2-Dimensional Attractor of* $x'(t) = -\mu x(t) + f(x(t-1))$ *(Mem. Am. Math. Soc. 113)* no 544 (Providence, RI: American Mathematical Sociecty)

Chapter 4

Creating stability out of instability

Christopher K R T Jones
Brown University

It is a compelling idea to conjure a stable, or robust, state out of an unstable state or combination of unstable states. In a natural way, this can be formulated as a problem in control theory: how can an instability be controlled to produce a stable state? A well-known example is the inverted pendulum. Kapitza (see [12]) showed that a vertically oscillating pivot can produce a stable pendulum that is upside down. This control strategy can be construed as placing the pendulum in a stable state (the vertical acceleration can overcome gravity on the way down) for part of the time. The pendulum is thus switching between a stable and an unstable state and the stable state wins. This suggests an interesting question: can switching between unstable states induce stability? This question can be answered positively with a system that jumps between two saddles, in a time-dependent way, in such a way that the saddles switch stable and unstable directions. As soon as the system moves off in the unstable direction, the switch makes that direction stable and so on. This can be implemented as a control strategy but it remains another question whether this mechanism is found in specific applications.

A related question occurs in wave propagation. A pulse can often be decomposed into phases: up and down. Suppose that either, or both, of the phases is intrinsically unstable, is it then possible that the full pulse itself is stable? In the same way as the system that oscillates between unstable states, the two phases would compensate for the instabilities in the other phase with a dampening effect that would come from their own stable directions.

Two quite different situations will be explored in this chapter. Both involve wave propagation and give examples of naturally occurring versions of this surprising phenomenon of switching between unstable states leading to a stable state. The first case comes from nonlinear optical pulse propagation. The current preferred technique of pulse propagation on optical fibres involves a technique called dispersion compensation through which chromatic dispersion is counteracted by reverse dispersion in a systematic fashion along the fibre. The mathematical explanation of the resulting stabilization is given in [25], where it is shown that this idea can be realized in the critical limit. This latter case can be construed as an example of switching between unstable states rendering a stable object. Despite its involving wave propagation, the stabilization due to dispersion compensation is actually an effect that involves switching in time. The second example will exhibit a case in which the spatial distribution of unstable phases, as described earlier, can lead to a stable pulse.

A number of fundamental questions are raised by the subjects covered here and *Aside* sections will be regularly placed throughout this chapter where these issues are explored.

4.1 Nonlinear optical fibres

The infrastructure of optical fibres is now the key component in the global communication system. Most systems run on a format based on the assumption that the fibres in use respond linearly to impinging light. If the amplitude is large enough, and the size needed depends on the material used in the fibre, then the response will become nonlinear. The inherent nonlinearity in fibres can be exploited to advantage through the use of optical solitons acting as carriers of 'bits' of information. This was first suggested by Hasegawa and Tappert [8] and later realized experimentally by Mollenauer *et al* [15]. The soliton exists by virtue of a balance between nonlinearity and dispersion. The fundamental equation is the nonlinear Schrödinger equation (NLS)

$$\mathrm{i}\frac{\partial q}{\partial z} = \alpha\frac{\partial^2 q}{\partial z^2} + |q|^2 q$$

where q is the amplitude of the envelope of the electric field. Note that the evolution is spatial and not temporal. Although this is, for a mathematician, counter-intuitive it is natural and correct from the physical standpoint. If a communication (e.g. a phone call) is being made then a time-trace is being imposed on the fibre at a fixed location. The communication is received at another location (down the fibre) as a time-trace that should faithfully reflect the input time-trace. The initialization is thus a function of time at a fixed spatial location and the output, at a place further down the fibre, is also a function of time. The evolution is therefore a function of time along the fibre.

4.1.1 Dispersion management

The current 'hot' approach to combating instabilities in optical fibres is through dispersion compensation. The idea is simple. The most significant destabilizing influence to any structure (pulse or front of on/off switch) is chromatic dispersion: the tendency of frequencies near to a carrier frequency to disperse and cause pulse broadening. Dispersion compensation counteracts this effect by a reversal of dispersion. This amounts to effectively running the equations backwards and is essentially a linear strategy. If a wave is evolving under the influence of the linear equation

$$\mathrm{i}u_z + u_{TT} = 0$$

then the idea is to subject it to the influence of the equation

$$\mathrm{i}u_z - u_{TT} = 0$$

to the extent that the first equation will make waves with nearby frequencies separate, the second equation will bring them back together.

The amazing fact is that fibres can actually be constructed that have juxtaposed pieces of opposite frequency. This was first suggested by Lin *et al* [13] as a strategy for stabilizing linear systems but was later adapted to nonlinear systems.

For nonlinear systems, i.e. those in which the amplitudes are large enough to excite the nonlinearities, the equations will have the form

$$\mathrm{i}u_z \pm u_{TT} + \varepsilon|u|^2 u = 0$$

where the $\pm$ will switch with the alternating fibre pieces. Smith *et al* [21] first found numerical pulses in the nonlinear regime. Gabitov and Turitsyn [5] developed an averaged equation for these dispersion-managed nonlinear fibres and Ablowitz and Biondini [1] numerically computed pulse solutions for the Gabitov–Turitsyn equations.

The dispersion management approach has met with astounding success. Leguen *et al* [9] reported at the 1999 Optical Fiber Communications Meeting that they had developed a dispersion-managed system with a capacity of over 1 Tbit/s over 1000 km. Subsequently, Mollenauer *et al* [16] reported system performance at 0.5 Tbit/s over 10 000 km. These involved increases of around 10^3 bit/s over previous systems.

The profile of varying dispersion along the fibre is called the dispersion map. It is usually taken to be piecewise constant with a fixed period. A key observation is that the intervals over which the map is positive and negative do not have to be equal, as long as the average dispersion over a period is balanced; see figure 4.1. The most advantageous situation is when the average (residual) dispersion is close to zero while the absolute dispersion of the individual parts remains large. A piece of positive dispersion fibre covering most of the period could then be balanced by a short piece of negative dispersion. This leads to the technologically attractive suggestion that an existing fibre could be converted to a dispersion-managed fibre by placing a piece of (strongly) negative dispersion fibre into the line at convenient locations, for instance, at amplifier stations; see again figure 4.1.

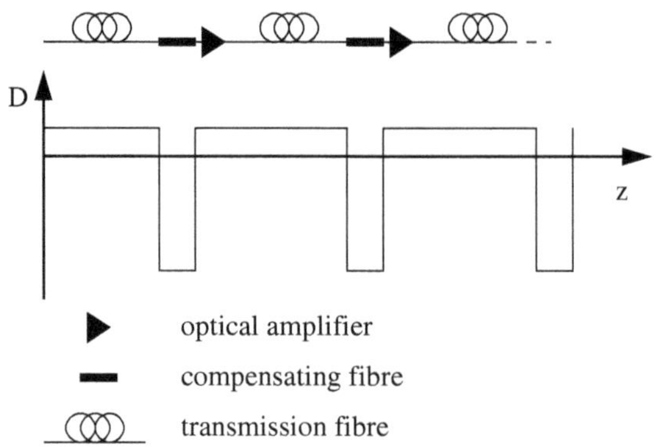

Figure 4.1. Dispersion management of fibre transmission.

4.1.2 Aside: linear versus nonlinear

The two main approaches to optical communications are identified respectively with linear and nonlinear theories. The dominant mode of information propagation on an optical fibre is the so-called Non-Return-to-Zero (NRZ) format. In this format a string of binary values is represented by On (1) or Off (0) and between two 1s the value does not return to 0. The string thus looks as in figure 4.2(*a*). It is assumed that the On state is of small amplitude so that the response of the medium can be assumed to be linear. It should be noted that the exact value of the amplitude is immaterial due to the linearity.

In this NRZ scenario, nonlinear effects are undesirable and thus the signal needs to be kept at as low a power as possible. This can cause a problem for long-haul transmission as losses on the fibre necessarily degrade the signal. Although this is compensated to some extent by periodic amplification, the case remains for taking nonlinearities in the fibre into account. Under a linear strategy, competing signals pass through each other unaffected by their interaction. Since multiple channels are carried by the same fibre, competing signals are interacting with great frequency and this gives a linear strategy a definite advantage. Ironically, however, during the interaction itself the amplitude of the signal will increase thus exciting the nonlinear properties of the fibre.

The nonlinear approach is based on a fundamentally different format in which information is encoded in pulses. The basic pulse is an optical soliton which is the soliton solution of the cubic nonlinear Schrödinger equation, as discovered by Zakharov and Shabat [24]. The idea of using optical solitons as information carriers was first suggested by Hasegawa and Tappert [8]. Each time window will either contain a pulse (1) or no pulse (0). Between 1 values the system returns to rest, as is sketched in figure 4.2(*b*), and the contrast with the NRZ format is thus clear. Pulse-based systems are now often referred to as Return-to-Zero (RZ). An advantage of the nonlinear system is that the amplitude

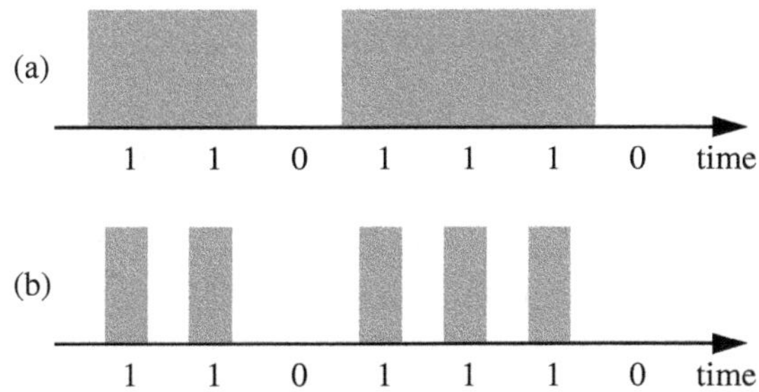

Figure 4.2. Non-Return-to-Zero (*a*) and Return-to-Zero (*b*) format for information propagation on an optical fibre.

is naturally restricted by the properties of the solitons. Since the nonlinearity is already being taken into account, the amplitude of the pulse can be larger. The interactions of pulses are less predictable. Provided the cubic NLS is a faithful approximation, the rules for soliton interactions coming from inverse scattering will hold. Perturbations of the cubic NLS will, however, be introduced by amplifiers and other adjustments to the basic fibre, such as those introduced to compensate the destabilizing effect of the amplification.

The area of optical communications presents an interesting context in which to evaluate the debate between linear and nonlinear strategies. Despite the excitement around nonlinear phenomena, such as those which are manifest in chaotic dynamics, I believe it is fair to say that most of the applied scientific and engineering world leans preferentially toward linear techniques. Nonlinear effects are generally viewed with some suspicion and the comfort of well-understood linear methods are sufficiently alluring that most users of mathematical techniques will try to mould a given system of interest into a linear regime. This raises a basic question which may seem simplistic from the theoreticians' point of view, but is, in view of current practice, quite important.

Question 4.1. *Should 'nonlinearity' be the object of damage control or a key component of the theory?*

There is a curious coming together of the linear and nonlinear approaches in optical communications. For long-distance transmission, the RZ format is becoming the preferred mode. The NRZ community refers to the pulses used in the RZ format as 'quasi-linear pulses' in reluctant recognition that nonlinearity is playing some role. Furthermore, with the advent of dispersion-managed systems, the possibility arises of converting existing systems, as described earlier, by inserting pieces of compensating fibre at convenient locations. It is then possible that erstwhile NRZ fibres could operate in a pulse format as RZ fibres and the nonlinearity be effectively harnessed.

4.1.3 Efficacy of dispersion management

On a dispersion-managed fibre, a pulse pulsates as it propagates. Over the negative dispersion piece, it will tend to broaden due to chromatic dispersion and then tighten in the focusing regime of the positive dispersion piece. This broadening and narrowing continues as it moves along the fibre giving the pulse an oscillating profile, with oscillations corresponding to the changes in dispersion characteristics of the fibre. If the dispersion map is periodic, then the expectation is that this pulse will also be periodic. This oscillating pulse, or its counterpart in the rapidly varying dispersion map limit, is called the Dispersion-Managed Soliton (DMS).

All of the evidence indicates that the DMS is a very stable object. Our goal in the work [25] was to explain this phenomenon mathematically. The natural approach is to use the averaged system as derived by Gabitov and Turitsyn [5] and to show that it carries a stable pulse solution. In this averaged limit, we would expect the pulse to be steady as the periodic variation of the DMS seen in the real systems corresponds to the different phases as it passes over the pieces of, respectively, negative and positive dispersion. It would be natural to show that the real system carried a periodic orbit that is a perturbation of the stable pulse of the averaged system. Since the underlying phase space is infinite-dimensional, this is not a realistic hope, as discussed later. The usual results that underscore the validity of the averaging procedure do, however, show that solutions of the full system starting near to the pulse will stay looking like it for long times. The two key points are thus the following elements:

(i) Validity of averaging. The main result shows that solutions of the dispersion-managed NLS (DMNLS) stay close to the corresponding solutions of the averaged equations for long times. This mimics the classical averaging result of ODEs.
(ii) Stable pulse. The averaged equation is shown to have a stable pulse by proving that a steady-state solution can be found that is a constrained minimum of the energy subject to constant charge (in the L^2-norm).

As a corollary of combining these two results, solutions of DMNLS with the appropriate initial shape stay pulse-like for long periods of time.

4.1.4 Averaging out dispersion variation

We start with the DMNLS written in the form

$$\mathrm{i}\varepsilon u_z + \left(d\left(\frac{z}{\varepsilon}\right) + \varepsilon\alpha\right) u_{tt} + \varepsilon|u|^2 u = 0 \tag{4.1}$$

or, scaling z,

$$\mathrm{i}u_\zeta + d(\zeta) + \varepsilon(\alpha u_{tt} + |u|^2 u) = 0. \tag{4.2}$$

Performing the Van der Pol transformation on (4.2) based on the evolution operator for the linear dispersion-managed equation

$$iu_\zeta + d(\zeta)u_{tt} = 0$$

which we write as $u(\zeta, t) = T(\zeta)u(0, t)$, i.e. setting $v = T(-\zeta)u$, we obtain

$$iv_\zeta + \varepsilon(\alpha v_{tt} + Q_\varepsilon(v, v, v)) = 0$$

where

$$Q_\varepsilon(v, v, v) = T(-\zeta)|T(\zeta)v|^2 T(-\zeta)v.$$

Now averaging over a dispersion map period (say 0 to 1)

$$iv_\zeta + \varepsilon(\alpha v_{tt} + Q(v, v, v)) = 0 \tag{4.3}$$

where

$$Q(v, v, v) = \int_0^1 T(-\zeta)|T(\zeta)v|^2 T(-\zeta)v \, d\zeta.$$

Or, in original z variables,

$$iv_z + \alpha v_{tt} + Q(v, v, v) = 0. \tag{4.4}$$

The averaging result should compare solutions of (4.1) to those of (4.4), or, equivalently, the solutions of (4.2) with those of (4.3). To that end, we have the following result.

Theorem 4.2. *If the residual dispersion $\alpha > 0$, v satisfies (4.1) and $\hat{v}$ satisfies (4.4), with the same initial data in H^s, $s \geqslant 4$, then*

$$\|v(\zeta, t) - \hat{v}(\zeta, t)\|_{H^{s-3}} = O(\varepsilon)$$

for $0 \leq \zeta \leq \frac{c}{\varepsilon}$, for some constant c.

This result can be viewed as a classical averaging result along the lines of the original work by Bogoluibov and Mitroploskii (see [7]), save for the fact that the underlying norms are appropriate for PDEs. It should be noted that a loss of derivatives occurs due to the presence of the Schrödinger evolution operator in the integral Q. The averaged equation (4.4) is equivalent to the Gabitov–Turitsyn equation but expressed in a new formulation in the real time domain, rather than the Fourier domain. The proof of this theorem, however, is carried out in the Fourier domain.

4.1.5 DMNLS pulse

To show that there is a base pulse for the DMNLS, we find a minimizer for the Hamiltonian associated with the averaged equation (4.3). In the following, it will be assumed throughout that the residual dispersion $\alpha > 0$. The case $\alpha = 0$ is particularly interesting as dispersion-managed systems are often run with small averaged dispersion, but serious technical complications prevent the proofs from going through. For the case of negative residual dispersion, it is shown in [25] that no minimum exists.

The equation (4.3) possesses a Hamiltonian given by

$$H_\alpha(v) = \int_0^1 \int_{-\infty}^{+\infty} (\alpha |v_t|^2 - |v^* Q|)\,\mathrm{d}t\,\mathrm{d}\zeta$$
$$= \int_0^1 \int_{-\infty}^{+\infty} (\alpha |v_t|^2 - |T(-\zeta)v|^4)\,\mathrm{d}t\,\mathrm{d}\zeta.$$

The charge (or power) is conserved, so that a 'ground state' is a minimizer of $H_\alpha(v)$ subject to fixed

$$P(v) = \int_{-\infty}^{+\infty} |v|^2\,\mathrm{d}t = \lambda.$$

If we set

$$\mathbf{C}_\lambda = \inf\{H_\alpha(v) : v \in H^1,\, P(v) = \lambda\}$$

it can be seen that $\lambda > 0$ implies that $\mathbf{C}_\lambda < 0$. This is a great help in the minimization procedure as it prevents a minimizing sequence from converging to 0. The standard strategy is to find a minimizing sequence $\{v_k\}$ and show that it converges strongly $v_k \to v$ with $H_\alpha(v) = \mathbf{C}_\lambda$ and $P(v) = \lambda$. From the Alaoglu theorem, it follows that $v_k \rightharpoonup v$ (converges weakly). The theory of concentration-compactness, due to Lions [14], tells us that there are now three possibilities:

(i) strong convergence: $v_k \to v$
(ii) vanishing: $v_k \to 0$ and
(iii) splitting: $v_k \to v^1 + v^2$.

If possibilities (ii) and (iii) can be ruled out then the sequence must converge strongly (possibility (i)). The key result is a lemma due to Zharnitsky that shows localization cannot occur in the linear Schrödinger equation. For instance, to rule out vanishing, from looking at the energy, and using the fact that $\mathbf{C}_\lambda < 0$, it suffices to show that $v_k \to 0$ implies that

$$\int_0^1 \int_{-\infty}^{+\infty} |T(-\zeta)v_k|^4\,\mathrm{d}t\,\mathrm{d}\zeta \to 0.$$

Zharnitsky's localization lemma shows that this cannot occur. It is a statement about the linear Schrödinger equation not being able to localize mass since $T(\zeta)$

is its evolution operator. It would seem that this fact should be well known but all the standard estimates for the linear Schrödinger equation fall short of implying it.

4.1.6 Aside: averaging in PDEs

Let us review what we can conclude from the previous analysis. We have the standard averaging theorem, albeit with some loss of derivatives, from which we know that the averaged and original solutions are $O(\varepsilon)$ close for times (or, in the optics case, distance down the fibre) of $O(\frac{1}{\varepsilon})$. Since we also know that there is a ground-state solution, we can conclude that with the ground-state set as initial data for the full DMNLS the resulting solutions will stay close for long periods of time (or distance down the fibre). The fact that the ground state is a steady state, let alone that it is stable, is not used at all in reaching this conclusion. There is information thus not being used that should strengthen the result that any pulse-like initial data will stay that way. In the context of ODEs there are standard results that afford conclusions about the existence of periodic orbits of the original system near to critical points of the averaged system; for instance, see [25]. It would be tempting to anticipate such a result in this context also. In particular, for the dispersion management problem it would follow that a pulsating orbit would exist for DMNLS (4.2) that tracks the ground state of the averaged equations. This basic result does not, however, go through for the PDE case. The impediment derives from the possible presence of infinitely many resonances. This raises the following question.

Question 4.3. *What is the analogue of the periodic orbit theorem for averaging in PDEs?*

Bambusi and Nekhoroshev [3] discuss this issue and suggest that the result should be an increased length of time, such as exponentially long, over which the averaged solution is a good approximation if that averaged solution is a stable state. Their work is in the context of nonlinear wave equations. An interesting and important issue is whether this can be extended to a wider class of problems.

4.1.7 Extending dispersion management stabilization of unstable states

Dispersion management has been shown to render stable pulses. This analysis is analogous to that for the NLS. In other words, the DMS has been shown to enjoy the same stability properties as the standard soliton of NLS. The expectation is, however, that the DMS is somehow more stable. This is corroborated by numerical calculations that show that even instabilities caused by numerical approximations (which might reasonably be viewed as reflecting instabilities due to losses in the system) are more effectively damped in DMNLS than in NLS; see [25].

A way to push dispersion management further is to apply it in cases that have only unstable steady states for the associated NLS, even in the case of positive (focusing) dispersion. A natural case is supplied by critical nonlinearities: quintic in one-dimension and cubic in two dimensions. The relevant equations in their pure NLS form are

$$\begin{aligned} \mathrm{i}u_z + \alpha u_{tt} + |u|^4 u &= 0 \\ \mathrm{i}u_z + \alpha \Delta_{\mathrm{T}} u + |u|^2 u &= 0 \end{aligned} \tag{4.5}$$

where

$$\Delta_{\mathrm{T}} = \frac{\partial^2}{\partial x^2} + \frac{\partial^2}{\partial y^2}.$$

We will not endow these equations with any particular physical interpretation in optics, but have written them in the usual optics form for the sake of consistency. In this interpretation Δ_{T} should be construed as the transverse Laplacian, i.e. where x and y are spatial variables transverse to the distance down the fibre z.

Even with positive dispersion ($\alpha > 0$) these equations possess no stable ground states, as was shown by Weinstein [22]. There are bound states, but they are all unstable, even the pulse-like ground state. Blow-up in finite time for some initial data in these critical NLS systems arbitrarily close to the pulse was proved by Weinstein [22]. This is clearly then a very unstable pulse and the question we addressed in [25] was whether this pulse could be stabilized by dispersion management. In other words, if we set

$$\alpha = d\left(\frac{z}{\varepsilon}\right)$$

in (4.5) and average, as in the previous section, do we obtain a ground-state pulse? It should be clearly noted that there is no particular physical problem associated with this procedure, it is just a mathematical test of the stabilization of dispersion management.

It turns out, as is shown in [25], that the analysis proceeds just as before: the averaging procedure is valid, i.e. theorem 4.2 holds, and there is a ground state. Given the drastic instability, as evidenced by the blow-up result of Weinstein, in these critical cases, this is a striking result. To return to the theme at the beginning of this chapter, namely the possibility of a stable state resulting from a jump between unstable states, we can see that exactly such an effect is occurring here. The jump is between a positive dispersion part and the negative dispersion part. In the latter part, there is no steady state at all, while in the former part, the state is unstable. Thus this dispersion-managed version of critical NLS (4.5) can reasonably be construed as jumping between a stable and an unstable state.

4.1.8 Aside: stability questions

There are some very important open stability questions related to nonlinear optical pulse propagation. One of the most fundamental of these concerns the possibility of multiple pulses that are stable. A multiple pulse is one in which the base pulse is repeated and the different pulses are well separated from each other. It has long been folklore in nonlinear optics that a multi-pulse will, of necessity, be unstable. The first full result is due to Yew *et al* [23] who exhibit instability of multiple pulses in coupled NLS systems. A novel mechanism for stable pulses that can be viewed as multi-pulses is given in Ostrovskaya *et al* [18]. These states become unstable, however, as the pulses separate from each other. The overall problem then remains open.

Question 4.4. *Are there stable multiple pulses in nonlinear optical systems?*

An affirmative answer to this question could have some important physical implications. For instance, it is conceivable that a multi-bit format could be used for message encoding. This would significantly multiply the amount that could be transmitted on a given fibre. This is explored in the context of phase-sensitive amplified fibres in [17].

Even if multiple pulses that have a stable spectrum, i.e. a spectrum lying entirely on the imaginary axis, are found, it is not known whether this implies stability relative to the full system, which is often called nonlinear stability. The main condition used to guarantee stability is the one used in this chapter, namely that a given state is a constrained minimum. The case that is likely to occur for multiple pulses with stable spectrum is that the Hamiltonian is indefinite.

Question 4.5. *Is a steady state with its linearization possessing an indefinite Hamiltonian but with no unstable eigenvalues, stable?*

This is not an easy question as it appears to be related to Arnol'd diffusion, which is not even well understood in finite dimensions.

The role of dispersion management, as can be seen clearly from the analysis of the preceding section, is as a stabilizing mechanism. Such mechanisms could be viewed in terms of wave control strategies. There are other interesting cases of wave stabilization as in, for instance, the work of Doelman *et al* [4]. In their work a strongly stable background shelf can be used to stabilize an unstable spike. They find this effect occurring in the Gray–Scott reaction–diffusion system.

Question 4.6. *Are wave stabilization mechanisms common? And can they be implemented as wave control strategies?*

An extremely important problem in optical pulse propagation concerns pulse–pulse interaction. A given fibre will usually be required to carry pulses of many frequencies. These are viewed as different channels that should co-exist on the fibre without interference. Due to the nonlinearity, however, they will inevitably interact and the effect of these interactions is not well understood.

Question 4.7. *Can we get an effective calculus of pulse–pulse interactions in nonlinear optical systems?*

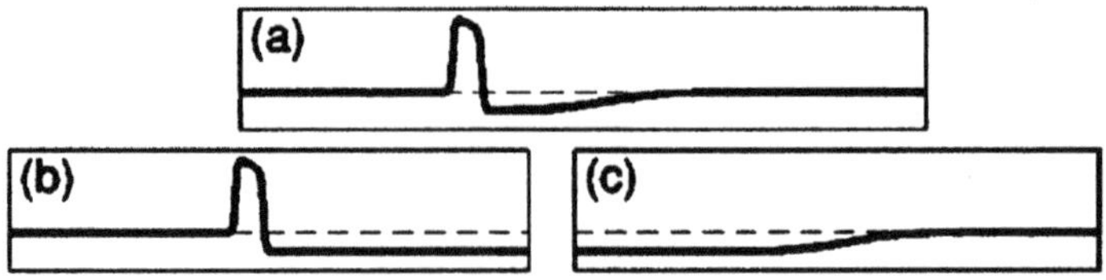

Figure 4.3. A pulse with a prolonged refractory period (*a*) can be viewed as a concatenation of a generalized front (*b*) and a generalized back (*c*). Reproduced from M M Romeo and C K R T Jones 2001 Stability of neuronal pulses composed of two concatenated unstable kinks *Phys. Rev.* E **63** 011904 ©2001 by the American Physical Society.

4.2 Can two unstable waves make a stable pulse?

The second scenario for creating stability out of instability was raised as a potential concatenation of unstable pulses to create a stable pulse. This is a spatial juxtaposition of unstable states. The idea has to be similar to the temporal juxtaposition of unstable states in that the stabilizing influence of each must act on the other to damp out their instabilities. How this will occur in this case of spatial distribution is mysterious.

A first observation is that the instability of the constituent waves cannot be through point spectrum. Indeed, according to Alexander *et al* [2], any unstable eigenvalues of these pieces would manifest themselves in corresponding unstable eigenvalues of the full pulse. It follows that any instability would have to be associated with the presence of the essential spectrum in the right half-plane. The essential spectrum in these wave problems comes from the far-field, or asymptotic, states. The concatenation of waves into a pulse involves a quiescent phase between the two constituent waves. This phase is an asymptotic state of each of the constituent waves, at minus and plus infinity for the front and back respectively. In the full pulse, it appears as a shelf that elongates as the separation between the waves increases; see figure 4.3. There will be an essential spectrum associated with this shelf that is manifest in the spectrum of the front and the back but not in the full wave itself. In the limit in which this intermediate, quiescent phase elongates, the question is whether there is any residual instability from the essential spectrum that pops out for the limiting profiles.

On the level of the travelling wave equations, the bifurcation of the pulse out of the two constituent waves is a homoclinic bifurcation that is called a T-point, a term originally coined by Glendinning and Sparrow [6].

We construct a specific example that exposes this extraordinary phenomenon in which a stable pulse bifurcates out of unstable fronts and backs.

4.2.1 The ultra-refractory FitzHugh–Nagumo equations

The FitzHugh–Nagumo equations were originally formulated as a simplification to the Hodgkin–Huxley model of nerve impulse propagation. They are formed by a diffusive equation for the voltage across a nerve membrane and an ODE for a lumped variable that governs the recovery of the nerve fibre.

$$\begin{aligned} u_t &= u_{xx} + f_1(u, w) \\ w_t &= f_2(u, w, \gamma). \end{aligned}$$

In the original FitzHugh–Nagumo system

$$\begin{aligned} f_1(u, w) &= f(u) - w \\ f_2(u, w, \gamma) &= \varepsilon(\gamma u - w) \end{aligned}$$

where $f(u) = u(u-a)(1-u)$ and $0 < a < \frac{1}{2}$, which is the usual cubic that occurs in the bistable system. We shall replace $f_2 = \varepsilon(\gamma u - w)$ by a more general term that we shall write as $g(u, w, p)$ where p is a parameter. We shall also focus on travelling waves, which are steady solutions in a moving frame. Setting $\xi = x - ct$, the reaction–diffusion system of interest will then be

$$u_t = u_{\xi\xi} + cu_\xi + f(u) - w \tag{4.6}$$

$$w_t = cw_\xi + g(u, w, p). \tag{4.7}$$

The ODEs for a travelling wave are then

$$u' = v \tag{4.8}$$

$$v' = -cv - f(u) + w \tag{4.9}$$

$$w' = -\frac{1}{c} g(u, w, p). \tag{4.10}$$

The scenario of interest is where a pulse bifurcates out of a concatenated front and back at a parameter value $p = p_0$; see figure 4.3. In the phase space of (4.8)–(4.10) this is a homoclinic orbit bifurcating out of a heteroclinic loop; see figure 4.4.

To study the stability of a given wave, we consider the linearization of the PDE, equations (4.6) and (4.7), at the wave: $(u(\xi), v(\xi))$, which is

$$\begin{aligned} q_t &= q_{\xi\xi} + cq_\xi + D_u f(u, w)q + D_w f(u, w)s \\ s_t &= cs_\xi + D_u g(u, w)q + D_w g(u, w)s. \end{aligned}$$

The right-hand side is a linear operator and so we write it as $L(q, s)$. This may be evaluated at the front, the back or the pulse.

From previous considerations, the scenario that will interest us is when L_{F} (the linearization at the front), and L_{B} (correspondingly at the back) have essential

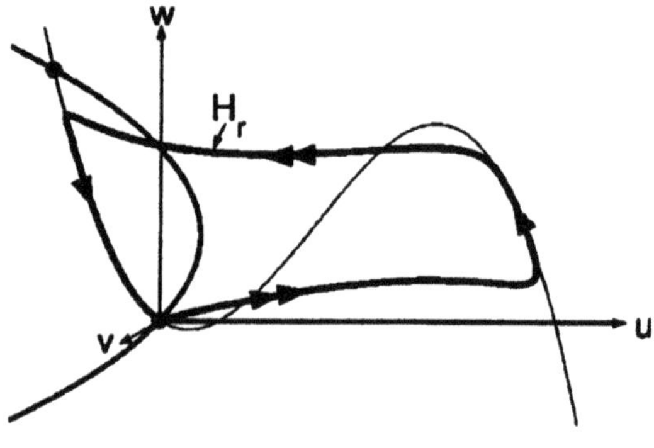

Figure 4.4. A homoclinic orbit corresponding to a pulse solution of the ultra-refractory FitzHugh–Nagumo system. Reproduced from M M Romeo and C K R T Jones 2001 Stability of neuronal pulses composed of two concatenated unstable kinks *Phys. Rev.* E **63** 011904 ©2001 by the American Physical Society.

spectrum in the right half-plane but for which L_{P} (at the pulse) has its essential spectrum contained entirely in the left half-plane. This situation pertains in the example considered later. The question raised by this situation is whether there is a residue from the essential spectrum in $\sigma(L_{\mathrm{P}})$.

The surprising fact was discovered simultaneously by Romeo and Jones [19] and Sandstede and Scheel [20] that there could, in fact, be no residue from the essential spectrum and the pulse could end up being stable, even though the constituent waves were unstable. Sandstede and Scheel [20] developed a general theory and numerically verified the hypotheses of their stability result for an example similar to that given here. In [19], we showed analytically that a stable pulse exists which bifurcates from a T-point.

The example in [19] is based on a quadratic representation of $g(u, w, p)$. In that representation, the null-cline $g = 0$ intersects $w = f(u)$ in more than one place to create a second rest state. As p varies, this rest state can interrupt the travelling wave; see figure 4.5. When ε is turned on, this leads to the bifurcation of a homoclinic orbit out of a heteroclinic loop. The heteroclinic loop is exactly of the T-point form.

What is shown in [19] is that for $\varepsilon > 0$ but sufficiently small, there is a $p^*(\varepsilon)$ for which a heteroclinic loop is present and that is $p > p^*(\varepsilon)$ there is an orbit homoclinic to rest. An analysis of the rest point shows that an unstable essential spectrum is present for the constituent waves (front and back). It can easily be shown, however, that the homoclinic orbit to rest is stable by invoking the result of [10] and observing that the changes in g do not affect the proof.

In [19], we also corroborate this stability result with careful numerical calculations. The computations are quite delicate in that the domain of attraction apparently shrinks as the bifurcation point is approached. The stability of the

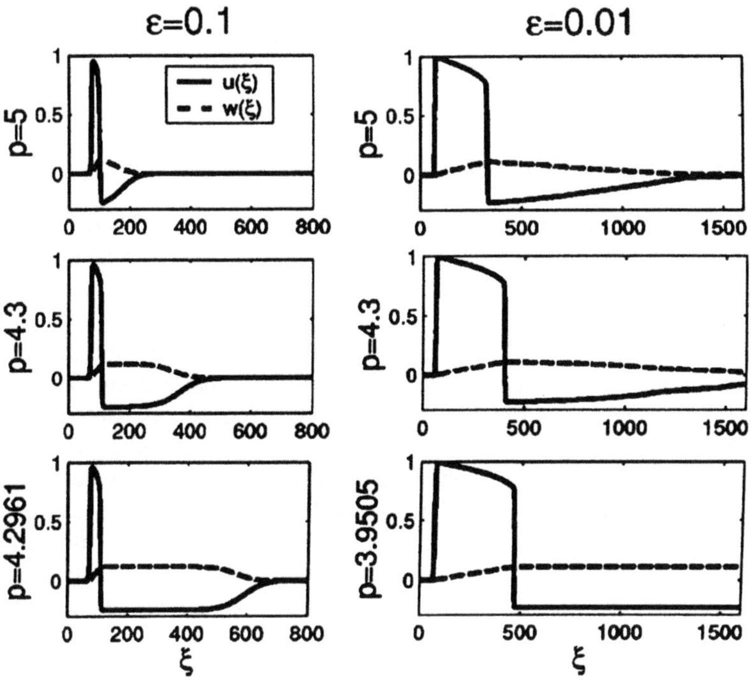

Figure 4.5. Lengthening of the quescient state with p for two fixed values of ε. As ε decreases, there is a longer disparity between slow and fast phases in addition to lengthening latent states. Reproduced from M M Romeo and C K R T Jones 2001 Stability of neuronal pulses composed of two concatenated unstable kinks *Phys. Rev.* E **63** 011904 ©2001 by the American Physical Society.

pulse is, however, still evident if the initial conditions are carefully set. A key point in the strategy is to continually reset the initial condition as the parameter is advanced. The new initial condition is taken to be the output pulse of the numerical experiment at the previous parameter setting. Progressively smaller steps in the parameter advance are also necessary. The results are rather convincing; see again figure 4.5.

As this work developed, I noticed that an interesting shift in my own perspective had occurred in the recent past. I really did feel that I needed the corroboration of the computational experiments to be convinced of this surprising phenomenon. As a mathematician, I know that a proof is the real arbiter of the truth, but as an applied mathematician, I crave numerical verification! As a tongue-in-cheek reflection of this dilemma, I pose the following question.

Question 4.8. *Do proofs validate numerical computations; or do numerical computations verify proofs?*

4.3 Some predictions, suggestions and questions

By way of conclusion, I would like to pose and discuss some basic questions concerning the future role of nonlinear dynamics.

Question 4.9. *Will the future of nonlinear dynamics be application driven?*

The subject of dynamical systems is, by now, mature. There is a certain mindset that is common to most practitioners of the subject, and many of the major problems have now been solved. It is, however, not yet clear how this material will be adapted to real-world applications. Much of the analytic development has been achieved in the context of simple models and the demands of realistic models take the entrenched ideas to their snapping point. The strength of dynamical systems lies in its adaptability, but much of the adaptation has not yet taken place. One area in which this can be seen is in the handling of numerical models. In areas involving geophysical fluid dynamics, such as oceanography or atmospheric science, any moderately realistic model is formulated as the numerical solution of a partial differential equation. Unless this model enjoys some magical source of periodicity, it will necessarily extend over only finite time and the concepts of dynamical systems, in so far as they are based on asymptotic conditions, need significant adaptation.

The relation of dynamical systems to computations and large data sets deserves singling out.

Question 4.10. *Will computations and data play a central role in the further development of dynamical systems?*

The answer to this seems to me to be inevitably: 'yes'. We cannot avoid the fact that data, both observational and computational, will proliferate in the future. Moreover, the sets of data will pertain to physical situations that cannot be captured by simple models and, hence, techniques will be needed that can incorporate ideas of dynamical systems into the analysis of data sets. This is dynamical systems without equations.

Question 4.11. *Is the main distinction between* linear *and* nonlinear *or between* local *and* global*?*

The fascination with the phenomena exposed by dynamical systems, such as chaotic motion, for the applied scientist has often been characterized as nonlinear. The techniques of physics and engineering are largely based on linear analysis and such exotic effects are prevented from occurring. It was therefore natural to view 'nonlinearity' as being responsible. However, this is misleading. The driving force of chaos, for instance, is the elementary linear effect of exponential stretching near a saddle-point. Nonlinearity merely supplies the recurrence needed to have the dynamics repeatedly experience the saddle effect. I address this point quite extensively in the article [11] and suggest that the Smale school were correct in using the term 'global dynamics' as it is the global effects that will be most challenging to us in the future.

References

[1] Ablowitz M J and Biondini G 1998 Multiscale pulse dynamics in communication systems with strong dispersion management *Opt. Lett.* **23** 1668–70

[2] Alexander J, Gardner R and Jones C K R T 1990 A topological invariant arising in the stability analysis of travelling waves *J. Reine Angew. Math.* **410** 167–212

[3] Bambusi D and Nekhoroshev N N 1998 A property of exponential stability in nonlinear wave equations near the fundamental linear mode *Physica* D **122** 73–104

[4] Doelman A, Gardner R A and Kaper T 1998 Stability of singular patterns in the 1-D Gray–Scott model *Physica* D **122** 1–36

[5] Gabitov I and Turitsyn S K 1996 Averaged pulse dynamics in the cascaded transmission system based on passive compensating technique *Opt. Lett.* **21** 327–9

[6] Glendinning P and Sparrow C 1986 T-points: A codimension two heteroclinic bifurcation *J. Stat. Phys.* **43** 479–88

[7] Hale J 1980 *Ordinary Differential Equations* (Florida: Krieger)

[8] Hasegawa A and Tappert F 1973 Transmission of stationary nonlinear optical pulses in dispersive dielectric bers: I Anomalous dispersion *Appl. Phys. Lett.* **23** 142–4

[9] Le Guen F, Del Burgo S, Moulinard M L, Grot D, Henry M, Favre F and Georges T 1999 Narrow band 1.02 Tbit/s (51 $\times$ 20 Gbit/s) soliton DWDM transmission over 1000 km of standard fiber with 100 km amplifier spans *Optical Fiber Communication Conference: OSA Technical Digest* (Washington, DC: Optical Society of America)

[10] Jones C K R T 1984 Stability of the travelling wave solution of the FitzHugh–Nagumo system *Trans. Am. Math. Soc.* **286** 431–69

[11] Jones C K R T 2001 Whither applied nonlinear dynamics? *Mathematics Unlimited: 2001 and Beyond: Part II* ed B Enquist and W Schmid (Berlin: Springer) pp 631–45

[12] Landau L D and Lifshitz E M 1960 *Mechanics* (Oxford: Pergamon)

[13] Lin C, Kogelnik H and Cohen L G 1980 Optical-pulse equalization of low-dispersion transmission in single-mode fibers in the 1.3–1.7 μm spectral region *Opt. Lett.* **5** 476–8

[14] Lions P-L 1985 The concentration compactness principle in the calculus of variations: The limit case: II *Rev. Mat. Iberoamericana* **1** 45–121

[15] Mollenauer L F, Stolen R H, and Gordon J P 1980 Experimental observation of picosecond pulse narrowing and solitons in optical fibers *Phys. Rev. Lett.* **45** 1095–8

[16] Mollenauer L F, Mamyshev P V, Gripp J, Neubelt M J, Mamysheva N, Gruner-Nielsen L and Veng T 2000 Demonstration of massive wavelength-division multiplexing over transoceanic distances by use of dispersion-managed solitons *Opt. Lett.* **25** 704–6

[17] Moore R, Sandstede B, Kath W, Jones C K R T and Alexander J 2001 Stability of multiple pulses in noisy optical fibers with phase-sensitive amplification *Opt. Commun.* **195** 127–39

[18] Ostrovskaya E A, Kivshar Y S, Scryabin D V and Firth W J 1999 Stability of multihump optical solitons *Phys. Rev. Lett.* **83** 296–9

[19] Romeo M M and Jones C K R T 2001 Stability of neuronal pulses composed of two concatenated unstable kinks *Phys. Rev.* E **63** 011904

[20] Sandstede B and Scheel A 2000 Gluing unstable fronts and backs together can produce stable pulses *Nonlinearity* **13** 1465–82

[21] Smith N J, Knox F M, Doran N J, Blow K J and Bennion I 1996 Enhanced power solitons in optical fibres with periodic dispersion management *Electron. Lett.* **32** 54–5

[22] Weinstein M 1986 Lyapunov stability of ground states of nonlinear dispersive evolution equations *Commun. Pure Appl. Math.* **39** 51–68

[23] Yew A, Sandstede B and Jones C K R T 2000 Instability of multiple pulses in coupled nonlinear Schrödinger equations *Phys. Rev.* E **61** 5886–92.

[24] Zakharov V E and Shabat A B 1972 Exact theory of two-dimensional self-focusing and one-dimensional nonlinear media *Sov. Phys.–JETP* **34** 62–9

[25] Zharnitsky V, Grenier E, Jones C K R T and Turitsyn S 2001 Stabilizing effects of dispersion management *Physica* D **152** 794–817

Chapter 5

Signal or noise? A nonlinear dynamics approach to spatio-temporal communication

Gregory D Van Wiggeren
Georgia Institute of Technology

Jordi Garcia-Ojalvo
Universitat Politecnica de Catalunya

Rajarshi Roy
University of Maryland

Communication with electromagnetic radiation has taken many guises and forms over the centuries. Recently, there have been efforts to utilize irregular waveforms output by chaotic lasers for encoding information. We explore the scientific basis for two such techniques that utilize polarization and spatio-temporal wavefronts and discuss possible future applications.

Our fascination with telecommunication has lead to the development of techniques that have increased the rate of information transmission at an exponential rate over the past century and a half [12]. Our focus has been on utilization of the amplitude or intensity of the radiation for this purpose, while harnessing the wavelength range available in the source (or sources) for multiplexing separate messages. At present we can communicate several Gbits/s over optical fibre communication channels, and have developed a sophisticated set of protocols for encoding and decoding the information. However there are still intriguing general questions: What is the maximum possible rate of communication of information (in bits/s) with a source of electromagnetic radiation, given the constraint that there is only a certain amount of power available? What would the spectrum of this radiation be?

On the basis of very general assumptions, Lachmann and colleagues [27] calculate the spectrum of this radiation source and arrive at an explicit expression for the maximum number of bits/s that could be transmitted as an information-carrying signal within prescribed constraints. They conclude that the spectrum of the source would be that of black body radiation, and emphasize that the radiation carrying the encoded message would appear as noise to anyone unable to decode the information. The characteristic temperature of the black body source would be determined by the power available, and the transmitter would utilize timing, energy, polarization and direction of the radiation; i.e. its spatio-temporal distribution, for encoding the message. If the spatio-temporal properties are not utilized, the information transmission rate possible drops by several orders of magnitude.

In this discussion, we focus on two new directions, attempting to utilize the properties of electromagnetic radiation more fully for communications. The first is the possibility of encoding and decoding information in the polarization fluctuations of a light source that is widespread in telecommunications—the erbium-doped fibre laser or EDFRL. Here, we use the vector property of light, its polarization, as a means for encoding and decoding information. We note that the polarization of light is possibly utilized for concealed intra-species communication by cuttlefish [40] and for inter-species communication by aliens [37]. The second direction concerns extending information coding and recovery to the spatio-temporal properties of the light waves. The emphasis will be on the use of wavefronts to transfer information in parallel, not in a bit-by-bit sequence.

Both ideas involve the use of nonlinear dynamical laser systems with time delays. Most laser cavities are of the order of or less than a metre in length and their operation can be described quite adequately by ordinary differential equations. The laser sources we discuss here are fibre oscillators with long cavities so that the time taken by light to go around is often several hundred nanoseconds—much longer than conventional systems. Such lasers generate signals with high-dimensional dynamics, and require delay-differential equations to model their behaviour. These models [1, 2, 48] will not be discussed in detail here; we will present their essential aspects and describe the main concepts involved and the experimental schemes through which they can be used for optical communication.

Traditional communication methods have used sine waves as carriers of information. A part of the receiver (the 'tuner') enables it to synchronize to the centre frequency of the carrier. The rest of the receiver is engaged with the task of recovering the information conveyed by the modulations of the amplitude or frequency of the carrier waveform.

It is natural to consider a generalization of the carrier to waveforms that are non-sinusoidal, and possibly irregular or chaotic. Over the past decade two developments have occurred that make it feasible to consider such methods of communication. The first is the realization that what are often considered unstable lasers, regarded as unsuitable for traditional communication techniques, emit light waves that are irregular and chaotic on very fast time scales and possess large bandwidths. The second is the extension of the concept of synchronization to signals that are spatio-temporally chaotic.

The study of chaos and nonlinear dynamics has made us realize that there are interesting deterministic dynamical phenomena to observe and measure that are highly irregular in time and space. Such systems have been discovered in every discipline and have lead to a plethora of studies that charted routes to chaos and developed quantitative mathematical measurements for characterizing the various signatures of nonlinear dynamical systems. The emphasis in the future will surely be to discover novel and interesting applications of such systems and to decipher how they may already be employed in biological contexts.

Though the synchronization of clocks (periodic systems) has been studied with great care over centuries, it was the surprising discovery of temporal synchronization between two chaotic systems [3, 16, 36] that initiated the field of 'chaotic communication'. Following the original implementation of this approach in electronic circuits [11, 23], a suggestion to use optical systems was made [10]. Optical chaotic systems offer the possibility of high-speed data transfer as shown in simulations of numerical models that include realistic operational characteristics of the transmitter, receiver and communication channel [34]. Recent work has shown that EDFRLs [31, 48] and semiconductor laser diodes [14] are capable of generating light waves that possess irregular and chaotic intensity fluctuations on very fast time scales. Abarbanel and coworkers [2] have developed a detailed delay-differential model for these EDFRL intensity dynamics. Numerical simulations of the model are able to reproduce many aspects of the experimentally observed dynamics. Also in recent work, a novel type of communication was demonstrated in which the chaotic intensity fluctuations generated by an EDFRL were used either to mask or carry a message [42–44]. Similar concepts were used to demonstrate optical communication using the chaotic wavelength fluctuation output from a diode laser [18]. Optical chaotic communication has also been demonstrated for semiconductor lasers with time-delayed optical feedback [15, 41].

In the rest of this chapter, we will outline two new directions:

(i) the exploration of the use of dynamical fluctuations of the polarization of light for fibre-optic communications (as detailed in section 1.2); and
(ii) a proposal for optical communications through free space using synchronization of broad area wavefronts that takes advantage of the parallelism of information transfer by spatio-temporal systems (as detailed in section 1.3).

These proposals utilize a general technique of time-delayed detection and recovery of signals, based on the fact that the transmitter itself is a dynamical system with a time delay (for a comprehensive introduction, see [44]). The message to be transmitted is encoded as a perturbation to the transmitter by a modulator that can change the amplitude, frequency or polarization of the light.

The signal from the transmitter is then sent over the communication channel, split into half and detected in two versions, one time delayed with respect to the other by the same time delay that is present in the transmitter. One of the halves may also be processed by passage through a generalized filter—a bandpass filter, polarization controller, nonlinear amplifier with specified gain spectrum, or other device, depending on the configuration and geometry of the transmitter. Dynamical changes introduced during one round-trip are compared to the signal from the previous round-trip and compared for synchronization. To recover the encoded message, the two halves are then combined through subtraction, division or other mathematical operation. Many examples of message encoding and recovery using amplitude modulation of the transmitter are given in our earlier work [44], which we extend here to dynamical modulation of the polarization state and spatio-temporal state of the field.

5.1 Communication with dynamically fluctuating states of light polarization

Information for communication can be encoded on the amplitude, frequency (phase) or polarization of a light wave. While the first two options have been explored extensively for decades [4], techniques to encode information on the polarization state of light have been developed only recently. Techniques in which the state-of-polarization (SOP) of light is used to carry information include multiplexing [13] and polarization-shift keying [7, 8]. In polarization multiplexing, a particular state of polarization is assigned to a given channel of information. The number of messages that can be multiplexed in this way depends on the precision with which one can identify different polarization states and maintain them in transmission through a fibre. Polarization-shift keying is the term used when one assigns a binary or multi-level digital coding to different states of polarization. In the case of a binary code, for example, left- and right-handed circularly polarized light could serve as zeros and ones.

The method described here is different [46]; there is no one-to-one correspondence between the SOP of the lightwave detected in the receiver and the value of the binary message bit that it carries. Instead, the binary message is used to modulate parameters of a dynamical laser system; in this case, an EDFRL laser. The modulation generates output light from the laser with fast, irregular polarization fluctuations. This light propagates through a communication channel to a suitable receiver, which is able to detect changes in the transmitter's polarization dynamics caused by the message signal and ultimately recover the message from the irregular polarization fluctuations of the transmitted light.

Here, we show that concepts related to those introduced by Van Wiggeren and Roy [42] and by Goedgebuer and colleagues [18] can be used to demonstrate communication using the irregular polarization-state fluctuations of light output by an EDFRL. When we began to study polarization fluctuations in the EDFRL

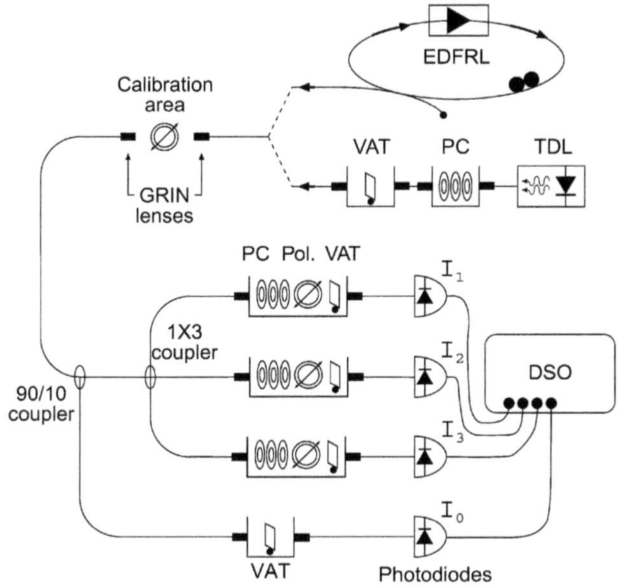

Figure 5.1. Polarization analyser apparatus. During the calibration process, light from the tunable diode laser (TDL) is transmitted to the apparatus. Once calibrated, light from an EDFRL is sent to the apparatus for measurement of the Stokes parameters. The variable attenuators (VATs) placed before the photodiodes prevent signal saturation. The polarization controllers (PCs) consist of a sequence of three waveplates and allow light from any input polarization to be adjusted to any output polarization state. The polarizers (Pol.) ensure that the photodiodes measure only one component of the light. All of these free space components are placed between gradient index (GRIN) lenses which allow the light to be coupled out of and back into the fibre. The digital sampling oscilloscope (DSO) records the intensities measured by the photodiodes. Reprinted from G D Van Wiggeren and R Roy High-speed fiber-optic polarization analyser: measurements of the polarization dynamics of an erbium doped fiber ring laser *Opt. Commun.* **164** 107–20 ©1999, with permission from Elsevier Science.

output some years ago, we did not find any available instruments to measure polarization-state fluctuations on a nanosecond time scale. To quantitatively specify the polarization state of light, one has to measure the Stokes parameters for the beam [6]. A novel high-speed fibre-optic polarization analyser shown in figure 5.1 was then constructed for this purpose by Greg Van Wiggeren [45]. The basic concept was to divide the light to be measured into four parts with matched time delays (accurate to a small fraction of a nanosecond) and send these beams through appropriate polarizing elements before measuring the intensities. We must then calibrate the system with suitable known sources of monochromatic polarized light and polarization controllers in order to compute

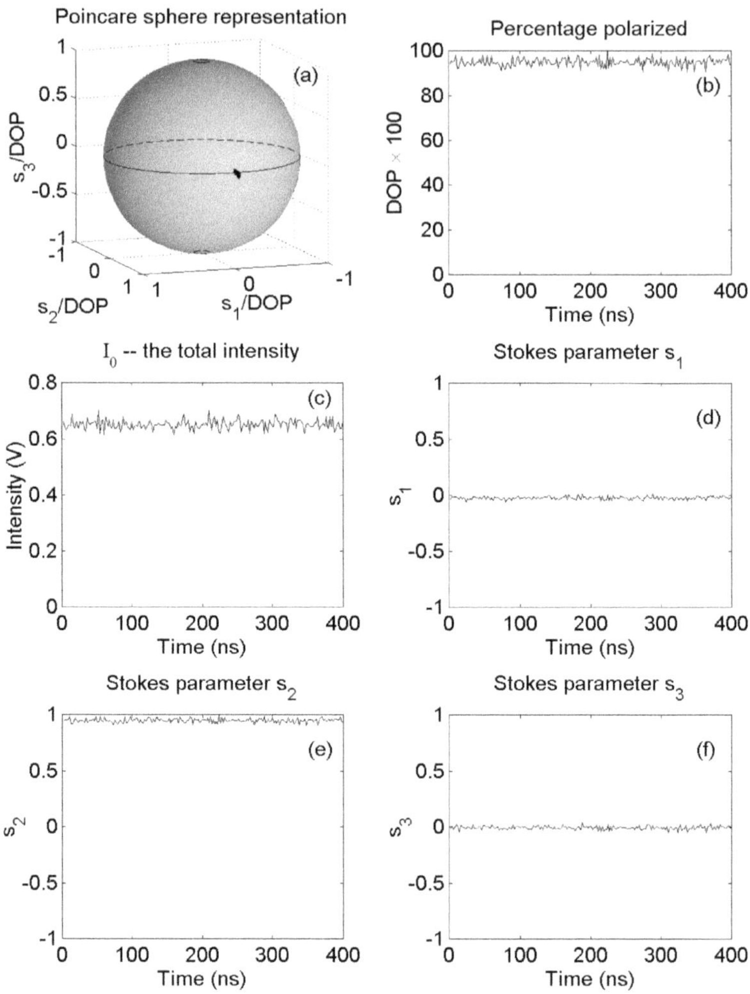

Figure 5.2. These measurements illustrate the accuracy of the polarization analyser in determining the state of polarization (SOP) of light produced by the EDFRL which possesses rapid intensity and polarization fluctuations. A polarizer at 45° has been placed in the calibration area. Thus the ideal measurement of the SOP should, in spite of the intensity fluctuations evident in panel (*c*), give $\vec{s} = (1, 0, 1, 0)$. The experimental measurement, as indicated by panels (*d*)–(*f*), is very close to this ideal. DOP represents the degree of polarization, which can be calculated according to the formula $\mathrm{DOP} = s_1^2 + s_2^2 + s_3^2$. The DOP shown in panel (*b*) is close to 100% as one would expect for light that passes through a polarizer. Reprinted from G D Van Wiggeren and R Roy High-speed fiber-optic polarization analyser: measurements of the polarization dynamics of an erbium doped fiber ring laser *Opt. Commun.* **164** 107–20 ©1999, with permission from Elsevier Science.

the polarization state of an unspecified light beam. Using this procedure, we were able to determine the Stokes parameters for the light beam from the EDFRL with a resolution of a couple of nanoseconds, limited in speed only by our detection electronics and digital sampling oscilloscope. The precision of the polarization analyser depends largely on the calibration process and electronics used. Figure 5.2 shows the Stokes parameters measured for light from the EDFRL after it was passed through a linear polarizer, and is therefore a calibration measurement for the apparatus. The size of the spot on the Poincaré sphere shows clearly the limits of precision for operation of the polarization analyser. The significance of the polarization fluctuations is shown in figure 5.3, which represents typical polarization fluctuations output from the EDFRL. The measurements were made in terms of normalized Stokes parameters which are plotted on the surface of the Poincaré sphere in figures 5.2 and 5.3. As is evident from figure 5.3, the polarization dynamics are quite irregular, and they are also localized on one side of the Poincaré sphere. The localization results from small levels of polarization-dependent loss (PDL) within the EDFRL itself, which tends to polarize the light output of the EDFRL due to differential net gain of certain polarization states in the system. The EDFRL model proposed by Abarbanel and colleagues [2] also displays similar behaviour for small levels of PDL.

The experimental apparatus used to demonstrate optical communication with dynamically fluctuating polarization states is shown in figure 5.4. The transmitter consists of a unidirectional EDFRL with a mandrel-type polarization controller (PC) and a phase modulator within the ring. The polarization controller consists of loops of fibre that can be twisted to alter their net birefringence. The phase modulator comprises a titanium in-diffused $LiNbO_3$ strip waveguide. Electrodes on either side of the waveguide induce a difference in the index of refraction between the TE and TM modes of the waveguide. In this way, it can also be used to alter the net birefringence in the ring. The length of the ring is about 50 m, which corresponds to a round-trip time for light in the ring of roughly 240 ns. This time delay in circulation of the light makes the dynamics observed in this type of laser quite different from those of more conventional cavities. A 70/30 output coupler directs roughly 30% of the light in the coupler into a fibre-optic communication channel while the remaining 70% continues circulating around the ring.

The communication channel, consisting of several metres of standard single mode fibre, transports the light to a receiver comprising two branches. Such a fibre does not maintain the polarization of the input light. Instead, due to random changes in birefringence along the length of the fibre, the polarization state of the input light evolves during its journey. The receiver is designed to divide the transmitted light into two branches. Light in the first branch of the receiver passes through a polarizer before being detected by photodiode A (125 MHz bandwidth). Light in the other branch passes through a polarization controller before it is incident on a polarizer. After passing through the polarizer, the light is measured by photodiode B (also 125 MHz bandwidth). Signals from these

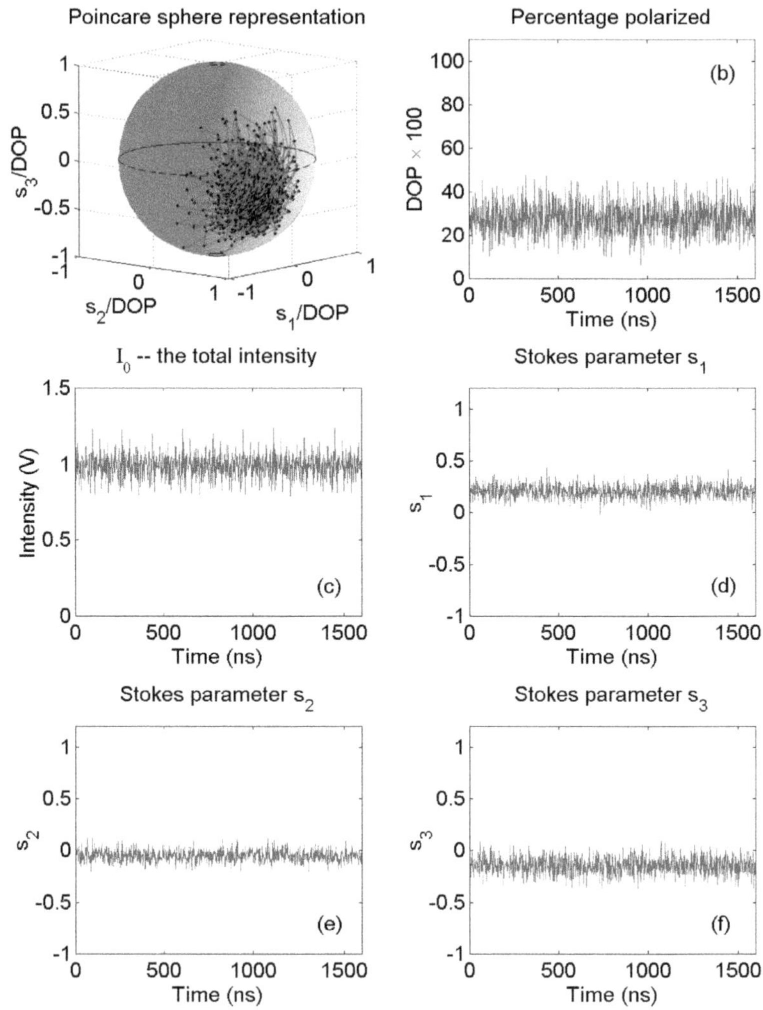

Figure 5.3. The polarization fluctuations of light generated by an EDFRL laser are shown plotted on the Poincaré sphere. (*a*) clearly shows that the fluctuations are somewhat localized about an elliptic SOP and the relatively low DOP in (*b*) indicates that these fluctuations are actually faster than can be observed even with the apparatus used here.

photodiodes are recorded by a digital oscilloscope at a 1 GS/s rate. A crucial element of the receiver operation is the time delay between the signals arriving at the photodiodes, which is equal to the round-trip time of the light in the transmitter laser. It is this dual detection with a time delay that allows us to differentially

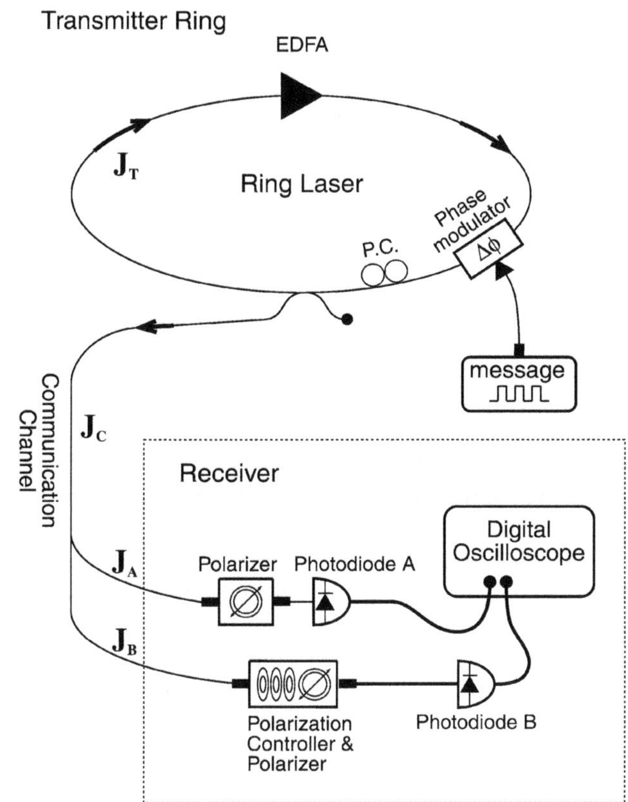

Figure 5.4. The apparatus for communication with dynamically fluctuating states of polarization. The transmitter consists of an EDFA, a polarization controller and a phase modulator. A time delay is present between the two photodiodes that is equal to the round-trip time for the light in the transmitter cavity. Reproduced from G D Van Wiggeren and R Roy 2002 Communication with dynamically fluctuating states of light polarization *Phys. Rev. Lett.* to appear ©2002 by the American Physical Society.

detect the dynamical changes that occur in the transmitter ring and recover the sequence of perturbations that constitute the digital message.

The electric field amplitude of a light wave located just before the output coupler in the transmitter EDFRL (see figure 5.4) at time t is given by the vector field $\mathbf{E}(t)$. As light propagates around the ring, the net action of the birefringence of the single mode fibre of the ring and other elements can be represented by a single unitary 2×2 $\mathbf{J}_T$, the so-called Jones matrix, with complex elements. Therefore,

$$\mathbf{E}(t + \tau_r) = \mathbf{J}_t \mathbf{E}(t) \tag{5.1}$$

where τ_r is the time for a single round-trip. It is assumed that the light field is not influenced significantly by noise or nonlinearities in the fibre or amplifier in a single round-trip, and that the PDL is negligible.

The elements of $\mathbf{J}_T$ can be changed by varying the voltage applied to the phase modulator in the EDFRL. In these experiments on communication, a data generator converts a binary message into a two-level voltage signal that is applied to the phase modulator. Thus, $\mathbf{J}_T$ can take on two different values,

$$\mathbf{J}_T = \mathbf{J}_0 \qquad \text{or} \qquad \mathbf{J}_T = \mathbf{J}_1 \tag{5.2}$$

depending on whether a '0' or a '1' bit is to be transmitted. In this way, the phase modulator drives the polarization dynamics of the EDFRL transmitter.

A fraction (30% in this experiment) of the light in the transmitter ring is coupled into a communication channel, which can be either free-space or fibre-optic. As mentioned previously, we chose a standard single mode fibre for our communication channel. There are variations of briefringence in this fibre which change the polarization state of the input light significantly even over short distances of a few metres. The effect of this birefringence is represented by a Jones matrix $\mathbf{J}_C$ which is assumed to be a unitary matrix with complex elements.

In the receiver, half of the light is directed toward photodiode B. Before it is measured by the photodiode, it propagates through the fibre, a polarization controller and a polarizer. The net effect of the birefringence in the fibre and the polarization controller can be represented by the unitary Jones matrix $\mathbf{J}_B$. The elements of $\mathbf{J}_B$ are completely controllable (with the constraint that the matrix is unitary) using the polarization controller. The effect of the polarizer is represented using another Jones matrix $\mathbf{P}$, though this matrix is non-unitary. Thus, the light actually measured by the photodiode in the arm B of the receiver can be written as a concatenation of Jones matrices:

$$\mathbf{P}\mathbf{J}_B\mathbf{J}_C\mathbf{E}(t). \tag{5.3}$$

However, the length of this arm is shorter to ensure that photodiode A measures a component of $\mathbf{E}(t+\tau_r)$ at the same time that photodiode B measures a component of $\mathbf{E}(t)$. Put another way, the light measured at photodiode A is the output from the transmitter one round-trip time after the light that is simultaneously measured by photodiode B. Thus, the field measured by photodiode A is

$$\mathbf{P}\mathbf{J}_A\mathbf{J}_C\mathbf{E}(t+\tau_r) = \mathbf{P}\mathbf{J}_A\mathbf{J}_C\mathbf{J}_T\mathbf{E}(t). \tag{5.4}$$

As mentioned earlier, when a '0' bit is to be communicated, the transmitter's Jones matrix, $\mathbf{J}_T$, is set equal to $\mathbf{J}_0$ by applying the proper voltage to the phase modulator. When $\mathbf{J}_T = \mathbf{J}_0$, synchronization of the signals measured by photodiodes A and B can be observed, provided that $\mathbf{J}_B$ is set properly using the polarization controller in that arm of the receiver. Synchronization in this situation occurs when

$$\mathbf{J}_B\mathbf{J}_C = \mathbf{J}_A\mathbf{J}_C\mathbf{J}_0. \tag{5.5}$$

It can be shown mathematically that this condition can always be satisfied by a proper adjustment of $\mathbf{J}_B$. For this proper setting of $\mathbf{J}_B$, synchronization is lost when the voltage applied to the phase modulator is switched so that $\mathbf{J}_T$ is set equal to $\mathbf{J}_1$ for communication of a '1' bit. Thus, by applying a binary voltage to the phase modulator in the transmitter, the signals measured by the photodiodes can be made to either synchronize or lose synchronization. The receiver interprets synchronized photodiode signals as '0' bits and unsynchronized signals as '1' bits.

Communication using this technique offers an unanticipated benefit compared with polarization shift keying, i.e. using one polarization direction for each bit. In fibre-optic communication channels, variations in temperature or stress in the channel induce fluctuations in the local birefringence of the channel that ultimately cause the channel Jones matrix, $\mathbf{J}_C$, to evolve in time. The evolution of $\mathbf{J}_C$ in typical commercial channels has been observed over tenths of seconds. Using the polarization shift keying technique, this evolution leads to ambiguity at the receiver which is harmful to accurate message recovery. In other words, the receiver cannot precisely determine the transmitted SOP from the received SOP. Complicated electronic tracking algorithms have been developed to compensate for this channel evolution in polarization-shift keying schemes.

In the technique presented here, no such complicated algorithms are necessary. If the transmitter Jones matrix for a '0' bit is chosen to be equal to the identity matrix, it can easily be shown that a proper choice of $\mathbf{J}_B$ can recover the message perfectly even when it is evolving in time. In this way, information can be transmitted through a channel with fluctuating birefringence, and a receiver can recover the information without any compensation for the channel's fluctuating birefringence. Conceptually, this is possible because, as previously mentioned, there is no one-to-one correspondence between the polarization state of the transmitted lightwave and the value of the binary message that it carries. In the method presented here, the message is encoded in the polarization dynamics (driven by the phase modulator) rather than in particular polarization states. Figure 5.5 shows experimental results demonstrating the communication technique described in this report. The message, the binary voltage signal shown in figure 5.5(*a*), drives the phase modulator as described earlier. It consists of a repeating sequence of 16 bits transmitted at 80 Mbits/s. At this rate, roughly 19 bits can be transmitted during the time light takes to complete one circuit around the ring. A completely random and non-repeating sequence of bits can also be transmitted successfully, but the repetitive message signal shown here provides some additional insights into the dynamics of the transmitter. Figure 5.5(*b*) shows the signal I_A measured by photodiode A in the receiver. Despite the repeating nature of the message, the measured signal does not show the same periodicity. Figure 5.5(*c*) shows the signal I_B measured simultaneously by photodiode B. As mentioned above, photodiode B measures the intensity, with a time delay, of a different polarization component of the same signal measured by photodiode A. A subtraction of I_B from I_A is shown

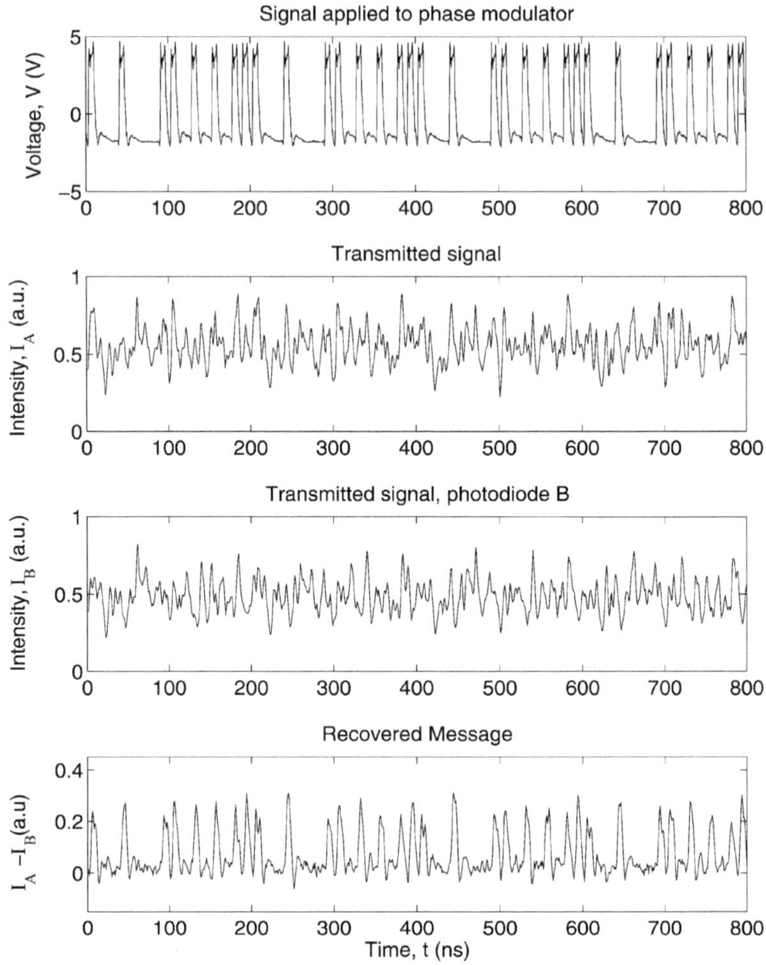

Figure 5.5. Demonstration of message recovery using dynamical fluctuations of SOP. Shown are the modulation voltage applied to the phase modulator (*a*); the transmitted signal measured by photodiode A (*b*); the time delayed signal measured by photodiode B (*c*); the message recovered (*d*) by subtraction of the data in (*c*) from that in (*b*). Loss of synchronization corresponds to '1' bits, while synchronization represents '0' bits. Reproduced from G D Van Wiggeren and R Roy 2002 Communication with dynamically fluctuating states of light polarization *Phys. Rev. Lett.* to appear ©2002 by the American Physical Society.

in figure 5.5(*d*). A '0' bit is interpreted when the subtracted signal is roughly zero because this value corresponds to synchronization of the signals measured by the

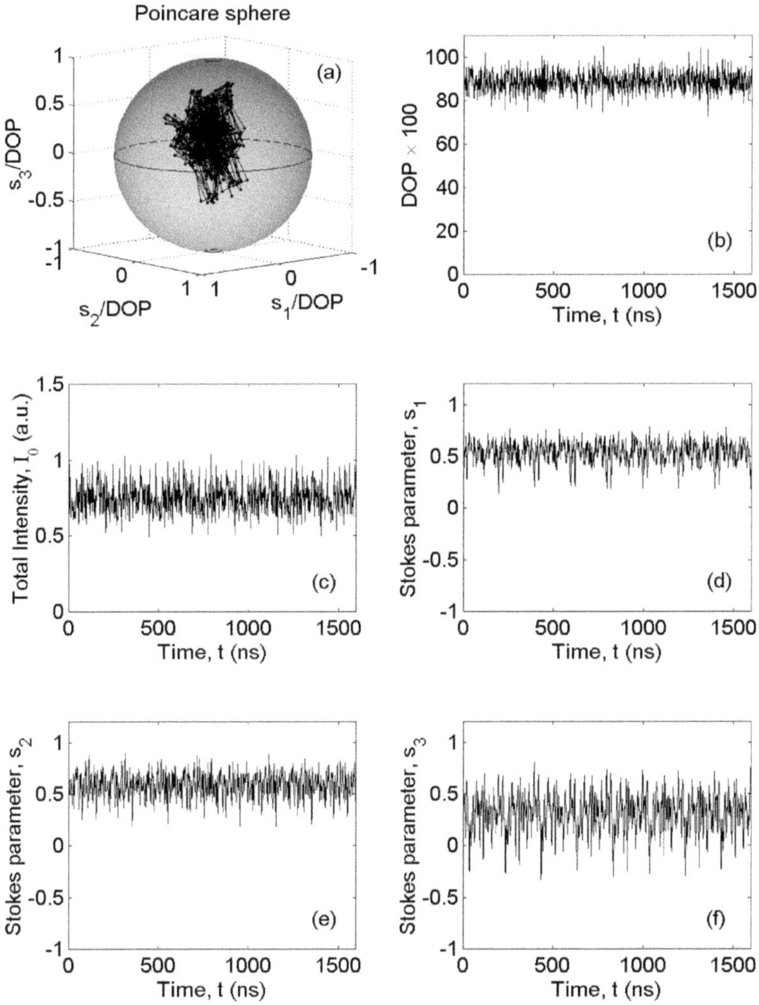

Figure 5.6. Measured polarization dynamics during message transmission are shown: the polarization dynamics of the transmitter are plotted on a Poincaré sphere (*a*); the degree-of-polarization (DOP) of the light output from the tranmsitter (*b*); the normalized Stokes parameters showing fluctuations in the SOP of the transmitted light (*d*)–(*f*). Reproduced from G D Van Wiggeren and R Roy 2002 Communication with dynamically fluctuating states of light polarization *Phys. Rev. Lett.* to appear ©2002 by the American Physical Society.

two photodiodes. A ‘1’ bit is interpreted when the difference signal is non-zero. We find that the transmitted message bits are accurately recovered by the receiver.

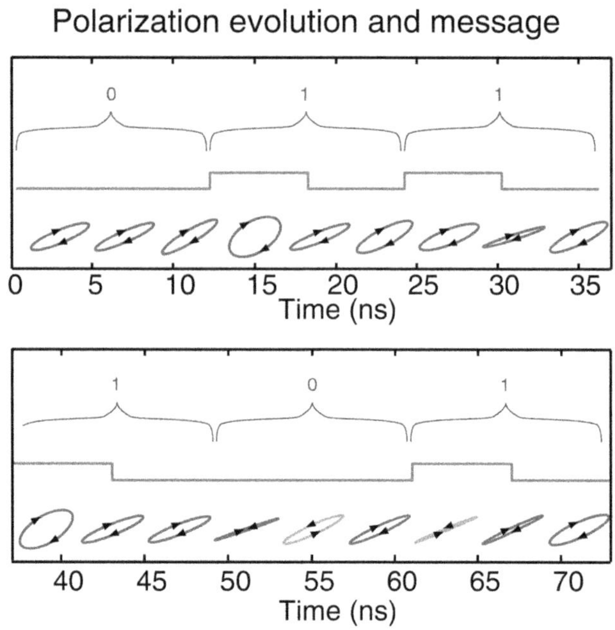

Figure 5.7. Schematic representation of the evolution of the SOP as message bits are encoded by the phase modulator in the transmitter. The ellipses display the SOP changes as measured every 4 ns by the polarization analyser. Flips from left- to right-handed polarization are seen to occur sometimes.

Figure 5.5(*d*) shows consistently positive values for the difference signal when a '1' is transmitted. This consistency can be attributed to the effect of the small amount of uncompensated PDL in the transmitter. With adjustments to the transmitter and receiver parameters, the difference signal could have been consistently negative, or even both positive and negative during a single '1' bit. While the message described in figure 5.5 was being communicated, the polarization analyser described earlier was used to measure the polarization dynamics of the transmitted signal. Figure 5.6 shows the results of this measurement. Figure 5.6(*a*) shows that the SOP of the transmitted light is localized in one region of the Poincaré sphere. This localization is, again, due to the small amount of PDL that remains in the transmitter. Figure 5.6(*b*) shows the calculated degree-of-polarization (DOP) of the light output from the transmitter. Figure 5.6(*c*) shows the total intensity of the light output from the transmitter. It shows a repetition with a period equal to the round-trip time for light in the ring. This same irregular, though nearly periodic, intensity behaviour is observed even when no message is being transmitted. Figure 5.6(*d*)–(*f*) shows the normalized Stokes parameters that were measured. The Stokes parameters reveal an imperfect periodicity of 200 ns. This is precisely the same period as the repeating 16-bit message. Although the message periodicity is revealed by the Stokes parameters, the message itself is not obviously disclosed. This periodicity may disappear if the PDL in the transmitter is further reduced. Figure 5.6 also illustrates one of the unique aspects of this communication technique. Previous discussion explained that in polarization-shift keying, a one-to-one correspondence exists between the polarization state of the light wave and the value of the message bit carried by the light wave. Figure 5.6 shows that this is certainly not true for the technique described here. It proves that the polarization state of the light is not simply alternating between two points on the Poincaré sphere as would be the case for polarization-shift keying. Instead, the polarization state fluctuates irregularly and evolves on time scales much shorter than the bit rate. This is schematically depicted in figure 5.7 where the ellipses show the state of polarization of the light as certain bits of information are transmitted. The polarization state occasionally flips from above the equator of the Poincaré sphere, to below it and vice versa. The direction of the arrows on the ellipse switch from clockwise to counterclockwise rotation of the field vector when this occurs.

The bit rate is limited only by the bandwidth of the detection apparatus. Measurements with a polarization analyser clearly show the irregular polarization and intensity dynamics of the transmitted light waves. This technique demonstrates the use of vector properties of a dynamical system, driven by the message, for optical communication.

5.2 Spatio-temporal communication with synchronized optical chaos

Our second example highlights the possibility of using chaotic signals as broadband carriers of spatio-temporal information. One of the most interesting recent realizations is that spatio-temporal waveforms display synchronization in nonlinear dynamical systems, and that the concept of synchronization needs to be generalized from the obvious one of identical synchronization to other less constrained forms [26], particularly in biological systems. We do not pretend to understand more than the most basic elements of the dynamics of coupled biological systems (in the eyes, ears, heart, brain or respiratory system, for example) at present; see also chapter 11 by Cowan and Bressloff, chapter 12 by Ermentrout and Osan and chapter 13 by Ditto. However, it is clear that experiments on physical spatio-temporal systems will be of great relevance to develop our understanding of spatio-temporal signal processing in biological ones.

In most optical realizations of communications with chaos, the message to be encoded drives the nonlinear transmitter, so that the message and carrier

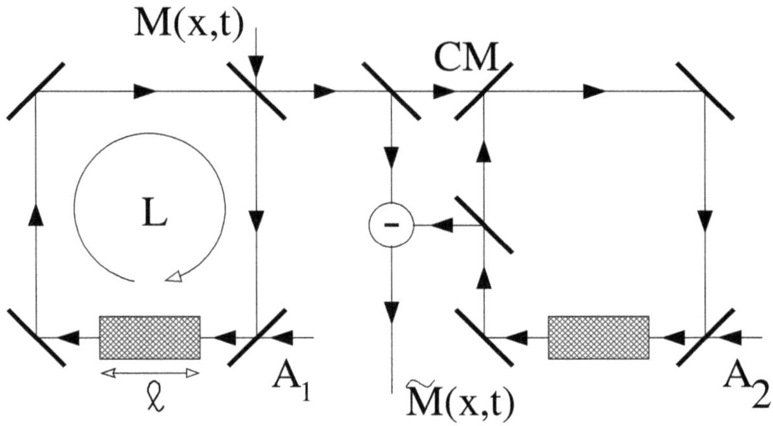

Figure 5.8. Scheme for communicating spatio-temporal information using optical chaos. CM is a coupling mirror. Reproduced from J Garcia-Ojalvo and R Roy 2001 Spatiotemporal communication with synchronized optical chaos *Phys. Rev. Lett.* **86** 5204–7 ©2001 by the American Physical Society.

become mixed in a non-trivial way. The resulting output is injected into a receiver, which, upon synchronization to the transmitter, allows for recovery of the original signal. The optical schemes developed so far have used purely temporal chaotic signals as information carriers. Here, we review a very recent proposal [17] for a nonlinear optical device exhibiting spatio-temporal chaos as the basis of a communication system capable of transmitting information in space and time. Spatio-temporal communication utilizes the inherent large-scale parallelism of information transfer that is possible with broad-area optical wave fronts.

Our scheme requires the existence of synchronization between transmitter and receiver and is a generalization of the model of synchronized ring cavities introduced by Abarbanel and Kennel [1]. Synchronization of spatio-temporal chaos was investigated extensively in previous years, but most studies were restricted to nonlinear oscillator arrays [24], coupled map lattices [49], and model partial differential equations [5, 9, 25]. Our system, however, is represented by an infinite-dimensional map spatially coupled in a continuous way by light diffraction. The experimental set-up proposed is shown schematically in figure 5.8. Two optical ring cavities are unidirectionally coupled by a light beam extracted from the left ring (the transmitter) and partially injected into the right ring (the receiver). Each cavity contains a broad-area nonlinear absorbing medium, and is subject to a continuously injected plane wave A_i. Light diffraction will be taken into account during propagation through the medium, in such a way that a non-uniform distribution of light in the plane transverse to the propagation direction may appear. In fact, an infinite number of transverse modes will, in principle, be allowed to oscillate within the cavity. A spatio-temporal message μ

can be introduced into the transmitter's cavity and recovered as μ' in the receiver, as explained later.

If no message is introduced and the receiver is absent, the transmitter is a standard nonlinear ring cavity, well known to exhibit temporal optical chaos [20, 21]. When transverse effects due to light diffraction are taken into account, a rich variety of spatio-temporal instabilities appear [33], giving rise to solitary waves [32,35], period-doubling bifurcations [19], spatial patterns [28], and spatio-temporal chaos [29, 38]. This latter behaviour is the one in which we are interested, since such chaotic waveforms will be used as information carriers in our set-up. The propagation of light through the nonlinear medium can be described by the following equation for the slowly varying complex envelope $E_n(x^r, t)$ of the electric field (assumed to be linearly polarized) in the nth passage through the resonator [38, 39]:

$$\frac{\partial E(x^r, z)}{\partial z} = \frac{\mathrm{i}}{2k}\nabla^2 E_n(x^r, z) - \frac{\alpha(1+\mathrm{i}\Delta)}{1+4|E_n|^2} E_n(x^r, z). \tag{5.6}$$

The first term on the right-hand side of (5.6) describes diffraction and the second describes saturable absorption. The propagation direction is denoted by z, whereas x^r is a vector in the plane orthogonal to the propagation direction. Equation (5.6) obeys the boundary condition

$$E_n(x^r, 0) = \sqrt{T}A + R\mathrm{e}^{\mathrm{i}kl} E_{n-1}(x^r, l) \tag{5.7}$$

which corresponds to the infinite-dimensional map that is the object of our analysis. The space point $z = 0$ in equation (5.7) denotes the input of the nonlinear medium, which has length l. The total length of the cavity is L. Other parameters of the model are the absorption coefficient α of the medium, the detuning Δ between the atomic transition and cavity resonance frequencies, the transmittivity T of the input mirror, and the total return coefficient R of the cavity (fraction of light intensity remaining in the cavity after one round-trip). The injected signal, with amplitude A and wavenumber k, is taken to be in resonance with a longitudinal cavity mode.

Previous studies have shown that for $\Delta < 0$, the model systems represented by equations (5.6) and (5.7) exhibit irregular dynamics in both space and time for large enough A [38, 39]. This spatio-temporally chaotic behaviour can become synchronized to that of a second cavity, also operating in a chaotic regime, coupled to the first one, as shown in figure 5.8. The coupling mechanism can be modelled in the following form [1]:

$$\begin{aligned} E_n^{(t)}(x^r, 0) &= M^{(t)}[E_{n-1}^{(t)}(x^r, 1)] \\ E_n^{(r)}(x^r, 0) &= M^{(r)}[(1-c)E_{n-1}^{(r)}(x^r, 1) + cE_{n-1}^{(t)}(x^r, 1)] \end{aligned} \tag{5.8}$$

where the application of $M^{(t,r)}$ represents the action of the map in equation (5.7) in every round-trip of the transmitter or receiver as specified by the superscript.

The coupling coefficient c is given by the transmittivity of the coupling mirror (CM) (figure 5.8). Earlier studies have shown that local sensor coupling is enough to achieve synchronization of spatio-temporal chaos in continuum models [22]. In our optical model, however, the whole spatial domain can be coupled to the receiver in a natural way. To estimate the synchronization efficiency of the earlier scheme, we have evaluated the synchronization error [25]

$$e_n = \sqrt{\frac{1}{S}\int_S |E_n^{(t)}(x^r,1) - E_n^{(r)}(x^r,0)|^2 \, \mathrm{d}x^r} \tag{5.9}$$

where S is the size of the system. This quantity has been computed for increasing values of the coupling coefficient c, by numerically integrating equation (5.6) with boundary condition (5.7) for both the transmitter and receiver operating in a regime of spatio-temporal chaos, and using the coupling scheme in equation (5.9). Simulations were performed in a 1D lattice of 1000 cells of size $\mathrm{d}x = 0.1$ spatial units, using a pseudospectral code for the propagation equation (5.6). Similar parameter values to those of Sauer and Kaiser [38, 39] are used here. The initially uncoupled systems evolve in time, starting from arbitrary initial conditions, and after 100 round-trips, when their unsynchronized chaotic dynamics is fully developed, the coupling is switched on. The synchronization error e_n is measured 100 round-trips later. The results are shown in figure 5.9(a), which plots the value of e_n for increasing coupling strengths. According to these results, a high degree of synchronization can be obtained for couplings as low as 40%. Another important issue to address at this point is how sensitive synchronization is to differences between the two cavities. We have extensively analysed the effect of different parameter mismatches on the synchronization error e_n. Our results indicate that parameters such as the absorption coefficient α, the detuning Δ, and the nonlinear medium length can be varied as much as 50% and still keep e_n below 10^{-2}. More sensitive parameters are the total length L of the cavity (due to its appearance in the phase-change term of equation (5.7)) and the amplitude A of the injected signal. Since the two cavity lengths can be matched experimentally, we now examine in detail the effect of a mismatch in A. This parameter could be controlled in real time if necessary and, hence, act as a control parameter for synchronization. The variation of e_n versus relative mismatch of A is shown in figure 5.9(b). It can be seen that the synchronization is quickly degraded as the two injected amplitudes differ, with e_n increasing well above 10^{-2} for mismatches of the order of 1%. Therefore, the value of A is critical for obtaining synchronization in the system.

We now use the synchronizing system described earlier to encode and decode information in space and time using the spatio-temporal chaotic carrier. We modify equations (5.9) according to figure 5.8 by inclusion of the message, which

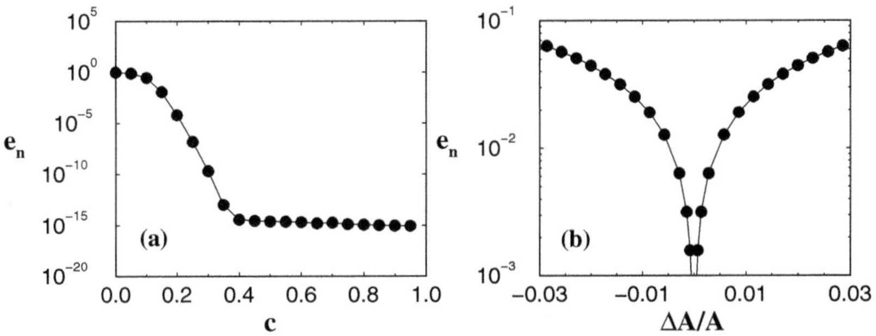

Figure 5.9. Synchronization error e_n versus coupling coefficient (*a*) and injected-amplitude mismatch (*b*). Parameters common to the two cavities are $\alpha = 100.0$, $\Delta = -10.0$, $R = 0.9$, $T = 0.1$, $k = 100.0$, $l = 0.01$, and $L = 0.015$. In (*a*), the common value of A is 7.0, which is also the value used for the transmitter in (*b*). Reproduced from J Garcia-Ojalvo and R Roy 2001 Spatiotemporal communication with synchronized optical chaos *Phys. Rev. Lett.* **86** 5204–7 ©2001 by the American Physical Society.

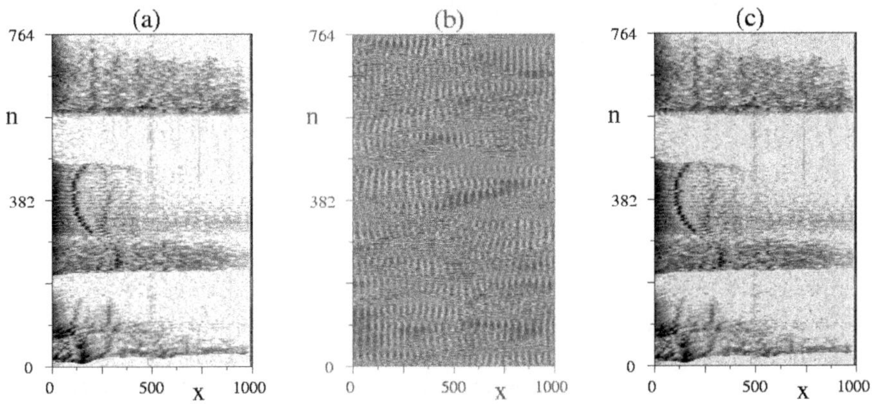

Figure 5.10. Transmission of 1D spatio-temporal data. Input spectrogram (*a*), real part of transmitted signal (*b*), and recovered data (*c*). Parameters are the same as shown in figure 5.9(*a*), and $c = 0.9$. Reproduced from J Garcia-Ojalvo and R Roy 2001 Spatiotemporal communication with synchronized optical chaos *Phys. Rev. Lett.* **86** 5204–7 ©2001 by the American Physical Society.

leads to

$$E_n^{(t)}(x^r, 0) = M^{(t)}[E_{n-1}^{(t)}(x^r, 1) + \mu_{n-1}(x^r)] \tag{5.10}$$
$$E_n^{(r)}(x^r, 0) = M^{(r)}[(1-c)E_{n-1}^{(r)}(x^r, 1) + c(E_{n-1}^{(t)}(x^r, 1) + \mu_{n-1}(x^r))].$$

If synchronization between transmitter and receiver is achieved, it will be possible to decode the signal by simply subtracting the transmitted signal and the one in the receiver:

$$\mu_n'(x^r) = E_n^{(t)}(x^r, 1) + \mu_n(x^r) - E_n^{(r)}(x^r, 1). \tag{5.11}$$

In the case of no mismatch, it can be seen analytically in a straightforward way that, as the coupling coefficient c tends to 1, the difference

$$|E_n^{(t)} - E_n^{(r)}| \to 0 \qquad \text{for all } x^r \tag{5.12}$$

which corresponds to perfect synchronization and, hence, to perfect spatio-temporal message recovery. It should be noted that the message is not merely added to the chaotic carrier, but rather the former *drives* the nonlinear transmitter itself. Therefore, as we will see in what follows, the amplitude of the message need not be much smaller than that of the chaotic signal to provide good masking of the information. The scheme described here was tested by encoding an analog 1D signal with complex evolution in space and time. The sample signal chosen is the spectrogram of a sample of speech. Chaotic systems have been used in the past to encode speech waveforms [1, 11], but the information that such signals provide is insufficient for voice-recognition purposes. Spectrograms, however, contain information on a broad range of frequencies as time evolves. Figure 5.10(*a*) shows a grey scale spectrogram of the word 'compute', with frequency components in the horizontal axis and time evolving from bottom to top. We will encode the frequency information in the 1D transverse direction of our set-up. The real part of the transmitted signal is shown in figure 5.10(*b*) for a message amplitude maximum of 0.5. This value should be compared to the maximum intensity of the chaotic carrier, which oscillates between 1 and 10, approximately, for the parameters chosen. The spatio-temporal chaotic state of the signal can be clearly observed. Finally, figure 5.10(*c*) shows the detected message, for a 90% coupling between transmitter and receiver. Figure 5.10 qualitatively shows that, even though coupling between transmitter and receiver is not complete, information varying in time *and* space can be successfully transmitted and recovered with the set-up sketched in figure 5.8. In order to have a quantitative measure of this effect, we have estimated the mutual information between input and output message signals, and its dependence on several system parameters. To that end, we discretize the values of μ and μ' in spacetime points, and compute the probability distributions $p(x)$, $p(y)$ and the joint probability $p(x, y)$ where x and y are the different values that μ and μ' may take, respectively. A measure of the mutual information between the two sets of data is given by

$$I = \sum_{x,y} p(x, y) \ln \left[\frac{p(x)p(y)}{p(x, y)} \right] \tag{5.13}$$

where the sums run over all possible values μ and μ'. This mutual information function I is 0 for completely independent data sets, and equal to the entropy of the common signal,

$$H = -\sum_{x} p(x) \ln p(x) \tag{5.14}$$

when the two messages are identical. Figure 5.11(a) shows the value of the mutual information I, for the message encoded in figure 5.10, versus the coupling coefficient c. It can be seen that, as c increases, I grows from 0 to perfect recovery, corresponding to the entropy of the input image, given by the horizontal dashed line in the figure. This result shows that, even though good synchronization is obtained for $c \lesssim 0.4$, satisfactory message recovery requires coupling coefficients closer to unity. This can also be seen by examining the behaviour of the entropy H of the recovered image, plotted as empty squares in figure 5.11(a): for values of c substantially smaller than 1, the entropy of the recovered data is appreciably larger than that of the input message, indicating a higher degree of randomness in the former. Finally, the behaviour of the mutual information in the presence of noise is shown as empty diamonds in figure 5.11(a). Uncorrelated, uniformly distributed noise is added continuously to the communication channel, with an amplitude of 1% that of the message. The results show that the scheme is reasonably robust, in agreement with previous studies [47]. We also examined the effect of parameter mismatch on the efficiency of message recovery. As in the synchronization characterization, we concentrated on the influence of the most sensitive parameter, namely, the amplitude A of the injected signal. The data plotted in figure 5.11(b) show that a slight mismatch in the value of A will degrade recovery, by leading to values of I much smaller that the entropy of the input message and leading to a recovered message with substantially larger entropy than the original. Finally, we should note that our set-up is also suitable for the transmission of two-dimensional information. To illustrate this, we have chosen to encode a static 2D image with the same mechanism discussed earlier. Figure 5.12 shows the results obtained in this case. As in figure 5.12, the left-hand plot depicts the input message, the middle plot the real part of the transmitted signal (a snapshot of it, in this case), and the right-hand plot the recovered data. The message amplitude maximum is now 0.01. Simulations are now performed on a square array with 256×256 pixels of width $\mathrm{d}x = 1.0$. The image is clearly recognizable even though the coupling coefficient is now as low as 0.7.

In summary, we have proposed a nonlinear optical model system that allows encoding and decoding information in space and time by means of spatio-temporal chaos synchronization. Synchronization occurs for a wide range of coupling values and system parameters. Spatiotemporal information can be successfully recovered for large enough coupling between the transmitter and receiver, and for small enough parameter mismatches. An infinite-dimensional map is required in this case for spatio-temporal communication with light wave fronts. The proposed set-up could be experimentally implemented upon identification of a suitable broad-area nonlinear medium such as a liquid crystal spatial light modulator.

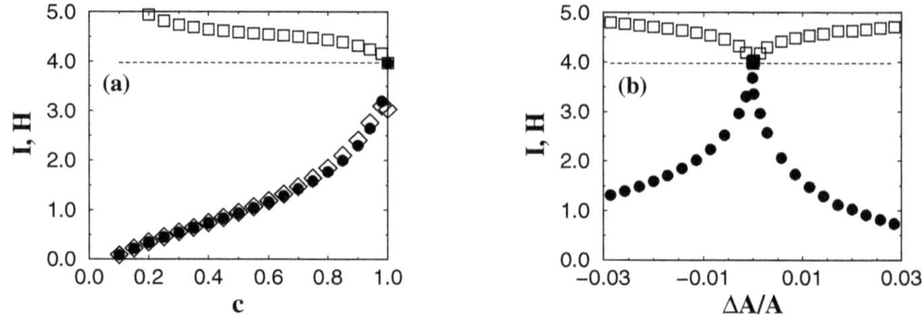

Figure 5.11. Information measures corresponding to the message encoding of figure 5.10: full circles, mutual information I; open squares, entropy H of the recovered data; horizontal dashed line, entropy of the original image. Empty diamonds are the values of I in the presence of noise (see text). Parameters are the same as in figure 5.9. Reproduced from J Garcia-Ojalvo and R Roy 2001 Spatiotemporal communication with synchronized optical chaos *Phys. Rev. Lett.* **86** 5204–7 ©2001 by the American Physical Society.

(*a*) (*b*) (*c*)

Figure 5.12. Transmission of a 2D static image. Input image (*a*), real part of the transmitted signal at a certain time (*b*), and recovered data (*c*). Parameters are the same as those in figure 5.9(*a*), and $c = 0.7$. Reproduced from J Garcia-Ojalvo and R Roy 2001 Spatiotemporal communication with synchronized optical chaos *Phys. Rev. Lett.* **86** 5204–7 ©2001 by the American Physical Society.

5.3 Conclusions

Some of the most challenging and interesting problems for scientists and engineers lie in the application of the concepts of nonlinear dynamics and the mathematical tools that have emerged over the last four decades to practical problems and to enhancing our understanding of the dynamics of biological systems. We are as yet nowhere near achieving the limits of communication theoretically possible with electromagnetic radiation. It would be of great interest to learn whether nonlinear dynamical concepts and tools can help us to approach these limits through suitable encoding processes and carrier waveforms and help us to develop more robust schemes for communication. It is clear that biological systems process vast amounts of information more efficiently and more flexibly than we have learned to do so far.

Acknowledgments

Financial support from the Office of Naval Research as well as from NATO (Project No CRG.971571) is acknowledged. JGO also acknowledges support from DGES (Spain). The spectrogram was provided by the Center for Spoken Language Understanding, Oregon Graduate Institute of Science and Technology. The two-dimensional image of the bird was provided by A+ Free Clip Art.

References

[1] Abarbanel H D I and Kennel M B 1998 Synchronizing high-dimensional chaotic optical ring dynamics *Phys. Rev. Lett.* **80** 3153–6

[2] Abarbanel H D I, Kennel M B and Lewis C T 1999 Chaotic dynamics in erbium-doped fibre ring lasers *Phys. Rev.* A **60** 2360–74

[3] Afraimovich V S, Verichev N N and Rabinovich M I 1986 General synchronization *Radiophys. Quantum Electron.* **29** 795–803

[4] Agrawal G P 1997 *Fiber-Optic Communication Systems* 2nd edn (New York: Wiley)

[5] Amengual A, Hernández-García E, Montagne R and San Miguel M 1997 Synchronization of spatio-temporal chaos: the regime of coupled spatio-temporal intermittency *Phys. Rev. Lett.* **78** 4379–82

[6] Azzam R M A and Bashara N M 1989 *Ellipsometry and Polarized Light* (Amsterdam: Elsevier)

[7] Betti S, De Marchis G and Iannone E 1992 Polarization modulated direct detection optical transmission systems *J. Lightwave Technol.* **10** 1985–97

[8] Benedetto S, Gaudino R and Poggiolini P 1997 Polarization recovery in optical polarization shift-keying systems *IEEE Trans. Commun.* **45** 1269–79

[9] Boccaletti S, Bragard J and Arecchi F T 1999 Controlling and synchronizing space time chaos *Phys. Rev.* E **59** 6574–8

[10] Colet P and Roy R 1994 Digital communication with synchronized chaotic lasers *Opt. Lett.* **19** 2056–8

[11] Cuomo K M and Oppenheim A V 1993 Circuit implementation of synchronized chaos with applications to communications *Phys. Rev. Lett.* **71** 65–8

[12] Davis C C 1996 *Lasers and Electro-Optics* (Cambridge: Cambridge University Press)

[13] Evangelides S G Jr, Mollenauer L F, Gordon J P and Bergano N S 1992 Polarization multiplexing with solitons *J. Lightwave Technol.* **10** 28–35

[14] Fischer I, van Tartwijk G H M, Levine A M, Elsäßer W, Göbel E and Lenstra D 1996 Fast pulsing and chaotic itinerancy with a drift in the coherence collapse of semiconductor lasers *Phys. Rev. Lett.* **76** 220–3

[15] Fischer I, Liu Y and Davis P 2000 Synchronization of chaotic semiconductor laser dynamics on subnanosecond time scales and its potential for chaos communication *Phys. Rev.* A **62** 011801(R)

[16] Fujisaka H and Yamada T 1983 Stability theory of synchronized motion in coupled-oscillator systems *Prog. Theor. Phys.* **69** 32–47

[17] Garcia-Ojalvo J and Roy R 2001 Spatiotemporal communication with synchronized optical chaos *Phys. Rev. Lett.* **86** 5204–7

[18] Goedgebuer J-P, Larger L and Porte H 1998 Optical cryptosystem based on synchronization of hyperchaos generated by a delayed feedback tunable laser diode *Phys. Rev. Lett.* **80** 2249–52

[19] Haelterman M 1993 Ikeda instability and transverse effects in nonlinear ring resonator *Opt. Commun.* **100** 389–98

[20] Ikeda K 1979 Muliple-valued stationary state and its instability of the transmitted light by a ring cavity system *Opt. Commun.* **30** 257–61

[21] Ikeda K, Daido H and Akimoto O 1980 Optical turbulence: chaotic behaviour of transmitted light from a ring cavity *Phys. Rev. Lett.* **45** 709–12

[22] Junge L and Parlitz U 2000 Synchronization and control of coupled Ginzburg–Landau equations using local coupling *Phys. Rev.* E **61** 3736–42

[23] Kocarev L, Halle K S, Eckert K, Chua L O and Parlitz U 1992 Experimental demonstration of secure communication via chaotic synchronization *Int. J. Bif. Chaos* **2** 709–13

[24] Kocarev L and Parlitz U 1996 Synchronizing spatio-temporal chaos in coupled nonlinear oscillators *Phys. Rev. Lett.* **77** 2206–9

[25] Kocarev L, Tasev Z and Parlitz U 1997 Synchronizing spatio-temporal chaos of partial differential equations *Phys. Rev. Lett.* **79** 51–4

[26] Kurths J, Pikovsky A and Rosenblum M G 2001 *Synchronization* (Cambridge: Cambridge University Press)

[27] Lachmann M, Newman M E J and Moore C 2001 The physical limits of communication *Santa Fe Institute Preprint* 99-07-054

[28] Le Berre M, Patrascu A S, Ressayre E and Tallet A 1996 Daisy patterns in the passive ring cavity with diffusion effects *Opt. Commun.* **123** 810

[29] Le Berre M, Patrascu A S, Ressayre E and Tallet A 1997 Localized structures in chaotic patterns: From disorder to ordering *Phys. Rev.* A **56** 3150–60

[30] Liu Y and Davis P 2000 Synchronization of chaotic mode hopping *Opt. Lett.* **25** 475–7

[31] Luo L, Tee T J and Chu P L 1998 Bistability of erbium-doped fibre laser *J. Opt. Soc. Am.* B **15**

[32] McLaughlin D W, Moloney J V and Newell A C 1983 Solitary waves as fixed points of infinite-dimensional maps in an optical bistable ring cavity *Phys. Rev. Lett.* **51** 75–8

[33] McLaughlin D W, Moloney J V and Newell A C 1985 New class of instabilities in passive optical cavities *Phys. Rev. Lett.* **54** 681–4

[34] Mirasso C R, Colet P and García-Fernández P 1996 Synchronization of chaotic semiconductor lasers: application to encoded communications *IEEE Photon. Technol. Lett.* **8** 299–301

[35] Moloney J V 1986 Many-parameter routes to optical turbulence *Phys. Rev.* A **33** 4061–78

[36] Pecora L M and Carroll T L 1990 Synchronization in chaotic systems *Phys. Rev. Lett.* **64** 821–4

[37] Sagan C 1997 *Contact* (New York: Pocket Books) p 83

[38] Sauer M and Kaiser F 1996 Crisis-induced intermittency in a spatially extended nonlinear optical system *Int. J. Bif. Chaos* **6** 1481–94

[39] Sauer M and Kaiser F 1996 Synchronized spatio-temporal chaos and spatio-temporal on-off intermittency in a nonlinear ring cavity *Phys. Rev.* E **54** 2468–73

[40] Shashar N, Rutledge P and Cronin T W 1996 Polarization vision in cuttlefish: a concealed communication channel *J. Exp. Biol.* **199** 999–1004

[41] Sivaprakasam S and Shore K A 1999 Demonstration of optical synchronization of chaotic external-cavity laser diodes *Opt. Lett.* **24** 466–8

[42] Van Wiggeren G D and Roy R 1998 Communication with chaotic lasers *Science* **279** 1198–200

[43] Van Wiggeren G D and Roy R 1998 Optical communication with chaotic waveforms *Phys. Rev. Lett.* **81** 3547–50

[44] Van Wiggeren G D and Roy R 1999 Chaotic communications using time delayed optical systems *Int. J. Bif. Chaos* **9** 2129–56

[45] Van Wiggeren G D and Roy R 1999 High-speed fiber-optic polarization analyser: measurements of the polarization dynamics of an erbium doped fiber ring laser *Opt. Commun.* **164** 107–20

[46] Van Wiggeren G D and Roy R 2002 Communication with dynamically fluctuating states of light polarization *Phys. Rev. Lett.* **88** 097903

[47] White J K and Moloney J V 1999 Multichannel communication using an infinite dimensional spatio-temporal chaotic system *Phys. Rev.* A **59** 2422–6

[48] Williams Q L, García-Ojalvo J and Roy R 1997 Fast intracavity polarization dynamics of an erbium-doped fibre ring laser: Inclusion of stochastic effects *Phys. Rev.* A **55** 2376–86

[49] Xiao J H, Hu G and Qu Z 1996 Synchronization of spatio-temporal chaos and its application to multichannel spread-spectrum communication *Phys. Rev. Lett.* **77** 4162–5

Chapter 6

Outstanding problems in the theory of pattern formation

Edgar Knobloch
University of Leeds

Many systems of interest in physics, chemistry and the biological sciences exhibit spontaneous symmetry-breaking instabilities that form structures that we may call *patterns* [14, 54]. The appearance of convection cells in a fluid layer heated from below [81] or of vortex structures in the flow between two independently rotating cylinders [22, 81] are familiar examples from fluid mechanics. But there are many other systems that form patterns. Recently studied examples of spatially periodic structures include the Turing instability [110] and the Faraday system [83, 101]. Vertically vibrated granular media exhibit similar pattern-forming instabilities [99, 124]. Related instabilities have been identified in the primary visual cortex and may be responsible for hallucinations (see chapter 11 by Bressloff and Cowan). Other systems form spiral waves or target patterns, emanating from apparently random locations, perhaps triggered by impurities or specks of dust [61]. A remarkable recent discovery is that of oscillons, or localized oscillations, in vertically vibrated granular media [156]. These oscillons may form 'bound states' that have been called dimers. Certain chemical systems break up into spots which grow, and fission into new spots or disappear depending on the density of spots around them [88, 112]. Such pattern-forming systems thus behave almost like a colony of living organisms.

The problems these experiments and simulations raise are fundamental. These range from the particular to the general. Typical of the former are studies of specific systems, which attempt to identify pattern-forming instabilities and their subsequent evolution using the (known) field equations describing the system. This approach may use analytical techniques (stability theory, perturbation methods) but often revolves around direct numerical simulation of the system. Significantly, the experiments and simulations often show that different systems behave in an almost identical fashion, regardless of the specific equations governing the system and the physical mechanism behind the pattern-forming instability. For this reason I focus here on the *universal* aspects of pattern formation—those aspects that are independent of the details of the governing equations. Indeed, even when the governing field equations are known, detailed quantitative comparisons with experiments have only rarely been attempted. Often a theory is only tested against measurements of linear stability thresholds. Such tests, while valuable, cannot be taken as confirming the theory. In general it is difficult to integrate the three-dimensional field equations describing a system, particularly if it is spatially extended. In such circumstances one resorts to simplifying assumptions which are believed to retain the essence of the system. These simplified formulations yield a great deal of valuable information, but often preclude detailed quantitative comparisons. This is regrettable, since a qualitative agreement between experiment and theory may in fact be fortuitous. The recent study of binary fluid convection by Batiste *et al* [8] shows just how difficult it is to perform quantitative comparisons even with very carefully documented experiments.

In the remainder of this section I mention a number of outstanding issues in the *theory* of pattern formation that are of particular personal interest. The list is not intended to be exhaustive, but rather to stimulate interest in certain topics, some longstanding, some new, where progress is needed, and to summarize both the state of the field and formulate the right questions. Some of these are then explored in greater detail in sections 6.1–6.5.

For simplicity the discussion is divided into two subsections, loosely called 'weakly nonlinear theory' and 'fully nonlinear theory'. To some extent this is an artificial division, and there is much overlap between the two sets of topics. A relatively recent review of the subject may be found in [32].

6.0.1 Weakly nonlinear theory

- Applicability and rigorous justification of envelope equations
- Boundary conditions for envelope equations
- Pattern formation via non-local envelope equations
- Dynamics of weakly damped driven systems
- Applicability of integrable PDEs to physical systems
- Quasi-patterns

Within weakly nonlinear theory we know very well how to deal with bounded systems undergoing a pattern-forming instability. Such systems have a spectral gap between any marginally unstable eigenvalue and the remaining stable eigenvalues. When there are no zero eigenvalues related to continuous symmetries such as rotational invariance, centre manifold reduction shows that the dynamics of the stable modes are slaved to the slow evolution of the near marginal mode (or modes). A systematic procedure has been developed for dealing with problems of this type. First, the critical parameter value(s) is (are) identified at which one or more modes are neutrally stable; these modes have eigenvalues on the imaginary axis and are called centre modes. Centre manifold reduction is performed on the system (ODEs or PDEs) to obtain the equations governing the (slow) evolution of the system on the centre manifold. Since the centre manifold is, in general, low-dimensional this procedure amounts to a dramatic reduction in the number of degrees of freedom of the system.

Next, near-identity nonlinear coordinate changes are performed to put the resulting equations into a simple form, called a normal form, subject to the requirement that they preserve the *dynamics* of the system. Consequently the stability properties of the various fixed points and limit cycles are unaffected by this simplification. The structure of the resulting normal form depends only on the linear problem and any symmetries (or degeneracies) that may also be present. Hence the same normal form finds application in a large variety of circumstances.

These normal forms must then be 'unfolded' by including the terms generated by small changes of the parameter(s) away from critical. In many cases these normal forms and their unfoldings have been worked out, and their dynamics are well understood. Different systems that reduce to the same normal form differ only in the values of the coefficients of the nonlinear terms in these equations. Separate techniques exist for the computation of these coefficients, and typically employ symbolic manipulation routines. It is sufficient usually to compute these coefficients for the critical parameter value(s). This approach has seen (and will continue to see) a large number of applications [29].

The main theoretical issue which remains unsolved concerns the truncation of the normal forms and conditions for the adequacy of the suggested unfoldings. In practice one does not know whether the complete (local) behaviour of the system is captured by any finite truncation of the normal form or how many unfolding parameters are required. In other words, one does not know whether the *codimension* of the problem is finite. These questions have only been answered for steady-state bifurcation problems where a general procedure for answering these questions exists [56], at least in simple cases. Most problems, however, involve dynamics and it appears unlikely that a corresponding theory can be developed for such cases. In practice, the pragmatic approach of truncating the unfolded normal form at various orders and comparing the results works well. Often the inclusion of higher order terms and/or the full parameter dependence of the nonlinear terms enlarges the range of applicability (in parameter space) of the theory, sometimes dramatically [161]. Liapunov–Schmidt reduction [56] can be used to establish rigorous results about solutions of simple type (steady states, relative equilibria or periodic orbits) even when other more complicated solutions are present.

Symmetry, if present, usually plays a paradoxical role. On the one hand it may increase the (geometric) multiplicity of the critical eigenvalue, leading to a higher-dimensional centre manifold and typically multiple solution branches. In particular if the centre manifold is at least three-dimensional, complex dynamics may be present near the bifurcation [2]. However, the presence of the symmetry can be exploited to solve these higher-dimensional equations, and hence allows one to understand behaviour that would be difficult to understand in generic systems [29]. For example, the action of the symmetry on an unstable eigenfunction may generate a linearly independent unstable eigenfunction with the same eigenvalue. This is the case if the instability breaks the symmetry of the problem and the symmetry acts absolutely irreducibly on the space of eigenfunctions. In contrast, continuous symmetries such as SO(2) which do not act absolutely irreducibly [57] do not generically admit steady-state symmetry-breaking bifurcations at all: in SO(2) invariant systems the generic symmetry-breaking bifurcation is a Hopf bifurcation producing a rotating wave [48].

In unbounded systems the situation is much less well understood. Here there is usually no spectral gap and the stable eigenvalues accumulate on the imaginary axis. In addition in two or more spatial dimensions there is often an orientational degeneracy. This is so, for example, in spatially isotropic systems. In such systems the linear stability theory predicts the wavenumber of the marginally stable modes but not their direction. Both problems can be traced to the presence of a non-compact symmetry group, the Euclidean group in two or three dimensions. Two techniques have been developed to deal with these problems. Envelope equations employ a slowly spatially varying envelope to take into account wavevectors near a particular marginal wavevector, but because they focus on the behaviour of modes of a certain type they are unable to capture significant changes in the magnitude and direction of the dominant wavevector [106]. The derivation of the corresponding equations is formal and, until recently, there were no rigorous results about the validity of the commonly used envelope equations. In one dimension the first such results were given by Van Harten [158] and Schneider [137]. These authors show that solutions of the Ginzburg–Landau equation track solutions of the original PDE on $O(\epsilon^{-1})$ length scales for $O(\epsilon^{-2})$ times. The most complete results on the validity and universality of the Ginzburg–Landau equation are given by Melbourne [97, 98] who discusses in detail the significance of higher order terms that depend on the fast spatial variable, and the way these terms lock the envelope to the underlying pattern.

Both these problems can be avoided by posing the PDE on a *lattice*, i.e. by imposing periodic boundary conditions in the unbounded directions. The translations then form a compact symmetry group (circle or torus group), and the allowed wavevectors become discrete. Problems of this type are amenable to straightforward analysis, and have now been worked out for both steady-state and Hopf bifurcations in two and, in some cases, three dimensions. This approach has had a number of successes, both in the theory of convection [58] and in the Faraday system [27, 140, 155]. The main reason is that the amplitude equations are easy to generate using group-theoretic techniques that utilize the symmetry properties of the chosen lattice, and that their solutions can be analysed and their stability properties determined in the abstract. Even without coefficient calculation the results from this approach often provide a qualitative understanding of the behaviour of the system. For some of the higher-dimensional representations the corresponding analysis may be quite involved and techniques from computer algebra (Gröbner bases etc) prove invaluable [16]. A brief review of these techniques is provided in section 6.1. Although there are still a number of cases that have not been worked out, the main theoretical interest undoubtedly centres on finding ways to relax the periodic boundary conditions and thereby make the analysis applicable to a broader class of problems. Attempts to do this include extensions of the Newell–Whitehead formalism to two- and three-dimensional patterns [18, 60, 69, 145] but are neither rigorous nor entirely convincing.

Although the use of envelope or amplitude equations is common they are typically used as model equations. This is the case when the equations are not derived via a rational expansion and hence still depend on the expansion parameter; in other cases the small parameter is retained in the boundary conditions imposed on the envelope equation. However, it is even more common to replace the boundary conditions by periodic ones, and to argue that the resulting solutions apply to systems in large domains. This approach is, in general, dangerous, and is no substitute for deriving the correct boundary conditions at the same time as the envelope equation. Unfortunately this often entails matching the solutions of the envelope equation to boundary layers computed from the original field equations, and is quite involved. In fact, the correct boundary conditions for the Hopf bifurcation in extended domains, hitherto treated heuristically [31], have only been derived recently [93, 94], while those for steady-state bifurcation are discussed in [122]. In this connection it is important to emphasize that, however large the domain sufficiently close to onset, the dynamics are inevitably strongly effected by the boundaries. This is simply because the primary instability corresponds to the excitation of a global eigenmode of the system, and centre manifold reduction guarantees that the behaviour near onset resembles this mode. Thus the system is *large*, in the sense that other modes participate in the dynamics in addition to the eigenmode, only when $\epsilon L \gg 1$. Here L is the dimensionless domain length and $\epsilon^2 \equiv |R - R_c|/R_c$ measures the distance from onset. This scaling is the reason why low-dimensional descriptions of nominally extended systems can still be successful—provided they are employed in the regime $\epsilon \ll L^{-1}$, cf [8].

In many cases the formal asymptotic methods lead to non-local amplitude or envelope equations. This is typically a consequence of the presence of some constraints such as mass conservation or the presence of multiple time scales. In catalysis, pressure effects are often responsible for the presence of non-local (global) coupling [72, 100]. Charge density waves in n-type semiconductors produced by the Gunn instability are described by a local equation subject to a non-local constraint due to an imposed voltage bias across the system [66]. In other cases, non-local terms are present due to the excitation of near marginal large scale modes, such as mean flows. These cases are examined in more detail in sections 6.3 and 6.4.

A great deal of effort has been devoted to the derivation of envelope or long-wave equations for ideal (i.e. non-dissipative) systems, such as ideal fluid flow. Water wave theory has led to several much studied equations of this type, the nonlinear Schrödinger equation, and the Korteweg–de Vries and Davey–Stewartson equations. All three are examples of *integrable* equations and admit localized structures called solitons that interact in a particle-like manner. However, real fluids are dissipative and the solitons must be sustained against dissipation by weak forcing. For example, in optical transmission lines solitons require periodic (spatial) pumping to restore their amplitude and compensate losses. The resulting system is no longer completely integrable; indeed, it is not even conservative. Equations of this type possess *attractors* which describe the asymptotically stable structures. These no longer interact in a particle-like manner and are called solitary waves. It should be mentioned that even solitons may be shown to be stable if small perturbations to their shape relax back to the original shape by radiating energy to infinity [113]. The analysis is delicate and requires coming to grips with eigenvalues embedded in a continuous spectrum, once again a consequence of an unbounded domain.

The relation between the behaviour of integrable systems and the dynamics of the corresponding weakly damped driven system constitutes a major challenge to research. While it is true that the dissipation and forcing 'select' a subset of the solutions of the integrable system, much more can (and does) happen, particularly in the regime in which the dissipation, forcing and nonlinearity are all comparable. A particularly simple example of what may happen is provided by recent work of Higuera *et al* [67] on the damped driven non-local Schrödinger equation. In this work several global bifurcations are identified that are responsible for the presence of complex dynamics. Different global bifurcations are responsible for multi-pulse homoclinic orbits when the damping is small relative to the other two terms, as discussed in detail by Haller and Wiggins [63–65].

The final topic under the heading of *weakly nonlinear theory* that I wish to mention is the issue of quasi-patterns. Such patterns are well known in physics, and have been generated in pattern-forming systems such as the Faraday system [49]. Of course a true quasi-pattern has structure over the whole plane, but the Fourier spectra of the observed patterns resemble those of true quasi-patterns. Such patterns are *not* periodic in space and, in general, form as a result of the presence of two or more incommensurate wavenumbers or wavevectors. Because these wavevectors are incommensurate there are, strictly speaking, no spatial resonances. As a result, formal asymptotic theory reveals no resonant terms and the amplitude equations that result contain to all orders only terms generated by self-interaction. The stability properties of the states that result are degenerate, since they possess zero eigenvalues that are not forced by symmetry. Most likely what is happening is the following: the near resonances lead to large values of the coefficients of (some of) the higher-order terms; and this fact reduces the range of validity of the amplitude equation: the higher the order of the equation the smaller the region of validity. To my knowledge even the simplest problem on the real line, the interaction between wavenumber 1 and an irrational wavenumber q is unsolved. What is necessary is to approximate q via its continued fraction expansion approximations, r_n/s_n, and examine the radius of convergence of the resulting amplitude equations as $s_n \to \infty$. One can think of this type of theory as an extension of KAM theory (see, e.g., [73] and references therein), well studied in the context of temporal resonance, into the spatial domain.

6.0.2 Fully nonlinear theory

- Dynamics of fronts and defects, and their interaction
- Existence and stability of spatially localized structures
- Classification of instabilities of nonlinear structures
- Wavelength selection close to and far from onset
- Relation between discrete and continuum systems
- Relation between unbounded and large but bounded systems
- Frequency selection in open systems
- Transition to spatio-temporal chaos close to and far from onset

In the context of *fully nonlinear theory* there is a number of interesting and important topics, many of which also arise in the context of the amplitude or envelope equations derived within weakly nonlinear theory, now considered as field theories in their own right. These concern the basic mechanism for the selection of the speed of a front (these may be linear [32] or nonlinear [9]), and existence and interaction of defects, such as spiral waves in reaction–diffusion systems. Many excitable systems exhibit spirals that are initiated by a finite amplitude perturbation. How such spiral structures are related to bifurcation from the trivial state remains unknown. The existence of spirals in a plane was proved by Scheel [132] but such theorems do not provide much information about the structure of the spiral (frequency, wavenumber and amplitude). It is of interest that observed spirals do not resemble the eigenfunctions associated with a Hopf bifurcation in planar systems described by polar coordinates. These either decay as $r^{-1/2}$ if the wavenumber is real or grow (or decay) exponentially if it is complex. Golubitsky *et al* [53] suggest that the latter case in fact provides a description of the *core* of the spiral, and that the constant amplitude wave seen outside the core is a consequence of nonlinearities that affect the solutions as $r \to \infty$ arbitrarily close to onset. In their view the core–spiral interface is a nonlinear front whose location is fixed by the imaginary part of the wavenumber selected by the Hopf frequency. This wavenumber is complex because in polar coordinates waves propagate preferentially in one direction (either outwards or inwards), so that the wavenumber is selected by the requirement that the instability be *absolute* [150]. This approach also appears to explain the breakup of spiral waves into 'chemical turbulence' far from the core [129, 150], as observed in some experiments [109]. These ideas await confirmation through quantitative studies of the relation between the spiral frequency and the core profile. It should be noted that spirals can also undergo a core instability [7, 129].

Localized structures come in a variety of forms. Perhaps the most interesting are the *oscillons* observed in granular media vibrated up and down [156]. These localized oscillations leave the rest of the system undisturbed, and may form 'bound' states that have been called dimers. These localized structures are believed to form as a consequence of a subcritical instability, essentially by the mechanism identified by Thual and Fauve [149]. However, there is no convincing explanation for their presence, largely because of an almost complete lack of understanding of the equations of motion governing granular media. There are many other systems where a subcritical bifurcation is responsible for the formation of steady localized states, including the recent discovery of 'convectons', that is, localized regions of field-free convection in a fluid layer with a uniform vertical magnetic field imposed across it [11], and related structures in reaction–diffusion systems [107]. These states can all be thought of as homoclinic connections from the trivial state back to itself. For the stability of the resulting state it is essential that the state at $x \to \pm\infty$ (i.e. the trivial state) be stable. It follows that such localized states can only be present for $\mu < 0$, where $\mu = 0$ denotes the primary bifurcation threshold. The situation is more interesting in systems undergoing a Hopf bifurcation since here homoclinic connections to the origin may be stable even for $\mu > 0$, cf [117, 138]. Homoclinic connections can also describe localized wavenumber and amplitude changes that may propagate, without change of profile, through a travelling wave-train, as in the examples computed by Spina *et al* [141]. Recent theory, based on the coupled complex Ginzburg–Landau equations (see section 6.4) indicates that such 'homoclinic holes' should be unstable [159], although this is clearly not always the case. Evidently the predictions of the theory, for example concerning the speed of propagation of such structures, should be checked against both numerical experiments using model equations and other field equations, and against actual experiments. It is likely that there are many such solitary wave solutions, although most are presumably unstable [131, 159], so that mechanisms that stabilize these structures need to be identified. Evans' function techniques are invaluable for studies of the stability properties of these states [51, 148].

The question of stability of finite amplitude structures, be they periodic or localized, and their bifurcation is a major topic that requires new insights. The theory for the former requires developments of a theory describing bifurcation from group orbits of solutions while the latter requires a theory for the bifurcation that results when an unstable eigenvalue emerges from the continuum. Bifurcations from a group orbit often produce drifts along the group orbit [22, 82, 126], resulting in unexpected *dynamics* arising from steady-state bifurcations [82]. The best known example of this bifurcation is the so-called parity-breaking bifurcation. This is a bifurcation from a circle of reflection-symmetric equilibria. At the bifurcation (at $\mu = 0$, say) there is a zero eigenvalue whose eigenvector breaks the reflection symmetry of the equilibrium. This zero eigenvalue is *in addition* to the zero eigenvalue due to rotation invariance, and together they are responsible for the ensuing drift of the solution along the group orbit, either clockwise or counterclockwise; the drift speed vanishes as $\mu^{1/2}$ as $\mu \downarrow 0$. Figure 6.1 shows some of the more complicated patterns that can be created in secondary bifurcations from a group orbit, this time from a pattern of standing hexagonal oscillations [127].

Bifurcations involving the continuum are also non-standard but in a different way, and lead to a saturation amplitude that scales like μ^2 instead of the more usual $\mu^{1/2}$, where $\mu > 0$ is now the growth rate of the instability. This is the case, for example, in the beam–plasma instability [26]. Related phenomena arise in ideal hydrodynamics. In particular, instability of a vorticity defect in inviscid plane Couette flow is described by a non-local evolution equation closely related to the Vlasov equation [5], with related non-local equations, obtained by matching across a critical layer, describing the evolution of long wavelength perturbations of marginally stable shear flows [6]. Such flows are prepared by deforming the shear profile until the necessary and sufficient conditions for instability are satisfied, in the same way that a beam–plasma instability is triggered by a bump on the tail of the particle distribution in plasma physics. Similar phenomena also arise in the theory of phase-coupled oscillators [25, 142]. Much remains to be learnt about these systems.

A major unsolved problem that has attracted attention at least since the 1960s is the problem of wavelength selection in extended or unbounded systems. Linear theory identifies the wavelength of the first unstable disturbance. However, in the nonlinear regime the observed wavelength usually differs. Analysis, initiated by Busse [13], revealed that above the neutral stability curve $\mu = \mu(k)$ there is usually an interval of stable wavenumbers k, limited by various secondary instabilities, such as the Eckhaus, skewed varicose, oscillatory, and zigzag instabilities. In most cases these instabilities leave a region of stable wavenumbers in the (μ, k)-plane, nowadays called the Busse balloon, but do not select a unique wavenumber. Yet experiments usually follow a unique path through this region, as discussed in the context of Rayleigh–Bénard convection by Koschmieder [81]. Thus although this problem has led to profound developments such as the introduction of phase equations (e.g., the Kuramoto–Sivashinsky equation) into the theory of pattern formation, and the notion of sideband (Eckhaus, Benjamin–Feir) instabilities, the original problem remains unsolved. In fact, the phase description is *the* appropriate description of *finite* amplitude patterns [105, 111]. This approach goes beyond envelope equations and has served to highlight the role played by focusing instabilities in wavelength selection. Unfortunately, the description breaks down at defects where the phase of the pattern is, by definition, undefined.

In many cases one supposes that a discretized version of a completely integrable PDE will behave the same way as the PDE. However, the discrete system may not be integrable and one needs to understand in detail how the integrals of motion appear as the continuum limit is approached; and indeed how the dynamics of the discrete system approach those of the integrable system. These issues have a profound significance for accurate numerical simulation of integrable PDEs.

One is often tempted to suppose that distant boundaries have a negligible

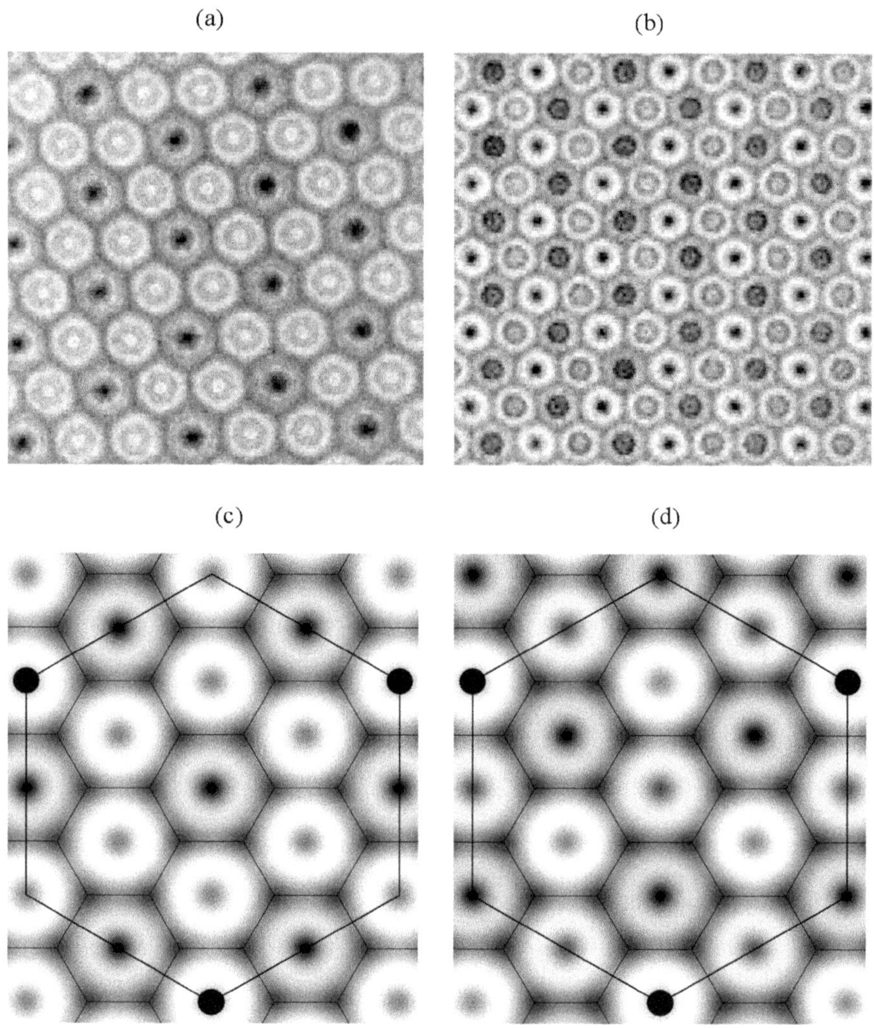

Figure 6.1. Experimental and reconstructed surface wave patterns in silicone oil with two-frequency temporal forcing ($f_1 : f_2 = 2 : 3$) arising from secondary instabilities of a small-scale hexagonal pattern: (*a*) hexagonal pattern on two scales; (*b*) pattern with instantaneous triangular symmetry; and (*c*), (*d*) theoretical reconstruction of (*a*), (*b*). Reproduced from A M Rucklidge, M Silber and J Fineberg Secondary instabilities of hexagons: a bifurcation analysis of experimentally observed Faraday wave patterns *Bifurcation, Symmetry and Patterns* ed J Buescu, S Castro, A P Dias and I Labouriau, to appear ©2002 by Birkhäuser Verlag.

effect on the process of pattern selection. While this may be so for steady-state pattern-forming instabilities for which the boundaries shift the threshold for primary instability by $O(L^{-2})$ where L measures the domain size [19], and modify the pattern substantially only near the boundary where matching to the boundary conditions is effected, systems supporting propagating waves behave quite differently. In such systems the wave is always in contact with the boundary, and the boundaries may exert profound influence. This is especially so when the waves have a preferred direction of propagation, as already mentioned in the context of our discussion of spiral waves. Since such waves cannot be reflected (all reflected waves are evanescent) the downstream boundary acts like an absorbing boundary. To overcome this dissipation the threshold for instability is shifted to $\mu = \mu_{\mathrm{f}} > 0$. It turns out that μ_{f} is related to the threshold μ_{a} for *absolute* instability [70] in the unbounded system: $\mu_{\mathrm{f}} = \mu_{\mathrm{a}} + O(L^{-2})$. For values of μ between the convective instability threshold in an unbounded system, $\mu = 0$, and μ_{f}, a disturbance originating at the upstream boundary grows as it propagates towards the downstream boundary, and piles up against it, increasing its wavenumber to such an extent that it ultimately decays. Thus in systems with broken reflection symmetry the presence of boundaries, however distant, changes the threshold for instability by an $O(1)$ amount! Of course an $O(L)$ transient is present before the presence of the downstream boundary manifests itself, a property of the system that can be traced to the non-normality of the linear stability problem [152, 153]. This system therefore demonstrates that the mere existence of boundaries may have a fundamentally important effect, and thus represents a situation in which the boundaries cannot be treated perturbatively, however far apart they may be. Additional consequences of boundaries in this system are discussed in section 6.5.

In problems of this type the upstream boundary serves as a ‘pacemaker’: this boundary selects the frequency which then determines the downstream amplitude and wavenumber from a nonlinear dispersion relation. If the amplitude at the upstream boundary is sufficiently small this frequency will be close to the linear theory frequency (and hence the frequency ω_{a} predicted by the global instability condition); in other cases the frequency solves a nonlinear eigenvalue problem and must be computed numerically [151]. As discussed in section 6.5 much remains poorly understood about these systems, particularly in cases with phase slips at the front that separates the upstream and downstream parts of the solution. Once again, we may think of the upstream part as the core of a spiral, and the downstream part as the fully developed (visible) spiral.

The final topic on the list is the transition to complex spatio-temporal behaviour in extended systems. While a certain number of routes to temporal chaos in low-dimensional dynamical systems have been identified and analysed [59, 108], the situation in spatially extended systems is much more complex. These systems are plagued by very long transients (in fluid systems one typically has to wait several *horizontal* diffusion times before transients die out) and there is even the possibility that the characteristic time on which the system relaxes may effectively diverge at some scale L, such that for scales $\ell > L$ the system 'never' finds a stable equilibrium even though one (or more) may be present. Loosely speaking one can think of an 'energy landscape' that has so many (steady) states, most of which are non-stable, that the system spends forever wandering in this 'landscape'. Analogies with annealing problems come to mind. Schmiegel and Eckhardt [134, 135] have explored some of the consequences of this picture in the context of shear flow instability.

6.1 Pattern selection on lattices

As an example of the type of results that may be obtained when pattern selection problems are posed on lattices, we describe briefly the results for the Hopf bifurcation on a square lattice [139]. We take an isotropic, spatially homogeneous system in a plane, with a trivial state $\Psi(x_1, x_2) = 0$, and suppose that this state loses stability to a symmetry-breaking Hopf bifurcation at $\mu = 0$. The linear stability theory predicts the associated Hopf frequency ω_c, and the critical wavenumber $k_c = |\mathbf{k}|$, assumed to be non-zero. In the following we impose periodic boundary conditions in two orthogonal directions, hereafter x_1, x_2, with period $2\pi/k_c$. This assumption reduces the symmetry group of the problem from $E(2)$, the Euclidean group of rotations and translations in two dimensions, to the group $\mathrm{D}_4\dot{+}T^2$, the semi-direct product of the symmetry of a square and a two-torus of translations, and selects the four wavevectors $\mathbf{k}_1 = k_c(1, 0)$, $\mathbf{k}_2 = k_c(0, 1)$, $\mathbf{k}_3 = k_c(-1, 0)$, $\mathbf{k}_4 = k_c(0, -1)$ from the circle of marginally stable wavevectors. We may therefore write the most general marginally stable eigenfunction in the form

$$\Psi(x_1, x_2) = (v_1(t)\mathrm{e}^{\mathrm{i}k_c x_1} + v_2(t)\mathrm{e}^{\mathrm{i}k_c x_2} + w_1(t)\mathrm{e}^{-\mathrm{i}k_c x_1} + w_2(t)\mathrm{e}^{-\mathrm{i}k_c x_2}) f(y) \quad (6.1)$$

where $y \equiv x_3$ denotes any transverse variables (if present). In the following we assume that the linear stability problem takes the form $\dot{\mathbf{z}} = \mu(\lambda)\mathbf{z}$, where $\mathbf{z} \equiv (v_1, v_2, w_1, w_2) \in \mathbb{C}^4$, and $\mu(0) = \mathrm{i}\omega_c$, $\mathrm{Re}(\mu'(0)) > 0$. Here λ is the bifurcation parameter. It follows that the quantities $(|v_1|, |w_1|)$ and $(|v_2|, |w_2|)$ represent amplitudes of left- and right-travelling waves in the (x_1, x_2) directions, respectively.

The group D_4 is generated by counterclockwise rotations by 90° (hereafter $\rho_{\pi/2}$) and reflections $x_1 \to -x_1$. These symmetries act on $\mathbf{z} \in \mathbb{C}^4$ as follows:

$$\rho_{\pi/2}: \quad (v_1, v_2, w_1, w_2) \to (w_2, v_1, v_2, w_1) \qquad \rho_{\pi/2} \in \mathrm{D}_4 \quad (6.2)$$

$$\kappa: \quad (v_1, v_2, w_1, w_2) \to (w_1, v_2, v_1, w_2) \qquad \kappa \in \mathrm{D}_4. \quad (6.3)$$

In addition spatial translations $(x_1, x_2) \to (x_1 + \theta_1/k_c, x_2 + \theta_2/k_c)$ act by

$$(\theta_1, \theta_2): \quad (v_1, v_2, w_1, w_2) \to (\mathrm{e}^{\mathrm{i}\theta_1} v_1, \mathrm{e}^{\mathrm{i}\theta_2} v_2, \mathrm{e}^{-\mathrm{i}\theta_1} w_1, \mathrm{e}^{-\mathrm{i}\theta_2} w_2)$$

$$(\theta_1, \theta_2) \in T^2. \quad (6.4)$$

Table 6.1. Fixed point subspaces corresponding to the different isotropy subgroups Σ of Γ. The coordinates specifying each Fix(Σ) are given in terms of $\mathbf{z} = (z_1, z_2, z_3, z_4)$ with each $v_j \in \mathbb{C}$, $j = 1, 2$.

	Name	Fix(Σ)	Σ
0.	Trivial solution (T)	$\mathbf{z} = (0, 0, 0, 0)$	Γ
I.	Travelling rolls (TR)	$\mathbf{z} = (v_1, 0, 0, 0)$	$\mathrm{SO}(2) \times \widetilde{\mathrm{SO}}(2)$
II.	Standing rolls (SR)	$\mathbf{z} = (v_1, 0, v_1, 0)$	$\mathrm{Z}_2 \times \mathrm{SO}(2)$
III.	Standing squares (SS)	$\mathbf{z} = (v_1, v_1, v_1, v_1)$	SO(2)
IV.	Alternating rolls (AR)	$\mathbf{z} = (v_1, iv_1, v_1, iv_1)$	SO(2)
V.	Standing cross-rolls (SCR)	$\mathbf{z} = (v_1, v_2, v_1, v_2)$	Z_2

Finally, in normal form the dynamical equations will commute with an S^1 phase shift symmetry in time acting by

$$\hat{\phi}: \quad \mathbf{z} \to \mathrm{e}^{\mathrm{i}\phi}\mathbf{z} \qquad \phi \in S^1. \tag{6.5}$$

As a result the full symmetry group of the dynamical equations near $\lambda = 0$ is $\Gamma \equiv \mathrm{D}_4 \dot{+} T^2 \times S^1$. An examination of the action of Γ on $\mathbf{z} \in \mathbb{C}^4$ shows that there are four axial isotropy subgroups of Γ with two-dimensional fixed points subspaces. The equivariant Hopf theorem [57] guarantees the existence of primary branches of solutions with the symmetries of these subgroups. These solutions, listed in table 6.1, are called standing squares (SS), and travelling (TR), standing (SR) and alternating rolls (AR). However, in the present example there is, in open regions of parameter space, a fifth primary solution branch whose existence is not revealed by the abstract theory. Such branches are sometimes called *submaximal*; in the present case the submaximal branch corresponds to standing cross-rolls (SCR). At present the only way of locating submaximal branches is by explicit calculation.

For this purpose we write down the most general set of equations commuting with these symmetries, truncated at third order [139]

$$\dot{v}_1 = \mu v_1 + (a|v_1|^2 + b|w_1|^2 + c|v_2|^2 + c|w_2|^2)v_1 + dv_2 w_2 \bar{w}_1 \tag{6.6}$$
$$\dot{v}_2 = \mu v_2 + (a|v_2|^2 + b|w_2|^2 + c|w_1|^2 + c|v_1|^2)v_2 + dv_1 w_1 \bar{w}_2 \tag{6.7}$$
$$\dot{w}_1 = \mu w_1 + (a|w_1|^2 + b|v_1|^2 + c|w_2|^2 + c|v_2|^2)w_1 + dv_2 w_2 \bar{v}_1 \tag{6.8}$$
$$\dot{w}_2 = \mu w_2 + (a|w_2|^2 + b|v_2|^2 + c|v_1|^2 + c|w_1|^2)w_2 + dv_1 w_1 \bar{v}_2. \tag{6.9}$$

The construction of these equations is algorithmic, and requires first the construction of the Hilbert basis of invariant functions and then the Hilbert basis of equivariant vector fields [16, 139]. These results can be used to generate the required amplitude equations to any desired order, via a procedure that can be automated. These equations can be used not only to compute the five non-trivial primary branches, but also to determine their stability properties with respect to perturbations on the chosen lattice, i.e. with respect to perturbations in $(v_1, v_2, w_1, w_2) \in \mathbb{C}^4$. At present there are no techniques for the *a priori* exclusion of submaximal primary branches, a fact that serves as an obstruction to a completely group-theoretic analysis of the primary bifurcation. In the present case the submaximal branch exists in an open region in coefficient space, and is always unstable [139, 146], although in other problems no restrictions on the existence and stability properties of the submaximal solutions are present [36].

The corresponding results for the steady-state bifurcation on a square lattice were first obtained by Swift [146], see also [55, 157], while those for the hexagonal lattice were obtained by Golubitsky *et al* [58]. The former case describes the competition between rolls and squares and shows that at most one of these states can be stable near onset and that the stable state is the one with the larger amplitude. On the hexagonal lattice the generic situation leads to a primary bifurcation with no (locally) stable branches. Hexagonal solutions bifurcate transcritically but are unstable on both sides of $\mu = 0$; rolls are either supercritical or subcritical but in either case are also unstable. This result is an example of a general result: primary branches in amplitude equations possessing quadratic equivariants are unstable whenever these equivariants do not vanish in the corresponding fixed point subspace. The hexagons on either side of $\mu = 0$ differ. In the convection context we speak of $H^{\pm}$, with H^{+} representing hexagons with rising fluid in the centre, while H^{-} denotes hexagons in which the fluid descends in the centre. Of these the subcritical one usually gains stability at a secondary saddle-node bifurcation. This, and secondary bifurcations to a branch of triangles, can be studied by looking at systems with a weakly broken mid-plane reflection symmetry [58]. It is important that hexagons and triangles not be confused. Unfortunately, these solutions are distinguished unambiguously only if the basic instability wavelength is known.

The Hopf bifurcation on a line is a special case of the Hopf bifurcation in a plane just discussed (set $v_2 = w_2 = 0$) and in amplitude-phase variables reduces to the steady-state bifurcation with D_4 symmetry [146], with travelling (rotating) waves taking the place of rolls, and standing waves taking the place of squares. Consequently here too at most one of these states can be stable near onset, and the stable state is the one with the larger amplitude. The corresponding results for the hexagonal lattice were worked out by Roberts *et al* [125], who computed the 11 primary branches guaranteed by the equivariant branching lemma together with their stability properties. Rotating lattices have also been analysed. A large number of applications of these results have now been worked out (e.g. [23]), but a major unsolved problem remains: there is no rigorous theory that would allow us to establish the stability of squares with respect to hexagonal perturbations and vice versa, essentially because there is no spatially periodic lattice that accommodates both solution types. Only in systems described by a potential can one compare the extrema of the potential corresponding to these states, and hence determine which of these states is stable (or metastable if both correspond to minima). It should be noted that certain sufficiently low-order truncations of the amplitude equations for steady-state bifurcations do yield gradient vector fields. This is, however, an artifact of the truncation and amplitude equations do not, in general, possess variational structure.

A possible approach to the vexed problem of the competition between squares and hexagons is based on higher-dimensional representations of the symmetry groups of planar lattices. In the case of the square lattice such a representation arises when periodic boundary conditions with a larger period are employed. For example, the reciprocal square lattice can intersect the circle of marginally stable wavevectors in eight instead of four places. These wavevectors are parametrized by two relatively prime integers $\alpha > 0$, $\beta > 0$, with the critical wavevectors given by $\mathbf{K}_1 = \alpha\mathbf{k}_1 + \beta\mathbf{k}_2$, $\mathbf{K}_2 = -\beta\mathbf{k}_1 + \alpha\mathbf{k}_2$, $\mathbf{K}_3 = \beta\mathbf{k}_1 + \alpha\mathbf{k}_2$, $\mathbf{K}_4 = -\alpha\mathbf{k}_1 + \beta\mathbf{k}_2$, where $\mathbf{k}_1 = (1, 0)k_c$, $\mathbf{k}_2 = (0, 1)k_c$, and produce a countably infinite number of eight-dimensional representations of $D_4\dot{+}T^2$. Group-theoretic results show that in this case there are six axial isotropy subgroups [40]; the new patterns that are possible include super-squares, anti-squares and two types of rectangular patterns. A similar theory for the 12-dimensional representation of $D_6\dot{+}T^2$ is available, and the stability properties of all the primary branches on both lattices have been worked out [41]. Of the potentially stable solutions super-triangles have been observed in the Faraday system [84, 140], while stable super-squares and anti-squares have been located in systems of reaction–diffusion equations [75], indicating that these higher-dimensional representations are indeed relevant to the problem of pattern formation. The abstract stability results enable one to compute the stability of hexagons with respect to rectangular patterns that are almost square [41]. However, the relevance of these results to the ultimate question of relative stability between hexagons and squares remains unclear. Related results for the Hopf bifurcation on the square super-lattice are given by Dawes [35].

Certain of the possible three-dimensional lattices have also been considered. The steady-state bifurcations on the simple cubic, face-centred cubic and body-centred cubic lattices are analysed by Callahan and Knobloch [16]. At $\mu = 0$ these bifurcations have a zero eigenvalue of multiplicity 6, 8 and 12, respectively, resulting from the selection of 6, 8 or 12 wavevectors from the sphere of marginally stable wavevectors. Except for the simple cubic case (which corresponds to a wreath product group) the remaining two cases are plagued by submaximal branches which greatly complicate the analysis of these bifurcations. The corresponding calculations for the Hopf bifurcation on the simple cubic lattice were performed by Dias [38] exploiting its wreath product structure, and for the face-centred cubic lattice by Callahan [15]. None of the higher-dimensional representations of the cubic lattices have been considered, and neither have the remaining lattices except to identify the axial isotropy subgroups [40], and hence the primary solutions guaranteed by the equivariant branching lemma [39].

6.2 Imperfection sensitivity

The results of equivariant bifurcation theory depend, of course, on the presence of the assumed symmetry. This symmetry may represent the result of an idealization of a physical system, or it may be the consequence of imposed periodic boundary conditions as in the examples described earlier. In either case a natural question arises concerning the robustness of the results when the assumed symmetry is broken. Forced symmetry breaking is almost inevitable in physical situations, and while the dynamics of highly symmetric systems may be of interest in their own right, applications demand that one attempts to identify those aspects of equivariant dynamics that persist under (small) perturbations of the assumed symmetry. This point of view is not only relevant to pattern formation problems, where boundary conditions may destroy the assumed spatial periodicity, or a small feed gradient may destroy the assumed homogeneity (and isotropy) of the system, but also in structural engineering, where attempts to build redundancy into a system require understanding the consequences of the loss of a particular strut or support beam [71]. At present there is no general theory that allows one to identify the consequences of such symmetry-breaking imperfections. The basic idea is simple: one seeks to embed the equivariant dynamical system in a larger class of vector fields with smaller symmetry. For the pitchfork bifurcation this notion leads to the universal unfolding of the pitchfork [56], and shows that the effects of all possible imperfections that break reflection symmetry are captured by just two unfolding parameters. A general formulation of problems of this type is given by Lauterbach and Roberts [87]; Callahan and Knobloch [17] introduce the notion of co-equivariance and use it to generate the most general isotropy-breaking contributions to the equations describing a steady-state bifurcation on the hexagonal lattice. Certain other examples have been worked out rigorously. Of these the best known is the behaviour of a steady-state symmetry-breaking bifurcation with D_4 symmetry under perturbations that reduce the symmetry to D_2. This situation arises naturally in pattern selection problems in containers of square and nearly square (rectangular) cross section [10], and in the normal form describing a symmetry-breaking Hopf bifurcation with O(2) symmetry when the O(2) symmetry is broken down to SO(2) [28, 157]. A related but distinct example of this type of problem arises in systems with Neumann boundary conditions on a square domain. Such problems can be embedded in the corresponding problem on the square *lattice*, an embedding that is destroyed when the shape of the domain becomes non-square even though it may retain D_4 symmetry. The resulting breakup of the primary bifurcation is discussed by Crawford [24] and confirmed experimentally by Crawford *et al* [27].

In the absence of a general theory a reasonable approach appears to be to include only the dominant symmetry-breaking terms. While this approach may not capture *all* possible effects of the loss of symmetry it seems plausible that it provides a good guide to the robustness of the results with the full symmetry. This approach is especially enlightening when it comes to problems with continuous symmetries. Continuous symmetries allow simple types of dynamics, such as rotating waves, because the phase of the wave decouples from the equation for its amplitude. When the translation invariance is broken (by spatial inhomogeneity or the presence of boundaries) the spatial phase couples to the amplitude, thereby raising the order of the dynamical system. This coupling in turn is often responsible for the introduction of global bifurcations into the dynamics [79], and these are often associated with the presence of chaotic dynamics. Thus symmetry-breaking imperfections can produce complex dynamics in systems that would otherwise behave in a simple fashion. Several examples of this type of behaviour have been worked out, for the Hopf bifurcation with O(2) symmetry broken down to Z_2 [33, 68], and for the Hopf bifurcation with D_4 symmetry broken down to D_2 [102, 103]. We describe here the latter case. This example is believed to be relevant to experiments on convection in ^{3}He/^{4}He mixtures in a finite but extended container in which the convective heat transport immediately above threshold ($\epsilon^2 \equiv (R - R_c)/R_c = 3 \times 10^{-4}$) may take place in a sequence of irregular bursts of large dynamic range despite constant heat input [144].

Numerical simulations of the two-dimensional equations in a container of aspect ratio $L = 16$ suggest that these bursts involve the interaction between the first two modes of the system [8, 74]. These have opposite parity and, because the neutral stability curve for the unbounded system has a parabolic minimum, set in in close succession as the bifurcation parameter is increased. Near threshold the perturbation from the trivial state then takes the form

$$\Psi(x, y, t) = \epsilon \operatorname{Re}\{z_+ f_+(x, y) + z_- f_-(x, y)\} + O(\epsilon^2) \tag{6.10}$$

where $\epsilon \ll 1$, $f_\pm(-x, y) = \pm f_\pm(x, y)$, and y again denotes transverse variables. The complex amplitudes $z_\pm(t)$ then satisfy the normal form equations [86]

$$\dot{z}_\pm = [\lambda \pm \Delta\lambda + \mathrm{i}(\omega \pm \Delta\omega)]z_\pm + A(|z_+|^2 + |z_-|^2)z_\pm + B|z_\pm|^2 z_\pm + C\bar{z}_\pm z_\mp^2. \tag{6.11}$$

In these equations the nonlinear terms have identical (complex) coefficients because of an approximate interchange symmetry between the odd and even modes when $L \gg 1$. When $\Delta\lambda = \Delta\omega = 0$ the resulting equations coincide with equations (6.6)–(6.9) in the *standing wave* subspace $v_1 = w_1$, $v_2 = w_2$, and have D_4 symmetry. This symmetry is weakly broken whenever $\Delta\lambda \neq 0$ and/or $\Delta\omega \neq 0$, a consequence of the finite aspect ratio of the system [86].

To identify the bursts we introduce the change of variables

$$z_\pm = \rho^{-1/2} \sin(\tfrac{1}{2}\theta + \tfrac{1}{4}\pi \pm \tfrac{1}{4}\pi) \mathrm{e}^{\mathrm{i}(\pm\phi+\psi)/2}$$

and a new time-like variable τ defined by $\mathrm{d}\tau/\mathrm{d}t = \rho^{-1}$. In terms of these variables equations (6.11) become

$$\frac{\mathrm{d}\rho}{\mathrm{d}\tau} = -\rho[2A_{\mathrm{R}} + B_{\mathrm{R}}(1+\cos^2\theta) + C_{\mathrm{R}}\sin^2\theta\cos 2\phi] - 2(\lambda + \Delta\lambda\cos\theta)\rho^2 \tag{6.12}$$

$$\frac{\mathrm{d}\theta}{\mathrm{d}\tau} = \sin\theta[\cos\theta(-B_{\mathrm{R}} + C_{\mathrm{R}}\cos 2\phi) - C_{\mathrm{I}}\sin 2\phi] - 2\Delta\lambda\rho\sin\theta \tag{6.13}$$

$$\frac{\mathrm{d}\phi}{\mathrm{d}\tau} = \cos\theta(B_{\mathrm{I}} - C_{\mathrm{I}}\cos 2\phi) - C_{\mathrm{R}}\sin 2\phi + 2\Delta\omega\rho \tag{6.14}$$

where $A = A_{\mathrm{R}} + \mathrm{i}A_{\mathrm{I}}$, etc. There is also a decoupled equation for $\psi(t)$ so that the fixed points and periodic solutions of equations (6.12)–(6.14) correspond, respectively, to periodic solutions and two-tori in equations (6.11). In the following we measure the amplitude of the disturbance by $r \equiv |z_+|^2 + |z_-|^2 = \rho^{-1}$; thus $\rho = 0$ corresponds to *infinite* amplitude states. Equations (6.12)–(6.14) show that the restriction to the invariant subspace $\Sigma \equiv \{\rho = 0\}$ is equivalent to taking $\Delta\lambda = \Delta\omega = 0$ in (6.13)–(6.14). Since the resulting D_4-symmetric problem is a special case of equations (6.6)–(6.9) we may use table 6.1 to conclude that there are three generic types of fixed points [147]: SS solutions with $\cos\theta = 0$, $\cos 2\phi = 1$; AR solutions with $\cos\theta = 0$, $\cos 2\phi = -1$; and SR solutions with $\sin\theta = 0$. In the following we refer to these solutions as u, v, w, respectively, to emphasize that their physical interpretation is now quite different. In fact, in the binary fluid context these solutions represent, respectively, mixed-parity travelling wave states localized near one of the container walls, mixed-parity chevron states and pure even ($\theta = 0$)- or odd ($\theta = \pi$)-parity chevron states. The chevron states consist of waves propagating outwards from the centre of the container (or inwards from the boundaries) that are either in phase at the boundaries (even parity) or out of phase (odd parity). In a finite container these are the only states that bifurcate from the trivial (conduction) state. Depending on A, B and C the subspace Σ may contain additional submaximal fixed points (SCR) as well as limit cycles [147]. In our scenario, a burst occurs for $\lambda > 0$ when a trajectory follows the stable manifold of a fixed point (or a limit cycle) $P_1 \in \Sigma$ that is *unstable* within Σ. The instability within Σ then kicks the trajectory towards another fixed point (or limit cycle) $P_2 \in \Sigma$. If this point has an unstable ρ eigenvalue the trajectory escapes from Σ towards a finite amplitude ($\rho > 0$) state, forming a burst. If $\Delta\lambda$ and/or $\Delta\omega \neq 0$ this state may itself be unstable to perturbations of type P_1 and the process then repeats. This bursting behaviour is thus associated with a codimension-one heteroclinic cycle between the *infinite amplitude* solutions P_1 and P_2 [103].

For the heteroclinic cycle to form it is necessary that at least one of the branches in the D_4-symmetric system be subcritical (P_1) and one be supercritical (P_2). For the parameters of figure 6.2 the u solutions are subcritical while v, w are supercritical when $\Delta\lambda = \Delta\omega = 0$ [102] and two of the resulting cycles are shown in figure 6.2. In each case the trajectory reaches infinity in finite time and the heteroclinic cycle therefore represents infinite amplitude bursts of *finite*

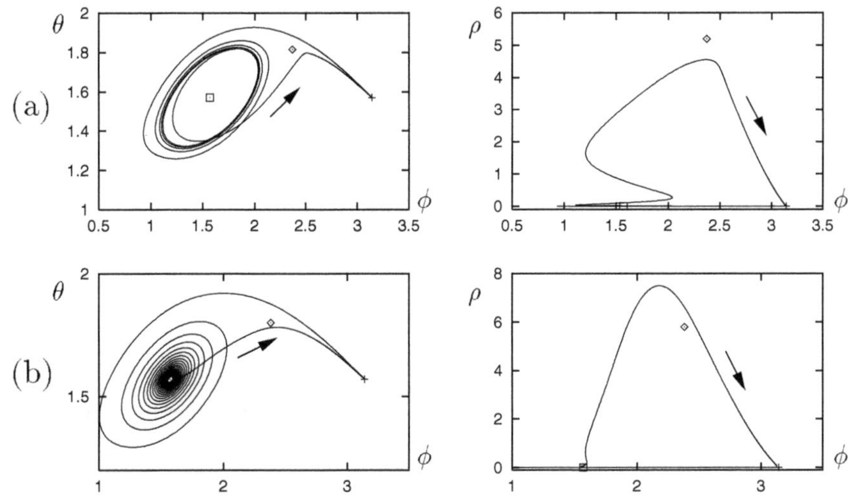

Figure 6.2. Numerically obtained approximate heteroclinic cycles for $\Delta\lambda = 0.03$, $\Delta\omega = 0.02$, $A = 1 - 1.5i$, $B = -2.8 + 5i$, and (*a*) $C = 1 + i$, (*b*) $C = 0.9 + i$ present at (*a*) $\lambda = 0.0974$ and (*b*) $\lambda = 0.08461$. The + signs indicate infinite amplitude u states responsible for the bursts, while the squares indicate infinite amplitude v states and the diamonds finite amplitude states. Reproduced from E Knobloch and J Moehlis Bursting mechanisms for hydrodynamical systems *Pattern Formation in Continuous and Coupled Systems* ed M Golubitsky, D Luss and S H Strogatz, pp 157–74 ©1999 by Springer-Verlag.

duration [103]. Consequently the time-averaged amplitude $\langle r \rangle$ may be dominated by the time spent near finite amplitude states, as in figure 6.3(*a*). Although neither of these solutions is in fact stable, they (and others like them) are responsible for the wealth of burst-like behaviour exhibited by this system. Figures 6.3(*a*) and (*b*) are an attempt to summarize some of this complexity in the form of bifurcation diagrams, but only the low period solutions have been followed. Much of the complexity revealed in these diagrams can be traced to the Shil'nikov-like properties of the dominant heteroclinic cycles [103].

The solutions shown in figure 6.2 are both (nearly) infinite period librations, characterized by bounded ϕ. But periodic trajectories in the form of rotations (ϕ increasing without bound) are also possible, and figure 6.4 shows an example of chaotic bursts generated by a stable chaotic rotation. Figure 6.5 shows the physical manifestation of the bursts arising from rotations and librations in the form of spacetime plots using the approximate eigenfunctions

$$f_{\pm}(x) = \{e^{-\gamma x + ix} \pm e^{\gamma x - ix}\} \cos \frac{\pi x}{L}$$

where $\gamma = 0.15 + 0.025i$, $L = 80$ and $-\frac{L}{2} \leq x \leq \frac{L}{2}$. The bursts in figure 6.5(*a*) are generated as a result of successive visits to *different* but symmetry-related

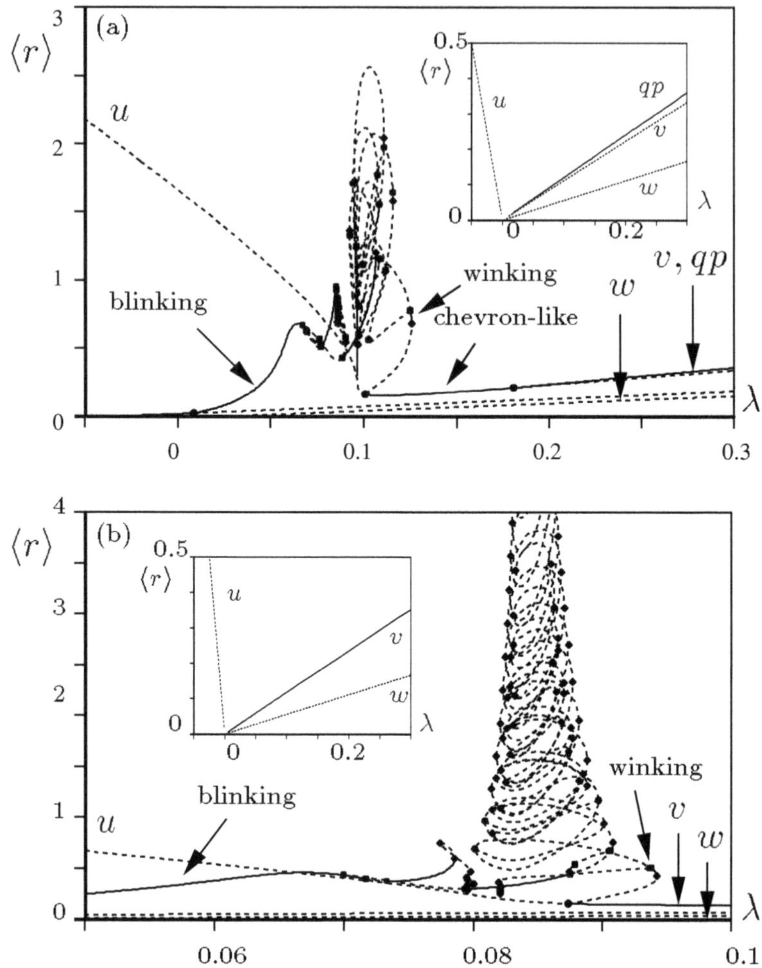

Figure 6.3. Partial bifurcation diagrams for (*a*) $C = 1 + i$ and (*b*) $C = 0.9 + i$ with the remaining parameters as in figure 6.2 showing the time-average of r for different solutions as a function of λ. Solid (dashed) lines indicate stable (unstable) solutions. The branches labelled u, v, w and qp (quasi-periodic) may be identified in the limit of large $|\lambda|$ with branches in the corresponding diagrams when $\Delta\lambda = \Delta\omega = 0$ (insets). All other branches correspond to bursting solutions which may be blinking or winking states (figure 6.5). Circles, squares, and diamonds in the diagram indicate Hopf, period-doubling, and saddle-node bifurcations, respectively. Reproduced from J Moehlis and E Knobloch Forced symmetry breaking as a mechanism for bursting *Phys. Rev. Lett.* **80** 5329–32 ©1998 by the American Physical Society.

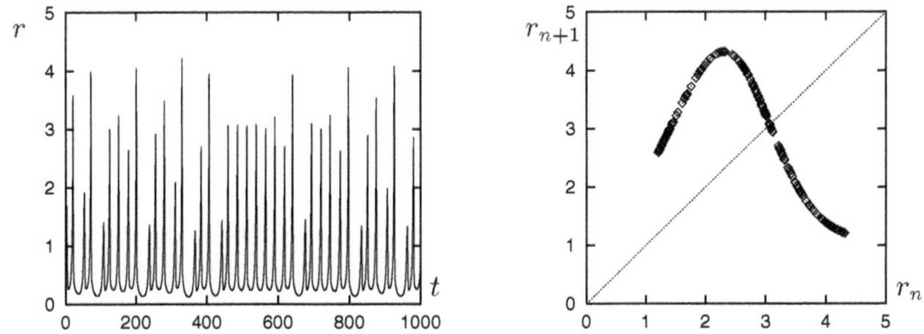

Figure 6.4. Time series and peak-to-peak plot showing bursts from chaotic rotations at $\lambda = 0.072$ for the parameters of figure 6.2(*a*). Reproduced from E Knobloch and J Moehlis Bursting mechanisms for hydrodynamical systems *Pattern Formation in Continuous and Coupled Systems* ed M Golubitsky, D Luss and S H Strogatz, pp 157–74 ©1999 by Springer-Verlag.

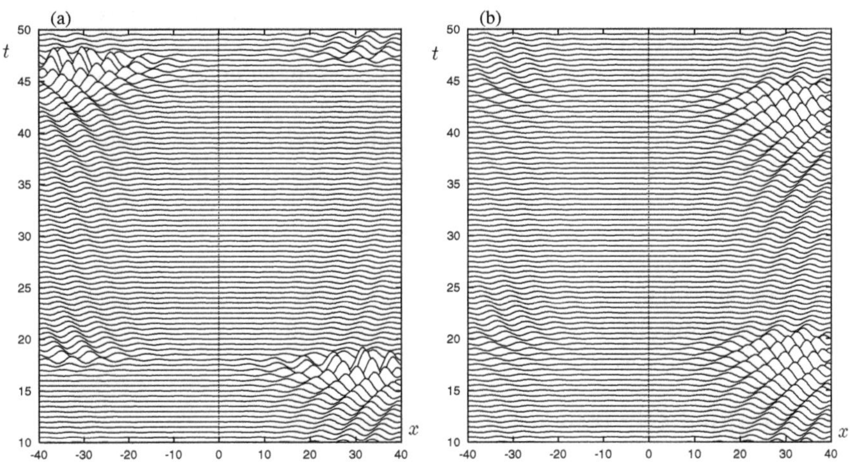

Figure 6.5. The perturbation Ψ from the trivial state represented in a spacetime plot showing (*a*) a periodic blinking state (in which successive bursts occur at opposite sides of the container) corresponding to a stable rotation at $\lambda = 0.1$, and (*b*) the periodic winking state (in which successive bursts occur at the same side of the container) corresponding to a stable libration at $\lambda = 0.1253$. Reproduced from J Moehlis and E Knobloch Forced symmetry breaking as a mechanism for bursting *Phys. Rev. Lett.* **80** 5329–32 ©1998 by the American Physical Society.

infinite amplitude *u* solutions; in figure 6.5(*b*) the generating trajectory makes repeated visits to the *same* infinite amplitude *u* solution. The former state is typical of the *blinking* state identified in binary fluid and doubly diffusive convection in rectangular containers [80, 118]. It is likely that the irregular bursts reported in [144] are due to such a state. The latter, a *winking* state, may be stable but often coexists with stable chevron-like states which are more likely to be observed in experiments in which the Rayleigh number is ramped upwards, cf figure 6.3(*a*).

The bursts described here are the result of oscillations in amplitude between two modes of opposite parity and 'frozen' spatial structure. Consequently this burst mechanism applies in systems in which bursts occur very close to threshold. This occurs not only in the convection experiments already mentioned but also in the mathematically identical Taylor–Couette system where counterpropagating spiral vortices play the same role as travelling waves in convection [1, 114]. In slender systems, such as the convection system described earlier or a long Taylor–Couette apparatus, a large aspect ratio L is required for the presence of the approximate D_4 symmetry. If the size of the D_4 symmetry-breaking terms $\Delta\lambda$, $\Delta\omega$ is increased too much the bursts fade away and are replaced by smaller amplitude, higher frequency states [103]. Indeed, if $\Delta\omega \gg \Delta\lambda$ averaging eliminates the C terms responsible for the bursts. From these considerations, we conclude that bursts will not be present if L is too small or ϵ too large. It is possible that the burst amplitude can become large enough that secondary instabilities not captured by ansatz (6.10) can be triggered. Such instabilities could occur on very different scales and result in *turbulent* rather than just large amplitude bursts. It should be emphasized that the physical amplitude of the bursts is $O(\epsilon)$ and so approaches zero as $\epsilon \downarrow 0$, cf (6.10). Thus despite their large dynamical range (i.e. the range of amplitudes during the bursts) the bursts are fully and correctly described by the asymptotic expansion that leads to equations (6.11). In particular, the mechanism is robust with respect to the inclusion of higher-order terms [103].

6.3 Coupled Ginzburg–Landau equations

We consider a translation-invariant system on the real line with periodic boundary conditions, undergoing a symmetry-breaking Hopf bifurcation from the trivial state, i.e. $k_c \neq 0$. Traditionally, one writes

$$\mathbf{u}(\mathbf{x}, t) = \epsilon \mathbf{u}(y)[A(X, T)e^{i(\omega t + k_c x)} + B(X, T)e^{i(\omega t - k_c x)} + \text{c.c.}] + \cdots \qquad (6.15)$$

where $\epsilon \ll 1$, $X = \epsilon x$, $T = \epsilon^2 t$ denote slow variables, and y denotes any transverse variables. We can think of the parameter ϵ as specifying the distance to the threshold of the primary instability, e.g., $(R - R_c)/R_c = \epsilon^2\mu$ as in section 6.2. The asymptotic theory applies on scales L such that $X = O(1)$, i.e. $L = O(\epsilon^{-1})$. It follows, therefore, that the resulting theory applies whenever $(R - R_c)/R_c = O(L^{-2}) \ll 1$. With the above ansatz substituted into the field equations, one recovers at leading order the linear stability problem for $\mathbf{u} = 0$, and hence determines k_c and ω_c, as well as R_c. From the remaining terms one obtains

$$A_T + \frac{c_g}{\epsilon} A_X = (\mu + \mathrm{i}\omega')A + (a|A|^2 + b|B|^2)A + cA_{XX} + O(\epsilon) \tag{6.16}$$

$$B_T - \frac{c_g}{\epsilon} B_X = (\mu + \mathrm{i}\omega')B + (a|B|^2 + b|A|^2)B + cB_{XX} + O(\epsilon) \tag{6.17}$$

where $c_g = \mathrm{d}\omega/\mathrm{d}k$ is the group velocity at $k = k_c$, $\omega' = \omega(\mu) - \omega_c$ and a, b and c are (complex) constants. These equations are often used as models of systems undergoing a symmetry-breaking Hopf bifurcation. However, when $c_g = O(1)$ (the usual case) they are clearly inconsistent as $\epsilon \to 0$. For this reason one typically has to assume that the system is near a codimension-two point such that $c_g = O(\epsilon)$. Only in this case do these equations make asymptotic sense. But there is no reason why one needs to restrict attention to this case. More generally, if $c_g = O(1)$ one sees that the advection term dominates the nonlinear terms, indicating that the time scale for advection at the group velocity is much faster than the time scale T on which the system equilibrates. We therefore introduce the intermediate time scale $\tau = \epsilon t$, and write $A = A(X, \tau, T; \epsilon) \equiv A_0 + \epsilon A_1 + \cdots$, $B = B(X, \tau, T; \epsilon) \equiv B_0 + \epsilon B_1 + \cdots$. At $O(\epsilon^2)$ one now finds

$$A_{0\tau} + c_g A_{0X} = 0 \qquad B_{0\tau} - c_g B_{0X} = 0 \tag{6.18}$$

indicating that $A_0 \equiv A_0(\xi, T)$, $B_0 \equiv B_0(\eta, T)$, where $\xi \equiv X - c_g\tau$, $\eta \equiv X + c_g\tau$, while at $O(\epsilon^3)$ one obtains

$$A_{1\tau} + c_g A_{1X} = -A_{0T} + (\mu + \mathrm{i}\omega')A_0 + (a|A_0|^2 + b|B_0|^2)A_0 + cA_{0\xi\xi} \tag{6.19}$$

$$B_{1\tau} - c_g B_{1X} = -B_{0T} + (\mu + \mathrm{i}\omega')B_0 + (a|B_0|^2 + b|A_0|^2)B_0 + cB_{0\eta\eta}. \tag{6.20}$$

The solvability condition for A_1, guaranteeing that ϵA_1 remains small relative to A_0 for $T = O(1)$, yields the required evolution equation for A_0:

$$A_{0T} = (\mu + \mathrm{i}\omega')A_0 + (a|A_0|^2 + b\langle|B_0|^2\rangle^{\eta})A_0 + cA_{0\xi\xi} \tag{6.21}$$

and, similarly,

$$B_{0T} = (\mu + \mathrm{i}\omega')B_0 + (a|B_0|^2 + b\langle|A_0|^2\rangle^{\xi})B_0 + cB_{0\eta\eta}. \tag{6.22}$$

Here

$$\langle|A_0|^2\rangle^{\xi} = \frac{1}{P}\int_0^P |A_0|^2\,\mathrm{d}\xi \qquad \langle|B_0|^2\rangle^{\eta} = \frac{1}{Q}\int_0^Q |B_0|^2\,\mathrm{d}\eta \tag{6.23}$$

where P and Q are the periods in ξ and η (perhaps infinite). Thus it is the presence of the intermediate advection time scale that is responsible for the non-local terms $\langle\ldots\rangle$ in the envelope equations (6.21)–(6.22). These equations, derived by Knobloch and De Luca [77] for dissipative systems, and by Chikwendu and Kevorkian [20], and Knobloch and Gibbon [78] for conservative systems, have been shown to be correct, in the sense that their solutions track those of the original field equations for $T = O(1)$ on spatial scales $X = O(1)$ [116]. Detailed stability results for these equations are given in [76, 96]. More recently, Riecke and Kramer [123] have explored the relation between equations (6.16)–(6.17) and the asymptotic equations (6.21)–(6.22), and showed that the solutions of the latter are indeed correct as a limit of the former, but that they may only apply in a small neighbourhood of $R = R_c$. Whether the local equations (6.16)–(6.17) have any validity beyond qualitative outside of this neighbourhood remains, at present, a matter of conjecture. Of course, non-local equations are harder to work with but it should be clear that they admit a larger class of solutions than local equations, and are of interest therefore for this reason alone.

Duan *et al* [47] have proved the existence of a finite-dimensional inertial manifold for the single non-local complex Ginzburg–Landau equation

$$C_T = \mu C + (a|C|^2 + b\langle|C|^2\rangle)C + cC_{XX} \tag{6.24}$$

that describes the dynamics of the system (6.21)–(6.22) in the invariant subspace $A_0 = B_0 \equiv C$. Equations of this type also arise as a result of pressure effects in incompressible flows [62] and in chemical kinetics [72] or of mass conservation [115], as well as describing ferromagnetic instabilities [50]. However, as already mentioned, existing derivations of such equations all ignore the presence of boundaries, and additional work must be carried out to derive the correct boundary conditions appropriate to specific applications. For the Hopf bifurcation in one spatial dimension two boundary conditions must be imposed at each boundary [93]. These boundaries do more than simply break translation invariance, since waves can also be reflected from them suffering a phase shift, in addition to losing energy.

It is important to remark that there is another way of rationalizing the expansion (6.16)–(6.17). This is to choose instead $(R - R_c)/R_c = \epsilon\mu = O(L^{-1})$, where L again measures the domain size. In this case the wave amplifies and interacts with the boundaries on the time scale τ and reaches a larger amplitude $\mathbf{u} = O(\epsilon^{1/2})$. One now obtains a pair of *hyperbolic* equations of the form

$$A_\tau + c_g A_X = (\mu + \mathrm{i}\omega')A + (a|A|^2 + b|B|^2)A + O(\epsilon) \tag{6.25}$$

$$B_\tau - c_g B_X = (\mu + \mathrm{i}\omega')B + (a|B|^2 + b|A|^2)B + O(\epsilon). \tag{6.26}$$

This time only two boundary conditions are required, and these are given by

$$A(0, t) = rB(0, t) \qquad B(L, t) = rA(L, t). \tag{6.27}$$

Here r is a (calculable) complex reflection coefficient [93]. The resulting equations can be written as a pair of coupled equations for the real amplitudes $|A|$, $|B|$; their solutions describe not only waves that propagate from one end of the container to the other, but also waves that 'blink', i.e. bounce back and forth between them [94]. Waves of this type have been seen in numerical simulations of doubly diffusive convection [37] and in experiments on both the oscillatory instability of convection rolls [30] and in binary fluid convection [80]. In the latter the subcritical nature of the instability complicates the dynamics substantially [8].

These equations represent the effects of a *global* non-locality: the global coupling does not fall off with distance. Yet in many cases (for example, in fluids that are not strictly incompressible) we would expect the finite speed of propagation of pressure waves to introduce a kernel with a cut-off into equations of this type. Equations with this weaker type of global coupling arise naturally in coupled oscillator systems as discussed in chapter 9 by Kuramoto. The introduction of the new length scale, the effective range of the global coupling, has interesting and non-trivial consequences for the dynamics of these systems. A related equation arises as a singular limit of a pair of coupled reaction–diffusion equations with disparate time scales [121].

6.4 The nearly-inviscid Faraday system

The nearly inviscid Faraday system provides another example of a system described by non-local amplitude equations. In this system a container of liquid is vibrated vertically; when the amplitude of the acceleration of the container exceeds a threshold value, the flat surface breaks up into a pattern of standing waves [52, 83, 101]. In the absence of forcing such surface gravity–capillary waves decay of an $O(C^{-1/2})$ time scale, where $C = \nu/(gh^3 + Th/\rho)^{1/2} \ll 1$, while hydrodynamic (i.e. viscous) modes decay yet more slowly, on an $O(C^{-1})$ time scale [91]. Here g is the gravitational acceleration, T is the coefficient of surface tension, ρ is the density and ν is the kinematic viscosity. Since the viscous modes decay so slowly they are easily excited by the vertical vibration; this excitation takes the form of a mean flow. In periodic domains in which the length of the domain is large relative to the wavelength of the instability this mean flow contains both viscous and inviscid contributions, and both couple to the amplitude equations for the surface waves. Traditionally such amplitude equations are derived using a velocity potential formulation. However, this formulation implicitly *excludes* large-scale streaming flows that may be driven by the time-averaged Reynolds stress in the oscillatory boundary layers at the container walls or at the free surface. Recently a systematic asymptotic technique has been developed that includes such flows in a self-consistent manner [160], and leads to a new class of (non-local) pattern-forming amplitude equations. These developments are summarized later.

We consider a container in the form of a right cylinder with horizontal cross section Σ filled level with the brim at $z = 0$. In this geometry the contact line is pinned at the lateral boundary and complications associated with contact line dynamics are reduced. The governing equations, non-dimensionalized using the unperturbed depth h as the unit of length and the gravity–capillary time

$[g/h + T/(\rho h^3)]^{-1/2}$ as the unit of time, are

$$\partial \mathbf{v}/\partial t - \mathbf{v} \times (\nabla \times \mathbf{v}) = -\nabla \Pi + C\Delta \mathbf{v} \qquad \nabla \cdot \mathbf{v} = 0 \text{ if } (x, y) \in \Sigma,\ -1 < z < f$$
$$\mathbf{v} = 0 \text{ if } z = -1 \text{ or } (x, y) \in \partial\Sigma \qquad f = 0 \text{ if } (x, y) \in \partial\Sigma$$
$$\mathbf{v} \cdot \mathbf{n} = (\partial f/\partial t)(\mathbf{e}_z \cdot \mathbf{n}) \qquad [(\nabla \mathbf{v} + \nabla \mathbf{v}^\top) \cdot \mathbf{n}] \times \mathbf{n} = 0 \qquad \text{at } z = f$$
$$\Pi - |\mathbf{v}|^2/2 - (1 - S) f + S\nabla \cdot [\nabla f/(1 + |\nabla f|^2)^{1/2}]$$
$$= C[(\nabla \mathbf{v} + \nabla \mathbf{v}^\top) \cdot \mathbf{n}] \cdot \mathbf{n} - 4\mu\omega^2 f \cos 2\omega t \qquad \text{at } z = f$$

where $\mathbf{v}$ is the velocity, f is the associated vertical deflection of the free surface (constrained by volume conservation), $\Pi = p + |\mathbf{v}|^2/2 + (1-S)z - 4\mu\omega^2 z \cos 2\omega t$ is the hydrostatic stagnation pressure, $\mathbf{n}$ is the outward unit normal to the free surface, $\mathbf{e}_z$ is the upward unit vector, and $\partial\Sigma$ denotes the boundary of the cross section Σ (i.e. the lateral walls). The real parameters $\mu > 0$ and 2ω denote the amplitude and frequency of the forcing, and $S = T/(T + \rho g h^2)$ is the gravity–capillary balance parameter. Thus $0 \leq S \leq 1$ with $S = 0$ and $S = 1$ corresponding to the purely gravitational limit ($T = 0$) and the purely capillary limit ($g = 0$), respectively.

In the (nearly inviscid, nearly resonant, weakly nonlinear) regime

$$C \ll 1 \qquad |\omega - \Omega| \ll 1 \qquad \mu \ll 1 \tag{6.28}$$

where Ω is an inviscid eigenfrequency of the linearized problem around the flat state, the vorticity contamination of the bulk from the boundary layers at the walls and the free surface remains negligible for times that are not too long, and the flow in the bulk is correctly described by an inviscid formulation but with boundary conditions determined by a boundary layer analysis. In general this flow consists of an *inviscid* part and a *viscous* part.

In this regime it is possible to perform a multi-scale analysis of the governing equations using C, L^{-1} and μ as unrelated small parameters. Here L is the dimensionless length of the container. The problem is simplest in two dimensions where we can use a stream function formulation, i.e. we write $\mathbf{v} = (-\psi_z, 0, \psi_x)$. We focus on two well-separated scales in both space ($x \sim 1$ and $x \gg 1$) and time ($t \sim 1$ and $t \gg 1$), and derive equations for small, slowly varying amplitudes A and B of left- and right-propagating waves defined by

$$f = \mathrm{e}^{\mathrm{i}\omega t}(A\mathrm{e}^{\mathrm{i}kx} + B\mathrm{e}^{-\mathrm{i}kx}) + \gamma_1 A\bar{B}\mathrm{e}^{2\mathrm{i}kx} + \gamma_2 \mathrm{e}^{2\mathrm{i}\omega t}(A^2\mathrm{e}^{2\mathrm{i}kx} + B^2\mathrm{e}^{-2\mathrm{i}kx})$$
$$+ f^+\mathrm{e}^{\mathrm{i}\omega t + \mathrm{i}kx} + f^-\mathrm{e}^{\mathrm{i}\omega t - \mathrm{i}kx} + \text{c.c.} + f^{\mathrm{m}} + NRT$$

with similar expressions for the remaining fields. The quantities $f^\pm$ and f^{m} represent resonant second-order terms, while NRT denotes non-resonant terms. The superscript m denotes terms associated with the mean flow; f^{m} depends weakly on time but may depend strongly on x. A systematic expansion procedure

[160] now leads to the equations

$$A_t - c_g A_x = \mathrm{i}\alpha A_{xx} - (\delta + \mathrm{i}d)A + \mathrm{i}(\alpha_3|A|^2 - \alpha_4|B|^2)A$$
$$+ \mathrm{i}\alpha_5\mu\bar{B} + \mathrm{i}\alpha_6\int_{-1}^{0} g(z)\langle\psi_z^{\mathrm{m}}\rangle^x\,\mathrm{d}z\,A + \mathrm{i}\alpha_7\langle f^{\mathrm{m}}\rangle^x A \quad (6.29)$$

$$B_t + c_g B_x = \mathrm{i}\alpha B_{xx} - (\delta + \mathrm{i}d)B + \mathrm{i}(\alpha_3|B|^2 - \alpha_4|A|^2)B$$
$$+ \mathrm{i}\alpha_5\mu\bar{A} - \mathrm{i}\alpha_6\int_{-1}^{0} g(z)\langle\psi_z^{\mathrm{m}}\rangle^x\,\mathrm{d}z\,B + i\alpha_7\langle f^{\mathrm{m}}\rangle^x B \quad (6.30)$$

$$A(x+L,t) \equiv A(x,t) \qquad B(x+L,t) \equiv B(x,t). \quad (6.31)$$

The first seven terms in these equations, accounting for inertia, propagation at the group velocity c_g, dispersion, damping, detuning, cubic nonlinearity and parametric forcing, are familiar from weakly nonlinear, nearly inviscid theories. These theories lead to the expressions

$$c_g = \omega'(k) \qquad \alpha = -\omega''(k)/2 \qquad \delta = \alpha_1 C_g^{1/2} + \alpha_2 C_g$$

$$\alpha_1 = \frac{k(\omega/2)^{1/2}}{\sinh 2k} \qquad \alpha_2 = \frac{k^2}{4\sigma^2}(1 + 8\sigma^2 - \sigma^4)$$

$$\alpha_3 = \frac{\omega k^2[(1-S)(9-\sigma^2)(1-\sigma^2) + Sk^2(7-\sigma^2)(3-\sigma^2)]}{4\sigma^2[(1-S)\sigma^2 - Sk^2(3-\sigma^2)]} + \frac{\omega k^2[8(1-S) + 5Sk^2]}{4(1-S+Sk^2)}$$

$$\alpha_4 = \frac{\omega k^2}{2}\left[\frac{(1-S+Sk^2)(1+\sigma^2)^2}{(1-S+4Sk^2)\sigma^2} + \frac{4(1-S)+7Sk^2}{1-S+Sk^2}\right] \qquad \alpha_5 = \omega k\sigma$$

where $\omega(k) = [(1-S+Sk^2)k\sigma]^{1/2}$ is the dispersion relation and $\sigma \equiv \tanh k$. The coefficient α_3 diverges at (excluded) resonant wavenumbers satisfying $\omega(2k) = 2\omega(k)$. The detuning d is given by

$$d = \alpha_1 C^{1/2} - (2\pi N L^{-1} - k)c_g \qquad N = \text{integer}$$

where the last term represents the mismatch between the wavelength $2\pi/k$ selected by the forcing frequency 2ω and the domain length L. The last two terms in equations (6.29)–(6.30) describe the coupling to the mean flow in the bulk (be it viscous or inviscid in origin) in terms of (a local average $\langle\cdot\rangle^x$ of) the stream function ψ^{m} for this flow and the associated free surface elevation f^{m}. The coefficients of these terms and the function g are given by

$$\alpha_6 = k\sigma/2\omega \qquad \alpha_7 = \omega k(1-\sigma^2)/2\sigma \qquad g(z) = 2\omega k\cosh[2k(z+1)]/\sinh^2 k \quad (6.32)$$

and are real. The new terms are therefore conservative, implying that at leading order the mean flow does not extract energy from the system.

The mean flow variables in the bulk depend weakly on time but strongly on both x and z, and evolve according to the equations

$$\Omega_t^{\mathrm{m}} - [\psi_z^{\mathrm{m}} + (|A|^2 - |B|^2)g(z)]\Omega_x^{\mathrm{m}} + \psi_x^{\mathrm{m}}\Omega_z^{\mathrm{m}} = C(\Omega_{xx}^{\mathrm{m}} + \Omega_{zz}^{\mathrm{m}}) \qquad (6.33)$$

where $\Omega^{\mathrm{m}} = \psi_{xx}^{\mathrm{m}} + \psi_{zz}^{\mathrm{m}}$, subject to

$$\begin{aligned} &\psi_x^{\mathrm{m}} - f_t^{\mathrm{m}} = \beta_1(|B|^2 - |A|^2)_x \qquad \psi_{zz}^{\mathrm{m}} = \beta_2(|A|^2 - |B|^2) \\ &(1-S)f_x^{\mathrm{m}} - Sf_{xxx}^{\mathrm{m}} - \psi_{zt}^{\mathrm{m}} + C(\psi_{zzz}^{\mathrm{m}} + 3\psi_{xxz}^{\mathrm{m}}) = -\beta_3(|A|^2 + |B|^2)_x \\ &\quad \text{at } z = 0 \end{aligned} \qquad (6.34)$$

and

$$\begin{aligned} \psi_z^{\mathrm{m}} = -\beta_4[\mathrm{i}A\bar{B}\mathrm{e}^{2\mathrm{i}kx} + \text{c.c.} + |B|^2 - |A|^2] \qquad \int_0^L \Omega_z^{\mathrm{m}}\,\mathrm{d}x = \psi^{\mathrm{m}} = 0 \\ \text{at } z = -1. \end{aligned} \qquad (6.35)$$

In addition $\psi^{\mathrm{m}}(x+L, z, t) \equiv \psi^{\mathrm{m}}(x, z, t)$, $f^{\mathrm{m}}(x+L, t) \equiv f^{\mathrm{m}}(x, t)$, subject to $\int_0^L f^{\mathrm{m}}(x, t)\,\mathrm{d}x = 0$, due to mass conservation. Here $\beta_1 = 2\omega/\sigma$, $\beta_2 = 8\omega k^2/\sigma$, $\beta_3 = (1-\sigma^2)\omega^2/\sigma^2$, $\beta_4 = 3(1-\sigma^2)\omega k/\sigma^2$. Thus the mean flow is forced by the surface waves in two ways. The right-hand sides of the first and third boundary conditions of (6.34) provide a *normal forcing mechanism*; this mechanism is the only one present in strictly inviscid theory [34, 115] and does not appear unless the aspect ratio is large. The right-hand sides of the second boundary condition of (6.34) and the first of (6.35) describe two *shear forcing mechanisms*, a tangential stress at the free surface [89] and a tangential velocity at the bottom wall [133]. Note that neither of these forcing terms vanishes in the limit of small viscosity (i.e. as $C \to 0$). The shear nature of these forcing terms leads us to retain the viscous term in equation (6.33) even when C is quite small. In fact, when C is very small, the effective Reynolds number of the mean flow is quite large. Thus the mean flow itself generates additional boundary layers near the top and bottom of the container, and these must be thicker than the original boundary layers for the validity of the analysis. This puts an additional restriction on the validity of the equations [160]. There is a third, less effective but inviscid, volumetric forcing mechanism associated with the second term in the vorticity equation (6.33), which looks like a horizontal force $(|A|^2 - |B|^2)g(z)\Omega^{\mathrm{m}}$ and is sometimes called the *vortex force*. Although this term vanishes in the absence of mean flow, it can change the stability properties of the flow and enhance or limit the effect of the remaining forcing terms.

Equations (6.29)–(6.31) and (6.33)–(6.35) may be referred to as the general coupled amplitude-mean-flow (GCAMF) equations. Like the equations derived in section 6.3 they are non-local, this time because the parametrically excited waves drive a whole spectrum of almost marginal large scale modes. The GCAMF equations differ from the exact equations forming the starting point for the analysis in three essential simplifications: the fast oscillations associated with the surface waves have been filtered out, the effect of the thin primary viscous boundary layers is replaced by effective boundary conditions on the flow in the bulk, namely the second condition of (6.34) and the first condition of (6.35), and the surface boundary conditions are applied at the unperturbed location of the free surface, namely $z = 0$. Thus only the much broader (secondary) boundary layers associated with the (slowly varying) mean flow need to be resolved in any numerical simulation.

The GCAMF equations describe small amplitude slowly varying wave-trains whenever the parameters C, L^{-1} and μ are small, but otherwise unrelated to one another. However, in the distinguished limit

$$\delta L^2/\alpha = \Delta \sim 1 \qquad dL^2/\alpha = D \sim 1 \qquad \mu L^2/\alpha \equiv M \sim 1 \tag{6.36}$$

with $k = O(1)$ and $|\ln C| = O(1)$ they can be simplified further, and their relation to the non-local equations of section 6.3 be brought out. These simplified equations are formally valid for $1 \ll L \ll C^{-1/2}$ if $k \sim 1$, assuming that $1 - S \sim 1$. Using x and t as *fast* variables and

$$X = x/L \qquad \tau = t/L \qquad T = t/L^2 \tag{6.37}$$

as *slow* variables, we write

$$\begin{aligned}
\psi^{\mathrm{m}}(x, z, X, \tau, T) &= \psi^{\mathrm{v}}(x, z, X, T) + \psi^{\mathrm{i}}(x, z, X, \tau, T) \\
\Omega^{\mathrm{m}}(x, z, X, \tau, T) &= \Omega^{\mathrm{v}}(x, z, X, T) + \Omega^{\mathrm{i}}(x, z, X, \tau, T) \\
f^{\mathrm{m}}(x, X, \tau, T) &= f^{\mathrm{v}}(x, X, T) + f^{\mathrm{i}}(x, X, \tau, T)
\end{aligned} \tag{6.38}$$

and demand that integrals over τ of ψ^{i}_x, ψ^{i}_X, ψ^{i}_z, Ω^{i} and f^{i} be bounded as $\tau \to \infty$, i.e. that the nearly inviscid mean flow is *purely oscillatory* on the time scale τ. In terms of these variables the local horizontal average $\langle\cdot\rangle^x$ becomes an average over the fast variable x. Since its frequency is of order L^{-1} (see equation (6.37)), which is large compared with C, the inertial term for this flow is large in comparison with the viscous terms (see equation (6.33)), except in two *secondary* boundary layers, of thickness of order $(CL)^{1/2}$ ($\ll 1$), attached to the bottom plate and the free surface. Note that, as required for the consistency of the analysis, these boundary layers are much thicker than the *primary* boundary layers associated with the surface waves, which provide the boundary conditions (6.34)–(6.35) for the mean flow. Moreover, the width of these secondary boundary layers remains small as $\tau \to \infty$ and (to leading order) the vorticity of this nearly inviscid mean flow remains confined to these boundary layers. It is possible to check that these boundary layers do not modify the boundary conditions (6.34)–(6.35) on the nearly inviscid bulk flow.

The complex amplitudes and the flow variables associated with the nearly inviscid bulk flow may be expanded as

$$(A, B) = L^{-1}(A_0, B_0) + L^{-2}(A_1, B_1) + \cdots$$
$$(\psi^{\rm v}, \Omega^{\rm v}, f^{\rm v}) = L^{-2}(\phi_0^{\rm v}, W_0^{\rm v}, 0) + L^{-3}(\phi_1^{\rm v}, W_1^{\rm v}, F_0^{\rm v}) + \cdots$$
$$(\psi^{\rm i}, \Omega^{\rm i}, f^{\rm i}) = L^{-2}(\phi_0^{\rm i}, 0, F_0^{\rm i}) + L^{-3}(\phi_1^{\rm i}, W_0^{\rm i}, F_1^{\rm i}) + \cdots . \tag{6.39}$$

Substitution of (6.36)–(6.39) into (6.29)–(6.35) leads to the following results:

(i) From (6.33)–(6.35), at leading order,

$$\phi_{0xx}^{\rm i} + \phi_{0zz}^{\rm i} = 0 \text{ in } -1 < z < 0 \qquad \phi_0^{\rm i} = 0 \text{ at } z = -1 \qquad \phi_{0x}^{\rm i} = 0 \text{ at } z = 0$$

together with $F_{0x}^{\rm i} = 0$. Thus

$$\phi_0^{\rm i} = (z+1)\Phi_0^{\rm i}(X, \tau, T) \qquad F_0^{\rm i} = F_0^{\rm i}(X, \tau, T).$$

At second order, the first and third boundary conditions of (6.34) yield

$$\phi_{1x}^{\rm i}(x, 0, X, \tau, T) = F_{0\tau}^{\rm i} - \Phi_{0X}^{\rm i} + \beta_1(|B_0|^2 - |A_0|^2)_X$$
$$(1-S)F_{1x}^{\rm i} - SF_{1xxx}^{\rm i} = \Phi_{0\tau}^{\rm i} - (1-S)F_{0X}^{\rm i} - \beta_3(|A_0|^2 + |B_0|^2)_X$$

at $z = 0$. Since the right-hand sides of these two equations are independent of the fast variable x and both $\phi_1^{\rm i}$ and $F_1^{\rm i}$ must be bounded in x, it follows that

$$\Phi_{0X}^{\rm i} - F_{0\tau}^{\rm i} = \beta_1(|B_0|^2 - |A_0|^2)_X \qquad \Phi_{0\tau}^{\rm i} - c_{\rm p}^2 F_{0X}^{\rm i} = \beta_3(|A_0|^2 + |B_0|^2)_X \tag{6.40}$$

where

$$c_{\rm p} = (1-S)^{1/2} \tag{6.41}$$

is the phase velocity of long wavelength surface gravity waves. Equations (6.40) must be integrated with the following additional conditions

$$\Phi_0^{\rm i}(X+1, \tau, T) \equiv \Phi_0^{\rm i}(X, \tau, T) \qquad F_0^{\rm i}(X+1, \tau, T) \equiv F_0^{\rm i}(X, \tau, T)$$

and the requirements that integrals over τ of $\Phi_{0X}^{\rm i}$ and $F_0^{\rm i}$ remain bounded as $\tau \to \infty$, with $\int_0^1 F_0^{\rm i}\,{\rm d}X = 0$.

(ii) The leading-order contributions to equations (6.29)–(6.30) yield $A_{0\tau} - c_{\rm g}A_{0X} = B_{0\tau} + c_{\rm g}B_{0X} = 0$. Thus $A_0 = A_0(\xi, T)$, $B_0 = B_0(\eta, T)$, where $\xi = X + c_{\rm g}\tau$ and $\eta = X - c_{\rm g}\tau$ are again the characteristic variables, and $A_0(\xi+1, T) \equiv A_0(\xi, T)$, $B_0(\eta+1, T) \equiv B_0(\eta, T)$. Substitution of these expressions into (6.40) followed by integration of the resulting equations yields

$$\Phi_0^{\rm i} = \frac{\beta_1 c_{\rm p}^2 + \beta_3 c_{\rm g}}{c_{\rm g}^2 - c_{\rm p}^2}[|A_0|^2 - |B_0|^2 - \langle |A_0|^2 - |B_0|^2\rangle^X]$$
$$+ c_{\rm p}[F^+(X + c_{\rm p}\tau, T) - F^-(X - c_{\rm p}\tau, T)] \tag{6.42}$$
$$F_0^{\rm i} = \frac{\beta_1 c_{\rm g} + \beta_3}{c_{\rm g}^2 - c_{\rm p}^2}[|A_0|^2 + |B_0|^2 - \langle |A_0|^2 + |B_0|^2\rangle^X]$$
$$+ F^+(X + c_{\rm p}\tau, T) + F^-(X - c_{\rm p}\tau, T)] \tag{6.43}$$

where $\langle\cdot\rangle^X$ denotes the mean value in the slow spatial variable X, and the functions $F^{\pm}$ are such that

$$F^{\pm}(X + 1 \pm c_{\mathrm{p}}\tau, T) \equiv F^{\pm}(X \pm c_{\mathrm{p}}\tau, T) \qquad \langle F^{\pm}\rangle^X = 0. \tag{6.44}$$

The particular solution (6.42)–(6.43) yields the usual inviscid mean flow included in nearly inviscid theories [115]; the averaged terms are a consequence of volume conservation [115] and the requirement that the nearly inviscid mean flow has a zero mean on the time scale τ; the latter condition is never imposed in strictly inviscid theories but is essential in the limit we are considering, as previously explained. To avoid breakdown of the solution (6.42)–(6.43) at $c_{\mathrm{p}} = c_{\mathrm{g}}$ we assume that

$$|c_{\mathrm{p}} - c_{\mathrm{g}}| \sim 1. \tag{6.45}$$

(iii) The evolution equations for A_0 and B_0 on the time scale T are readily obtained from equations (6.29)–(6.31), invoking (6.36), (6.38) and (6.42)–(6.44), and eliminating secular terms (i.e. requiring $|A_1|$ and $|B_1|$ to be bounded on the time scale τ):

$$\begin{aligned} A_{0T} &= \mathrm{i}\alpha A_{0\xi\xi} - (\Delta + \mathrm{i}D)A_0 + \mathrm{i}[(\alpha_3 + \alpha_8)|A_0|^2 - \alpha_8\langle|A_0|^2\rangle^{\xi} \\ &\quad - \alpha_4\langle|B_0|^2\rangle^{\eta}]A_0 + \mathrm{i}\alpha_5 M\langle\overline{B}_0\rangle^{\eta} + \mathrm{i}\alpha_6\int_{-1}^{0} g(z)\langle\langle\phi_{0z}^{\mathrm{v}}\rangle^x\rangle^X\,\mathrm{d}z\,A_0 \\ B_{0T} &= \mathrm{i}\alpha B_{0\eta\eta} - (\Delta + \mathrm{i}D)B_0 + i[(\alpha_3 + \alpha_8)|B_0|^2 - \alpha_8\langle|B_0|^2\rangle^{\eta} \\ &\quad - \alpha_4\langle|A_0|^2\rangle^{\xi}]B_0 + \mathrm{i}\alpha_5 M\langle\overline{A}_0\rangle^{\xi} - \mathrm{i}\alpha_6\int_{-1}^{0} g(z)\langle\langle\phi_{0z}^{\mathrm{v}}\rangle^x\rangle^X\,\mathrm{d}z\,B_0. \end{aligned} \tag{6.46}$$

Here $\langle\cdot\rangle^x$, $\langle\cdot\rangle^X$, $\langle\cdot\rangle^{\xi}$ and $\langle\cdot\rangle^{\eta}$ denote mean values over the variables x, X, ξ and η, respectively, and

$$\alpha_8 = [\alpha_6(2\omega/\sigma)(\beta_1 c_{\mathrm{p}}^2 + \beta_3 c_{\mathrm{g}}) + \alpha_7(\beta_1 c_{\mathrm{g}} + \beta_3)]/(c_{\mathrm{g}}^2 - c_{\mathrm{p}}^2).$$

Equations (6.46) are independent of $F^{\pm}$ because of the second condition in (6.44).

When $\Delta > 0$, $\langle|A_0|^2 - |B_0|^2\rangle^{\tau} = \langle|A_0|^2\rangle^{\xi} - \langle|B_0|^2\rangle^{\eta} \to 0$ as $T \to \infty$, and equations (6.33)–(6.35) become

$$W_{0T}^{\mathrm{v}} - \phi_{0z}^{\mathrm{v}} W_{0x}^{\mathrm{v}} + \phi_{0x}^{\mathrm{v}} W_{0z}^{\mathrm{v}} = Re^{-1}(W_{0xx}^{\mathrm{v}} + W_{0zz}^{\mathrm{v}}) \qquad W_0^{\mathrm{v}} = \phi_{0xx}^{\mathrm{v}} + \phi_{0zz}^{\mathrm{v}} \quad \text{in } -1 < z < 0 \tag{6.47}$$

$$\phi_{0x}^{\mathrm{v}} = \phi_{0zz}^{\mathrm{v}} = 0 \qquad \text{at } z = 0 \tag{6.48}$$

$$\langle\langle W_{0z}^{\mathrm{v}}\rangle^x\rangle^X = \phi_0^{\mathrm{v}} = 0 \qquad \phi_{0z}^{\mathrm{v}} = -\beta_4[\mathrm{i}\langle A_0\overline{B}_0\rangle^{\tau}\mathrm{e}^{2\mathrm{i}kx} + \mathrm{c.c.}] \qquad \text{at } z = -1 \tag{6.49}$$

$$\phi_0^{\mathrm{v}}(x + L, X + 1, z, T) \equiv \phi_0^{\mathrm{v}}(x, X, z, T) \tag{6.50}$$

where

$$Re = 1/(CL^2) \tag{6.51}$$

is the *effective Reynolds number* of the viscous mean flow. Note that $Re = O(C^{-1/2})$ if $k \sim 1$.

Some remarks about these equations and boundary conditions are now in order.

First, the viscous mean flow is driven by the short gravity–capillary waves through the inhomogeneous term in the third boundary condition of (6.49), i.e. at the *lower* boundary. Since $\langle A_0\overline{B}_0\rangle^\tau$ depends on both X and T (unless either A_0 or B_0 is spatially uniform) the boundary condition implies that ϕ_0^{v} (and hence W_0^{v}) depends on both the fast and slow horizontal spatial variables x and X. This dependence cannot be obtained in closed form and one must therefore resort to numerical computations for realistically large values of L. Note, however, that in fully three-dimensional situations [46] in which lateral walls are included a viscous mean flow will be present even when $k \gg 1$ because the forcing of the mean flow in the oscillatory boundary layers along these walls remains.

Second, the change of variables

$$A_0 = \tilde{A}_0 \mathrm{e}^{-\mathrm{i}k\theta} \qquad B_0 = \tilde{B}_0 \mathrm{e}^{\mathrm{i}k\theta} \tag{6.52}$$

where

$$\theta'(T) = -\alpha_6 k^{-1} \int_{-1}^{0} g(z)\langle\langle \phi_{0z}^{\mathrm{v}}\rangle^x\rangle^X \,\mathrm{d}z \tag{6.53}$$

reduces equations (6.46) to the much simpler form

$$\begin{aligned}
\tilde{A}_{0T} &= \mathrm{i}\alpha\tilde{A}_{0\xi\xi} - (\Delta + \mathrm{i}D)\tilde{A}_0 + \mathrm{i}[(\alpha_3+\alpha_8)|\tilde{A}_0|^2 \\
&\quad - (\alpha_4+\alpha_8)\langle|\tilde{B}_0|^2\rangle^\eta]\tilde{A}_0 + \mathrm{i}\alpha_5 M\langle\overline{\tilde{B}_0}\rangle^\eta \\
\tilde{B}_{0T} &= \mathrm{i}\alpha\tilde{B}_{0\eta\eta} - (\Delta + \mathrm{i}D)\tilde{B}_0 + \mathrm{i}[(\alpha_3+\alpha_8)|\tilde{B}_0|^2 \\
&\quad - (\alpha_4+\alpha_8)\langle|\tilde{A}_0|^2\rangle^\xi]\tilde{B}_0 + \mathrm{i}\alpha_5 M\langle\overline{\tilde{A}_0}\rangle^\xi \\
\tilde{A}_0(\xi+1, T) &\equiv \tilde{A}_0(\xi, T) \qquad \tilde{B}_0(\eta+1, T) \equiv \tilde{B}_0(\eta, T)
\end{aligned} \tag{6.54}$$

from which the mean flow is absent. This *decoupling* is a special property of the regime (6.36), but is not unique to it. The resulting equations provide perhaps the simplest description of the Faraday system at large aspect ratio and it is for this reason that they have been extensively studied [67,92]. Except for the presence of the parametric forcing term M they are of the same form as the coupled non-local CGL equations derived in section 6.3. Tracing back through the derivation one can see that the non-local terms in both sets of equations have identical origin: dynamics occuring on two distinct slow time scales.

In short domains, however, with $L = O(1)$ only two time scales are present, and the resulting coupled amplitude–mean flow equations are considerably simpler:

$$\begin{aligned}
A_t = & -(\delta + \mathrm{i}d)A + \mathrm{i}(\alpha_3|A|^2 - \alpha_4|B|^2)A + \mathrm{i}\alpha_5\mu\bar{B} \\
& - \mathrm{i}\alpha_6 L^{-1}\int_{-1}^{0}\int_{0}^{L} g(z)u^{\mathrm{v}}\,\mathrm{d}x\,\mathrm{d}z\,A \\
B_t = & -(\delta + \mathrm{i}d)B + \mathrm{i}(\alpha_3|B|^2 - \alpha_4|A|^2)B + \mathrm{i}\alpha_5\mu\bar{A} \\
& + \mathrm{i}\alpha_6 L^{-1}\int_{-1}^{0}\int_{0}^{L} g(z)u^{\mathrm{v}}\,\mathrm{d}x\,\mathrm{d}z\,B
\end{aligned} \tag{6.55}$$

with A and B spatially constant and the coefficients given by expressions that are identical to those in (6.29)–(6.30). Here t denotes a slow time whose magnitude is determined by the damping $\delta > 0$ and the detuning d, both assumed to be of the same order as the forcing amplitude μ; in the long time limit $|A|^2 = |B|^2 \equiv R^2$. It follows that the mean flow $(u^{\mathrm{v}}(x, z, t), w^{\mathrm{v}}(x, z, t))$ is now entirely viscous in origin, and obeys a two-dimensional Navier–Stokes equation of the form (6.47). If we absorb the standing wave amplitude R (and some other constants) in the definition of the Reynolds number,

$$Re \equiv 2\beta_4 R^2/C_{\mathrm{g}} \tag{6.56}$$

this equation is to be solved subject to the boundary conditions

$$\begin{aligned}
& u^{\mathrm{v}} = -\sin[2k(x-\theta)] \qquad w^{\mathrm{v}} = 0 \qquad \text{at } z = -1 \\
& u_z^{\mathrm{v}} = 0 \qquad w^{\mathrm{v}} = 0 \qquad \text{at } z = 0.
\end{aligned} \tag{6.57}$$

Because the structure of equations (6.55) is identical to that of equations (6.46) the change of variables (6.52) leads to a decoupling of the amplitudes from the spatial phase θ which now satisfies the equation

$$\theta_t = \frac{\alpha_6}{kL}\int_{-1}^{0}\int_{0}^{L} g(z)u^{\mathrm{v}}(x, z, t)\,\mathrm{d}x\,\mathrm{d}z. \tag{6.58}$$

Martín *et al* [95] solve the resulting equations numerically with periodic boundary conditions (corresponding to an annular domain of length L) and demonstrate that a basic pattern of subharmonic standing waves can indeed lose stability at finite amplitude through the excitation of a viscous mean flow. This may happen in one of two ways, either at a parity-breaking steady-state bifurcation or at a reflection symmetry-breaking Hopf bifurcation. The former results in a steady drift of the pattern while the latter produces a state that has been called a direction-reversing wave [85].

6.5 Nonlinear waves in extended systems with broken reflection symmetry

The effects of distant boundaries on the onset of travelling wave instabilities in systems with broken reflection symmetry differ remarkably from the corresponding steady-state situation in reflection-symmetric systems. In the latter case the imposition of boundary conditions at the ends of a domain of aspect ratio L leads to a correction to the instability threshold of $O(L^{-2})$ when L is large. However, as already mentioned in the introduction, in systems that lack reflection symmetry the primary instability is always a Hopf bifurcation to travelling waves with a preferred direction of propagation, and in this case the imposition of similar boundary conditions results in an $O(1)$ change to the threshold for instability. Moreover, the initial eigenfunction at the onset is a wall mode rather than wave-like, and the frequency of this mode differs substantially from that of the most unstable, spatially periodic solution. With increasing values of the instability parameter μ a front forms, separating an exponentially small wave-train near the upstream boundary from a fully developed one downstream, with a well-defined wavenumber, frequency and amplitude, whose location moves further and further upstream as μ continues to increase. Both the spatial wavenumber and the amplitude are controlled by the temporal frequency, which, in turn, is controlled by μ, and in the fully nonlinear regime by the boundary conditions as well. The resulting changes in the frequency may trigger secondary transitions to quasi-periodic and/or chaotic states. These phenomena have been described in a recent paper by Tobias *et al* [151] (see also [21]), and the criterion for the presence of a global unstable mode was found to be closely related to that for absolute instability of the basic state in an unbounded domain [128, 151]. The secondary transitions are likewise related to the absolute instability of the primary wave-train.

The simplest model of this set of phenomena is provided by the complex Ginzburg–Landau equation with drift:

$$\frac{\partial A}{\partial t} = c_{\mathrm{g}}\frac{\partial A}{\partial x} + \mu A + a|A|^2 A + \lambda \frac{\partial^2 A}{\partial x^2} \qquad 0 \le x \le L \tag{6.59}$$

subject to the boundary conditions

$$A(0) = A(L) = 0. \tag{6.60}$$

Here $c_{\mathrm{g}} > 0$ represents the drift term (and is the group velocity $\mathrm{d}\omega/\mathrm{d}k$ if $A(x, t)$ is the complex amplitude of a wave-train with frequency $\omega(k)$) and $\lambda = 1 + \mathrm{i}\lambda_{\mathrm{I}}$, $a = -1 + \mathrm{i}a_{\mathrm{I}}$. In the following we take c_{g} to be of order unity, so that reflection symmetry is broken strongly. In figure 6.6 we show a solution of equations (6.59)–(6.60) obtained numerically for $\mu > \mu_{\mathrm{a}} \equiv c_{\mathrm{g}}^2/[4(1 + \lambda_{\mathrm{I}}^2)]$ and $c_{\mathrm{g}} = 1$. The figure is in the form of a spacetime diagram, with time increasing upwards. Infinitesimal disturbances travel to the left from $x = L$ and grow as they do so, evolving into a stationary primary front, separating a small amplitude upstream wave (barely visible) from a fully developed downstream wave with $O(1)$ amplitude. For μ not too large this wave-train persists until almost $x = 0$, where there is a boundary layer within which A drops to zero. However, for larger μ a remarkable phenomenon appears. The wave-train develops a secondary

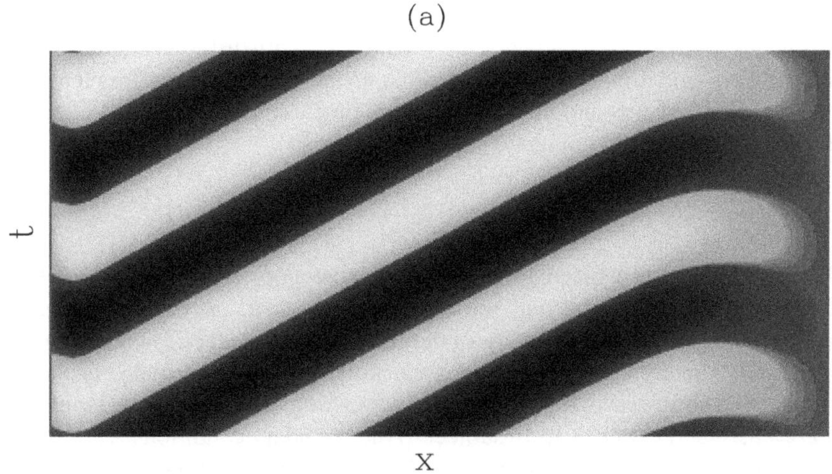

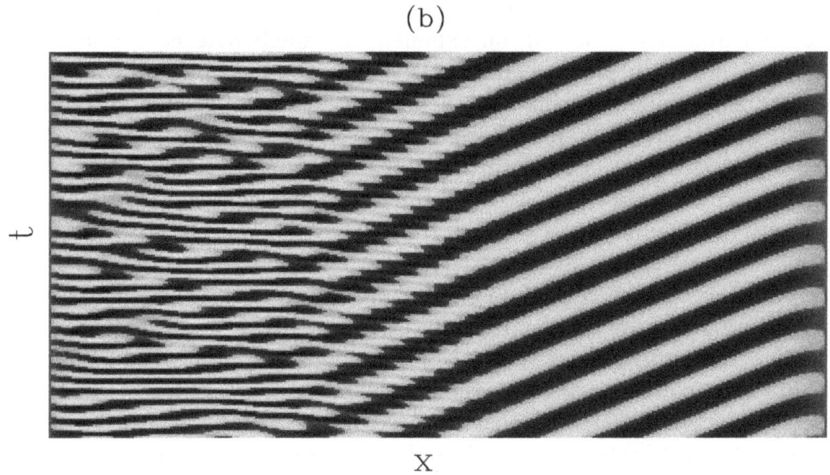

Figure 6.6. Grey scale spacetime plots obtained for the CGL equation (6.59) subject to the boundary conditions (6.60) with $L = 60.0$, $c_g = 1.0$, $\lambda_I = 0.45$, $A_I = 2.0$ and (*a*) $\mu = 0.3$, (*b*) $\mu = 1.5$. The domain $(0, L)$ lies along the horizontal axis, with time increasing upward. In (*a*) the waves change their direction of propagation as they pass through the primary front; in (*b*) a secondary front signals the transition to spatio-temporal chaos. Reproduced from S Tobias, M R E Proctor and E Knobloch Convective and absolute instabilities of fluid flows in finite geometry *Physica* D **113** 43–72 ©1998 by Elsevier Science.

front prior to $x = 0$ separating the laminar $O(1)$ wave-train from a new wave-train in the region adjacent to $x = 0$. In the example shown this wave-train is chaotic or 'turbulent'. As a result the position of the secondary front fluctuates but nonetheless has a well-defined mean. For other parameter values one may find a secondary front between two periodic wave-trains, with different amplitudes and wavenumbers, and if there are phase slips along the front, different frequencies as well. Figure 6.7 shows another example, this time computed from the equations describing the mean field dynamo [119],

$$\frac{\partial A}{\partial t} = \frac{DB}{1+B^2} + \frac{\partial^2 A}{\partial x^2} - A \qquad \frac{\partial B}{\partial t} = \frac{\partial A}{\partial x} + \frac{\partial^2 B}{\partial x^2} - B. \tag{6.61}$$

Here $D > 0$ is a bifurcation parameter analogous to μ, and A and B are the poloidal field potential and the toroidal field itself. Both are real-valued functions. This model has the merit of describing a real magnetic field as opposed to the CGL equation, which only describes the evolution of a slowly varying envelope of a short-wave, high-frequency wave-train. The figure is computed for (*a*) $D = 8.75$ and (*b*) $D = 15.0$, subject to the boundary conditions $A_x = B = 0$ at $x = 0$ and $A = B = 0$ at $x = L$. Figure 6.7(*a*) shows a single front near the right-hand boundary separating a stationary fully developed but laminar wave-train from a region with an exponentially small state near $x = L$. In contrast, figure 6.7(*b*) shows two fully developed wave-trains with well-defined wavenumbers, amplitudes and frequencies separated by a secondary front in the left half of the container. Both wave-trains are laminar and are separated by a more-or-less stationary secondary front. It is tempting to think of these fronts as shock-like structures, across which there is finite jump in three quantities: amplitude, wavenumber and perhaps frequency as well. These shocks are of considerable mathematical interest since the underlying equations are neither hyperbolic nor conservative. As a result there are no obvious candidates for the type of jump conditions familiar from gas dynamics. The *a priori* prediction of the location of these shocks, and of the associated jump conditions as a function of μ constitutes an interesting and important problem for future research. Analysis of the CGL equation (6.59) shows that these secondary fronts form once the primary wave train loses stability to a *secondary* absolute Benjamin–Feir instability [151]. As already mentioned the resulting scenario also arises in attempts to explain the far-field breakup of spiral waves in reaction–diffusion systems [150].

As is well known, systems that are only convectively unstable (but absolutely stable) act as efficient noise amplifiers [42]. This property manifests itself in extended but finite domains as well, as discussed by Proctor *et al* [120]. Figure 6.8 summarizes the effects of injecting small amplitude noise at the upstream boundary in the two cases of interest: for Benjamin–Feir stable ($\lambda_I a_I < 1$) and Benjamin–Feir unstable ($\lambda_I a_I > 1$) regimes. In the former case the spatially uniform Stokes solution of (6.59) is *stable* with respect to long wavelength perturbations; this is not the case in the Benjamin–Feir unstable regime [143]. As a result any non-trivial dynamics present in the system must be a consequence of the boundaries. Little changes if the noise is injected instead throughout the domain $0 < x < L$, since the dynamics are primarily determined by the perturbations that have travelled farthest, and hence grew the most.

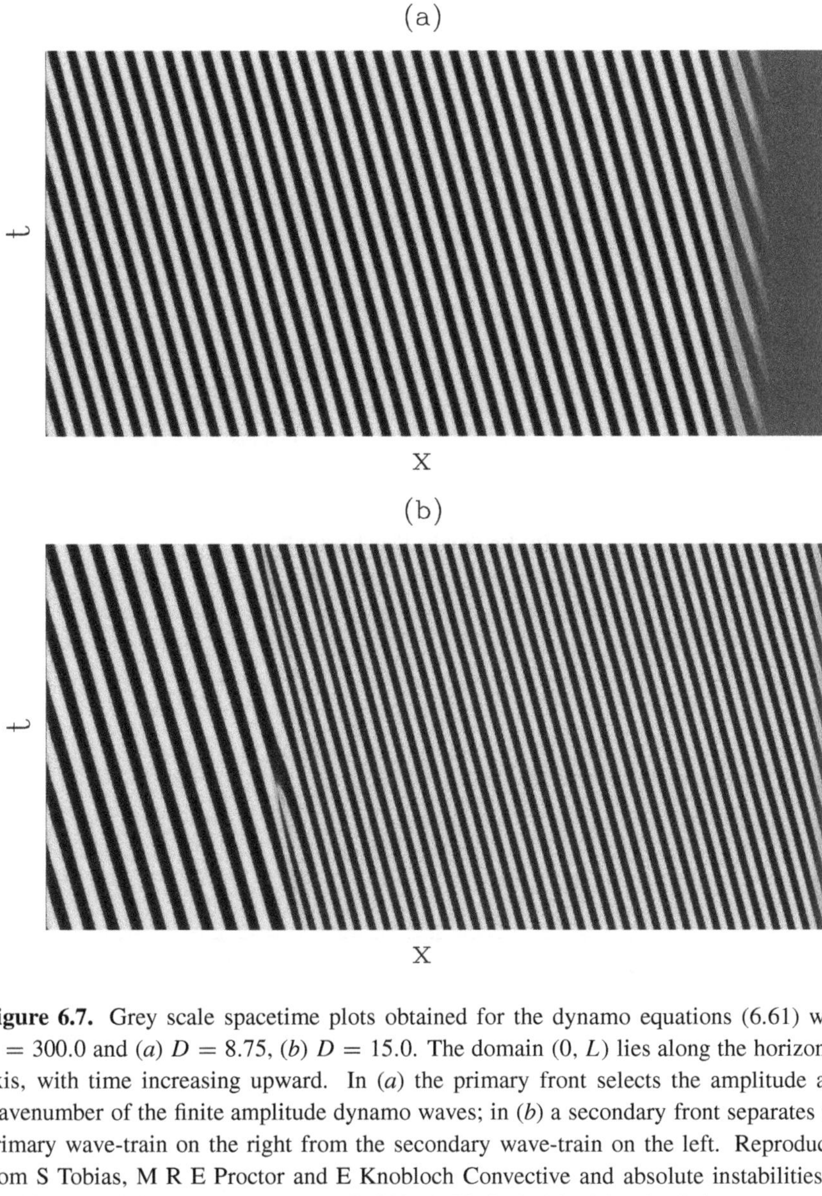

Figure 6.7. Grey scale spacetime plots obtained for the dynamo equations (6.61) with $L = 300.0$ and (*a*) $D = 8.75$, (*b*) $D = 15.0$. The domain $(0, L)$ lies along the horizontal axis, with time increasing upward. In (*a*) the primary front selects the amplitude and wavenumber of the finite amplitude dynamo waves; in (*b*) a secondary front separates the primary wave-train on the right from the secondary wave-train on the left. Reproduced from S Tobias, M R E Proctor and E Knobloch Convective and absolute instabilities of fluid flows in finite geometry *Physica* D **113** 43–72 ©1998 by Elsevier Science.

If the trivial state $A = 0$ is completely stable, then all such disturbances decay at every location, and so have little effect. However, if the trivial state is convectively unstable, there are disturbances that grow in an appropriately moving reference frame, even though at fixed location all disturbances eventually decay.

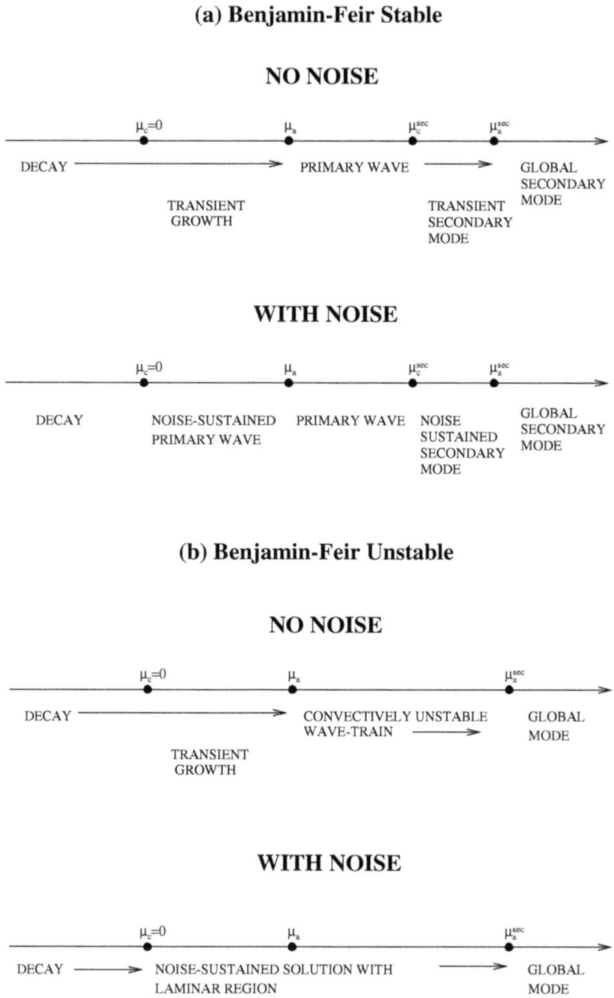

Figure 6.8. Schematic diagram of the possible roles of noise in the CGL equation (6.59). The two scenarios presented are for (*a*) Benjamin–Feir stable parameter values ($\lambda_I a_I < 1$) and (*b*) Benjamin–Feir unstable parameter values ($\lambda_I a_I > 1$). In (*a*) the role of the noise depends critically on the value of the bifurcation parameter, while in (*b*) the noise-induced primary wave-train is always susceptible to noise-induced disruption as discussed by Deissler [43]. Reproduced from M R E Proctor, S M Tobias and E Knobloch Noise-sustained structures due to convective instability in finite domains *Physica* D **145**191–206 ©2000 by Elsevier Science.

In these circumstances the continued injection of disturbances at an upstream location can produce persistent noise-sustained structures downstream [42]. Such structures have been studied in experiments [3, 4, 154], and in various model problems based on the CGL equation [12, 42–45, 90, 104, 130]. In fact, the addition of noise can lead to the destabilization not only of the trivial state but also of the primary wave-train (figure 6.8). Moreover the travelling wave nature of the transient instability that is sustained by noise injection leads to a powerful frequency selection effect, which determines the spatio-temporal properties of the resulting noise-sustained structures. This mechanism, identified in [120] in the context of the CGL equation (6.59), is believed to have general applicability. Briefly, the system selects the frequency $\omega_{\max}$ that *maximizes the smallest spatial growth rate in the downstream direction*. This frequency selection mechanism differs from the intuitive (but incorrect) idea that one wishes to maximize the downstream spatial growth rate on the grounds that the mode growing most rapidly to the left would be the one observed. This is a consequence of the downstream boundary condition. Because of this boundary the solution of the linear problem on $0 < x < L$ consists of a linear combination of exponentials involving the (complex) roots of the dispersion relation. At the left-hand (or downstream) boundary the amplitudes of all the component modes are in general of the same order; for each value of the frequency ω the mode seen far from the boundary is then the one that decays least rapidly to the right. One seeks therefore the frequency $\omega_{\max}$ that maximizes this least decay rate. Figure 6.9 provides a confirmation of this idea for the dynamo equations (6.61). Thus once again the mere fact that there *is* a downstream boundary has a profound effect on the behaviour of the system: open (i.e. semi-infinite) systems without a downstream boundary behave very differently, and for these the intuitive argument *is* correct. It is worth mentioning that in the absence of noise a semi-infinite system without a downstream boundary only generates structure above the absolute instability threshold. In contrast a semi-infinite system without an *upstream* boundary is already unstable above the convective instability threshold $\mu = 0$ [162].

The linear theory just described predicts [120] that the noise-sustained structure will be present in $x < x_{\text{front}}$, where

$$x_{\text{front}} \equiv L - \kappa_{\max}^{-1} |\ln \epsilon|. \tag{6.62}$$

Here $\kappa_{\max}$ is the maximum value of the least spatial growth rate determined by the previously described procedure, and ϵ is a measure of the amplitude of the noise injected at $x = L$, assumed to be uniformly distributed in $[-\epsilon, \epsilon]$. For (6.59) $\omega_{\max} = \mu\lambda_{\mathrm{I}}$ and the resulting noise-sustained structure has a well-defined wavenumber and amplitude, given by

$$\begin{aligned} k &= \frac{1}{2(a_{\mathrm{I}} + \lambda_{\mathrm{I}})} \left(c_{\mathrm{g}} - \sqrt{c_{\mathrm{g}}^2 - 4\mu(\lambda_{\mathrm{I}}^2 - a_{\mathrm{I}}^2)} \right) \\ |A_0|^2 &= \frac{1}{2(a_{\mathrm{I}} + \lambda_{\mathrm{I}})^2} \left(4\mu\lambda_{\mathrm{I}}(a_{\mathrm{I}} + \lambda_{\mathrm{I}}) - c_{\mathrm{g}}^2 + c_{\mathrm{g}}\sqrt{c_{\mathrm{g}}^2 - 4\mu(\lambda_{\mathrm{I}}^2 - a_{\mathrm{I}}^2)} \right). \end{aligned} \tag{6.63}$$

As shown in [120], this solution can, in turn, be convectively unstable, and

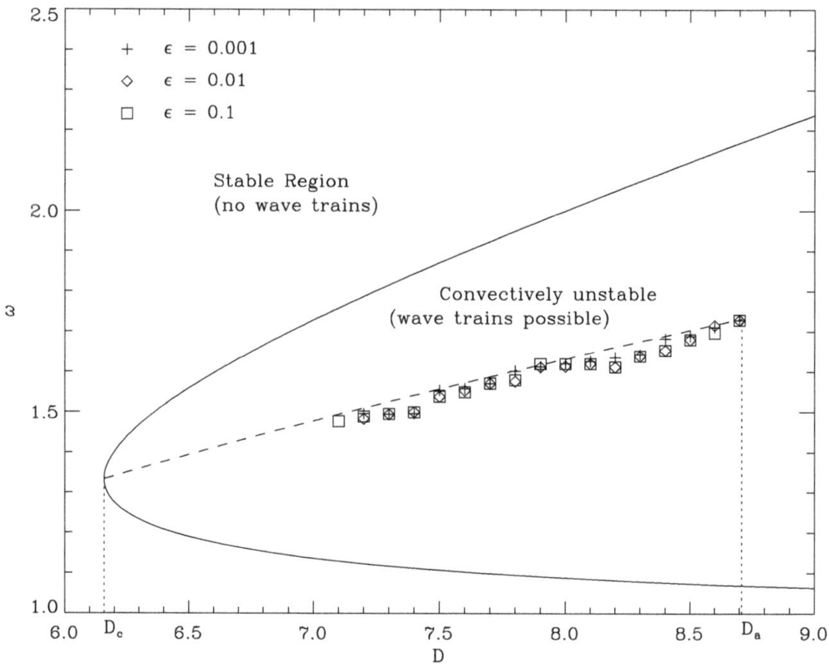

Figure 6.9. Noise-sustained dynamo waves computed from equations (6.61): selected frequency ω as a function of the dynamo number D in the convectively unstable regime $D_{\mathrm{c}} < D < D_{\mathrm{a}}$. The full curve is the boundary of the region in which wave-trains occur. The broken line shows the theoretical prediction. The symbols show the results of numerical simulations with different values of D and noise level ϵ. Dotted lines show thresholds for convective and absolute instability of the trivial state. Reproduced from M R E Proctor, S M Tobias and E Knobloch Noise-sustained structures due to convective instability in finite domains *Physica* D **145**191–206 ©2000 by Elsevier Science.

so support secondary noise-induced structures or be absolutely unstable (cf figure 6.8).

The phenomenon of a noise-sustained secondary instability is perhaps of greatest interest. We consider therefore the case of a primary wave-train when $\mu_{\mathrm{f}} < \mu_{\mathrm{c}}^{\mathrm{sec}} < \mu < \mu_{\mathrm{a}}^{\mathrm{sec}}$, i.e. when this wave-train is convectively unstable. The primary wave-train is unaffected by noise injection at the upstream boundary because it is a global mode ($\mu > \mu_{\mathrm{f}} = \mu_{\mathrm{a}} + O(L^{-2})$). In the absence of noise the secondary convective instability has no effect, provided that the group velocity associated with this instability is leftward. The argument is the same as for the primary instability: perturbations are advected towards the boundary at $x = 0$ where they are dissipated, and the instability will produce secondary

Figure 6.10. Noise-sustained secondary structures in the CGL equation (6.59). Spacetime plot of Re(A) for $\mu = 1.0$, $\epsilon = 10^{-4}$, $\lambda_{\mathrm{I}} = 0.45$ and $a_{\mathrm{I}} = 2.0$ with x increasing to the right and time increasing upwards. The irregular phase at the *left* of the picture is sustained by noise injection at the *right* boundary, and decays away in its absence. Reproduced from M R E Proctor, S M Tobias and E Knobloch Noise-sustained structures due to convective instability in finite domains *Physica* D **145** 191–206 ©2000 by Elsevier Science.

structures only when the threshold for secondary absolute instability is exceeded. However, in the presence of noise a noise-sustained secondary structure will form. This structure appears first at the left-hand boundary, and with increasing μ extends further and further to the right, separated from the primary wave-train by a secondary front. Figure 6.10 shows an example computed from equation (6.59) for $\mu = 1.0$, $\epsilon = 10^{-4}$, $\lambda_{\mathrm{I}} = 0.45$, $a_{\mathrm{I}} = 2.0$ and $c_{\mathrm{g}} = 1.0$. For these parameters the secondary instability boundary in the absence of noise is at $\mu_{\mathrm{a}}^{\mathrm{sec}} \approx 1.5$, and indeed the chaotic structure present near the left boundary in figure 6.10 decays away when ϵ is set to zero. Guided by the discussion of the primary instability we expect that, in its linear phase, the secondary mode has a well-defined spatial growth rate, wavenumber and temporal frequency and that these are determined by a criterion similar to that for the primary noise-sustained mode. This expectation is borne out by an investigation of the linear instability of the primary wave-train along the lines of [151].

It should be noted that even in systems that are 'small' the breaking of reflection symmetry may have quite unexpected consequences, such as the appearance of a Hopf bifurcation 'by frustrated drifts' in the Rayleigh–Bénard problem between out-of-phase wavy boundaries [136].

6.6 Summary and conclusions

In this chapter I have pointed to a number of issues in the general area of pattern formation which remain unsolved, and attempted to indicate why I believe they are important. The problems mentioned range from the mathematical (extension of KAM theory into the spatial domain) to the practical, such as confronting the predictions of model equations with experimental data, or at least data obtained from accurate numerical simulations of the governing field equations. I discussed various techniques for generating such model equations, including both rigorous derivations and symmetry-based methods. In each case I discussed the assumptions required and their appropriateness for the physical problem at hand. In many cases these equations provide us with a good but partial understanding; in some problems, such as the oscillons in granular media, this may be as much as we can expect, while in others much more can, in principle, be done. This is because in the former case we know only the microscopic interactions but lack effective constitutive relations for a macroscopic theory, while in other areas the correct field equations are well known. Many pattern-forming systems, for example chemical systems, fall in between these two extremes. These systems are governed by an ever growing variety of model equations whose structure depends sensitively on the assumptions made about the spatial and temporal scales in the system, both intrinsic and externally imposed. In this chapter I have emphasized systems that are described by fundamentally non-local equations and explained the reasons for this fact. These equations provide new classes of pattern-forming equations and their behaviour is only just beginning to be understood. Traditional equivariant theory is of much less use in these systems because of its implicit assumptions that all interactions are local, that is, local in both space and time, and in phase space. As a result the derivations may appear to be involved, but at present this is unavoidable. I have also emphasized the importance of considering the applicability or relevance of the solutions of these model equations to the underlying physical system. These questions raise additional mathematical issues, some well known from the theory of averaging, but others such as imperfection sensitivity of equivariant dynamics still poorly understood.

I expect substantial progress on a number of the issues raised here over ensuing several years, although I will not risk making any predictions. The field of pattern formations is a lively one, and this article will have succeeded if it stimulates both new work and new ideas in this area.

Acknowledgments

I am grateful to Alastair Rucklidge for a careful reading of this article. This work was supported in part by the NASA Microgravity Program under grant NAG3-2152 and by the National Science Foundation under grant DMS-0072444, and represents results obtained in collaboration with a number of colleagues: T Callahan, M Higuera, C Martel, J Moehlis, M Proctor, S Tobias and J M Vega.

References

[1] Andereck C D, Liu S S and Swinney H L 1986 Flow regimes in a circular Couette system with independently rotating cylinders *J. Fluid Mech.* **164** 155–83

[2] Arnéodo A, Coullet P, Tresser C and Spiegel E A 1985 Asymptotic chaos *Physica* D **14** 327–47

[3] Babcock K L, Ahlers G and Cannell D S 1994 Noise amplification in open Taylor–Couette flow *Phys. Rev.* E **50** 3670–92

[4] Babcock K L, Cannell D S and Ahlers G 1992 Stability and noise in Taylor–Couette flow with through-flow *Physica* D **61** 40–6

[5] Balmforth N J, Del-Castillo-Negrete D and Young W R 1997 Dynamics of vorticity defects in shear *J. Fluid Mech.* **333** 197–230

[6] Balmforth N J and Young W R 1997 Long-wave instability of marginally stable shear flows *Phys. Rev. Lett.* **79** 4155–8

[7] Bär M and Or-Guil M 1999 Alternative scenarios of spiral breakup in a reaction–diffusion model with excitable and oscillatory dynamics *Phys. Rev. Lett.* **82** 1160–3

[8] Batiste O, Knobloch E, Mercader I and Net M 2001 Simulations of oscillatory binary fluid convection in large aspect ratio containers *Phys. Rev.* E **65** 016303 (19pp)

[9] Benguria R D and Depassier M C 1994 Validity of the linear speed selection mechanism for fronts of the nonlinear diffusion equation *Phys. Rev. Lett.* **73** 2272–4

[10] Bergeon A, Henry D and Knobloch E 2001 Three-dimensional Marangoni-Bénard flows in square and nearly square containers *Phys. Fluids* **13** 92–8

[11] Blanchflower S 1999 Magnetohydrodynamic convectons *Phys. Lett.* A **261** 74–81

[12] Brand H R, Deissler R J and Ahlers G 1991 Simple model for the Bénard instability with horizontal flow near threshold *Phys. Rev.* A **43** 4262–8

[13] Busse F H 1967 On the stability of of two-dimensional convection in a layer heated from below *J. Math. Phys.* **46** 140–9

[14] Busse F H and Müller S C (ed) 1998 *Evolution of Spontaneous Structures in Dissipative Continuous Systems* (Berlin: Springer)

[15] Callahan T K 2000 Hopf bifurcation on the FCC lattice *EQUADIFF 99 (Proc. Int. Conf. on Diff. Eqs.)* vol 1, ed B Fiedler, K Gröger and J Sprekels (Singapore: World Scientific) pp 154–6

[16] Callahan T and Knobloch E 1997 Symmetry-breaking bifurcations on cubic lattices *Nonlinearity* **10** 1179–216
[17] Callahan T and Knobloch E 2000 Pattern formation in weakly anisotropic systems *EQUADIFF 99 (Proc. Int. Conf. on Diff. Eqs.)* vol 1, ed B Fiedler, K Gröger and J Sprekels (Singapore: World Scientific) pp 157–62
[18] Callahan T and Knobloch E 2001 Long wavelength instabilities of three-dimensional patterns *Phys. Rev.* E **64** 036214 (25pp)
[19] Chen Y-Y 1992 Boundary conditions and linear analysis of finite-cell Rayleigh–Bénard convection *J. Fluid Mech.* **241** 549–85
[20] Chikwendu S C and Kevorkian J 1972 A perturbation method for hyperbolic equations with small nonlinearities *SIAM J. Appl. Math.* **22** 235–58
[21] Chomaz J-M and Couairon A 1999 Against the wind *Phys. Fluids* **11** 2977–83
[22] Chossat P and Iooss G 1994 *The Couette–Taylor Problem* (Berlin: Springer)
[23] Clune T and Knobloch E 1994 Pattern selection in three-dimensional magnetoconvection *Physica* D **74** 151–76
[24] Crawford J D 1991 Surface waves in non-square containers with square symmetry *Phys. Rev. Lett.* **67** 441–4
[25] Crawford J D 1994 Amplitude expansions for instabilities in populations of globally-coupled oscillators *J. Stat. Phys.* **74** 1047–84
[26] Crawford J D 1994 Universal trapping scaling on the unstable manifold of a collisionless electrostatic mode *Phys. Rev. Lett.* **73** 656–9
[27] Crawford J D, Gollub J P and Lane D 1993 Hidden symmetries of parametrically forced waves *Nonlinearity* **6** 119–64
[28] Crawford J D and Knobloch E 1988 On degenerate Hopf bifurcation with broken O(2) symmetry *Nonlinearity* **1** 617–52
[29] Crawford J D and Knobloch E 1991 Symmetry and symmetry-breaking bifurcations in fluid mechanics *Ann. Rev. Fluid Mech.* **23** 341–87
[30] Croquette V and Williams H 1989 Nonlinear competition between waves on convective rolls *Phys. Rev.* A **39** 2765–8
[31] Cross M C 1988 Structure of nonlinear travelling waves in finite geometries *Phys. Rev.* A **38** 3593–600
[32] Cross M C and Hohenberg P C 1993 Pattern formation outside of equilibrium *Rev. Mod. Phys.* **65** 851–1112
[33] Dangelmayr G and Knobloch E 1991 Hopf bifurcation with broken circular symmetry *Nonlinearity* **4** 399–427
[34] Davey A and Stewartson K 1974 On three-dimensional packets of surface waves *Proc. R. Soc.* A **338** 101–10
[35] Dawes J H P 2001 Hopf bifurcation on a square superlattice *Nonlinearity* **14** 491–511
[36] Dawes J H P, Matthews P C and Rucklidge A M, Reducible representations of $D_4\dot{+}T^2$: superlattice patterns and hidden symmetries *Nonlinearity* submitted
[37] Deane A E, Knobloch E and Toomre J 1988 Travelling waves in large-aspect-ratio thermosolutal convection *Phys. Rev.* A **37** 1817–20
[38] Dias A 1998 Hopf bifurcation for wreath products *Nonlinearity* **11** 247–64
[39] Dionne B 1993 Planforms in three dimensions *ZAMP* **44** 673–94
[40] Dionne B and Golubitsky M 1992 Planforms in two and three dimensions *ZAMP* **43** 36–62

[41] Dionne B, Silber M and Skeldon A 1997 Stability results for the steady, spatially periodic planforms *Nonlinearity* **10** 321–53
[42] Deissler R J 1985 Noise-sustained structures, intermittency, and the Ginzburg–Landau equation *J. Stat. Phys.* **40** 371–95
[43] Deissler R J 1987 Spatially growing waves, intermittency, and convective chaos in an open-flow system *Physica* D **25** 233–60
[44] Deissler R J 1989 External noise and the origin and dynamics of structure in convectively unstable systems *J. Stat. Phys.* **54** 1459–88
[45] Deissler R J 1989 Thermally sustained structure in convectively unstable systems *Phys. Rev.* E **49** R31–4
[46] Douady S, Fauve S and Thual O 1989 Oscillatory phase modulation of parametrically forced surface waves *Europhys. Lett.* **10** 309–15
[47] Duan J, Ly H V and Titi E S 1996 The effects of non-local interactions on the dynamics of the Ginzburg–Landau equation *ZAMP* **47** 432–55
[48] Ecke R E, Zhong F and Knobloch E 1992 Hopf bifurcation with broken reflection symmetry in rotating Rayleigh–Bénard convection *Europhys. Lett.* **19** 177–82
[49] Edwards W S and Fauve S 1994 Patterns and quasi-patterns in the Faraday experiment *J. Fluid Mech.* **278** 123–48
[50] Elmer F J 1988 Nonlinear and non-local dynamics of spatially extended systems: stationary states, bifurcations and stability *Physica* D **30** 321–42
[51] Evans J W 1972 Nerve axon equations. III. Stability of nerve impulse *Indiana Univ. Math. J.* **22** 577–93
[52] Fauve S 1995 Parametric instabilities *Dynamics of Nonlinear and Disordered Systems* ed G Martínez Mekler and T H Seligman (Singapore: World Scientific) pp 67–115
[53] Golubitsky M, Knobloch E and Stewart I 2000 Target patterns and spirals in planar reaction–diffusion systems *J. Nonlin. Sci.* **10** 333–54
[54] Golubitsky M, Luss D and Strogatz S H (ed) 1999 *Pattern Formation in Continuous and Coupled Systems* (New York: Springer)
[55] Golubitsky M and Roberts M 1987 A classification of degenerate Hopf bifurcations with O(2) symmetry *J. Diff. Eqns* **69** 216–64
[56] Golubitsky M and Schaeffer D G 1985 *Singularities and Groups in Bifurcation Theory* vol I (New York: Springer)
[57] Golubitsky M, Stewart I and Schaeffer D G 1988 *Singularities and Groups in Bifurcation Theory* vol II (New York: Springer)
[58] Golubitsky M, Swift J W and Knobloch E 1984 Symmetries and pattern selection in Rayleigh–Bénard convection *Physica* D **10** 249–76
[59] Guckenheimer J and Holmes P 1984 *Nonlinear Oscillations, Dynamical Systems and Bifurcations of Vector Fields* (New York: Springer)
[60] Gunaratne G H 1993 Complex spatial patterns on planar continua *Phys. Rev. Lett.* **71** 1367–70
[61] Hagan P S 1981 Target patterns in reaction–diffusion systems *Adv. Appl. Math.* **2** 400–16
[62] Hall P 1984 Evolution equations for Taylor vortices in the small-gap limit *Phys. Rev.* A **29** 2921–3
[63] Haller G and Wiggins S 1995 N-pulse homoclinic orbits in perturbations of resonant Hamiltonian systems *Arch. Rat. Mech. Anal.* **130** 25–101

[64] Haller G and Wiggins S 1995 Multi-pulse jumping orbits and homoclinic trees in a modal truncation of the damped-forced nonlinear Schrödinger equation *Physica* D **85** 311–47

[65] Haller G 1998 Homoclinic jumping in the perturbed nonlinear Schrödinger equation *Commun. Pure Appl. Math.* **52** 1–47

[66] Higuera F J and Bonilla L L 1992 Gunn instability in finite samples of GaAs II. Oscillatory states in long samples *Physica* D **57** 161–84

[67] Higuera M, Porter J and Knobloch E 2002 Heteroclinic dynamics in the non-local parametrically driven nonlinear Schrödinger equation *Physica* D **162** 155–87

[68] Hirschberg P and Knobloch E 1996 Complex dynamics in the Hopf bifurcation with broken translation symmetry *Physica* D **90** 56–78

[69] Hoyle R B 1993 Long-wavelength instabilities of square patterns *Physica* D **67** 198–223

[70] Huerre P and Monkewitz P A 1990 Local and global instabilities in spatially developing flows *Ann. Rev. Fluid Mech.* **22** 473–537

[71] Ikeda Y and Murota K 2001 *Imperfection Bifurcation in Structures and Materials* (New York: Springer) in press

[72] Imbihl R, Engel H and Eiswirth M 1998 Dynamics of patterns of chemical reactions *Evolution of Spontaneous Structures in Dissipative Continuous Systems* ed F H Busse and S C Müller S C (New York: Springer) pp 384–410

[73] Iooss G and Los J 1990 Bifurcation of spatially quasi-periodic solutions in hydrodynamic stability problems *Nonlinearity* **3** 851–71

[74] Jacqmin D and Heminger J 1994 Double-diffusion with Soret effect in a rectangular geometry: Linear and nonlinear travelling wave instabilities *Preprint*

[75] Judd S L and Silber M 2000 Simple and superlattice Turing patterns in reaction–diffusion systems: bifurcation, bistability, and parameter collapse *Physica* D **136** 45–65

[76] Knobloch E 1992 Non-local amplitude equations *Pattern Formation in Complex Dissipative Systems* ed S Kai (Singapore: World Scientific) pp 263–74

[77] Knobloch E and De Luca J 1990 Amplitude equations for travelling wave convection *Nonlinearity* **3** 975–80

[78] Knobloch E and Gibbon J D 1991 Coupled NLS equations for counter-propagating waves in systems with reflection symmetry *Phys. Lett.* A **154** 353–6

[79] Knobloch E, Hettel J and Dangelmayr G 1995 Parity breaking bifurcation in inhomogeneous systems *Phys. Rev. Lett.* **74** 4839–42

[80] Kolodner P, Surko C M and Williams H 1989 Dynamics of travelling waves near the onset of convection in binary fluid mixtures *Physica* D **37** 319–33

[81] Koschmieder E L 1993 *Bénard Cells and Taylor Cells* (Cambridge: Cambridge University Press)

[82] Krupa M 1990 Bifurcations from relative equilibria *SIAM J. Math. Anal.* **21** 1453–86

[83] Kudrolli A and Gollub J P 1997 Patterns and spatio-temporal chaos in parametrically forced surface waves: A systematic survey at large aspect ratio *Physica* D **97** 133–54

[84] Kudrolli A, Pier B and Gollub J P 1998 Superlattice patterns in surface waves *Physica* D **123** 99–111

[85] Landsberg A S and Knobloch E 1991 Direction-reversing travelling waves *Phys. Lett.* A **159** 17–20

[86] Landsberg A S and Knobloch E 1996 Oscillatory bifurcation with broken translation symmetry *Phys. Rev.* E **53** 3579–600

[87] Lauterbach R and Roberts M 1992 Heteroclinic cycles in dynamical systems with broken spherical symmetry *J. Diff. Eqns* **100** 428–48

[88] Lee K J, McCormick W D, Swinney H L and Pearson J E 1994 Experimental observation of self-replicating chemical spots in a reaction–diffusion system *Nature* **369** 215–18

[89] Longuet-Higgins M S 1953 Mass transport in water waves *Phil. Trans. R. Soc.* A **245** 535–81

[90] Lücke M L and Szprynger A 1997 Noise sustained pattern growth: bulk versus boundary effects *Phys. Rev.* E **55** 5509–21

[91] Martel C and Knobloch E 1997 Damping of nearly inviscid water waves *Phys. Rev.* E **56** 5544–8

[92] Martel C, Knobloch E and Vega J M 2000 Dynamics of counterpropagating waves in parametrically forced systems *Physica* DS **137** 94–123

[93] Martel C and Vega J M 1996 Finite size effects near the onset of the oscillatory instability *Nonlinearity* **9** 1129–71

[94] Martel C and Vega J M 1998 Dynamics of a hyperbolic system that applies at the onset of the oscillatory instability *Nonlinearity* **11** 105–42

[95] Martín E, Martel C and Vega J M 2001 Drift instability of standing Faraday waves *Preprint*

[96] Matkowsky B J and Volpert V 1993 Stability of plane wave solutions of complex Ginzburg–Landau equations *Q. J. App. Math.* **51** 265–81

[97] Melbourne I 1998 Derivation of the time-dependent Ginzburg–Landau equation on the line *J. Nonlin. Sci.* **8** 1–15

[98] Melbourne I 1999 Steady-state bifurcation with Euclidean symmetry *Trans. Am. Math. Soc.* **351** 1575–603

[99] Melo F, Umbanhowar P B and Swinney H L 1995 Hexagons, kinks, and disorder in oscillated granular layers *Phys. Rev. Lett.* **75** 3838–41

[100] Middya U, Sheintuch M, Graham M D and Luss D 1993 Patterns of temperature pulses on electrically heated catalytic ribbons *Physica* D **63** 393–409

[101] Miles J and Henderson D 1990 Parametrically forced surface waves *Ann. Rev. Fluid Mech.* **22** 143–65

[102] Moehlis J and Knobloch E 1998 Forced symmetry breaking as a mechanism for bursting *Phys. Rev. Lett.* **80** 5329–32

[103] Moehlis J and Knobloch E 2000 Bursts in oscillatory systems with broken D_4 symmetry *Physica* D **135** 263–304

[104] Neufeld M, Walgraef D and San Miguel M, 1996 Noise-sustained structures in coupled complex Ginzburg–Landau equations for a convectively unstable system *Phys. Rev.* E **54** 6344–55

[105] Newell A C, Passot T and Lega J 1993 Order parameter equations for patterns *Ann. Rev. Fluid Mech.* **25** 399–453

[106] Newell A C and Whitehead J A 1969 Finite bandwidth, finite amplitude convection *J. Fluid Mech.* **38** 279–303

[107] Ohta T, Mimura M and Kobayashi R 1989 Higher-dimensional localized patterns in excitable media *Physica* D **34** 115–44

[108] Ott E 1993 *Chaos in Dynamical Systems* (Cambridge: Cambridge University Press)

[109] Ouyang Q and Flesselles J-M 1996 Transition from spirals to defect turbulence driven by a convective instability *Nature* **379** 143–6

[110] Ouyang Q and Swinney H L 1991 Transition from a uniform state to hexagonal and striped Turing patterns *Nature* **352** 610–12

[111] Passot T and Newell A C 1994 Towards a universal theory for natural patterns *Physica* D **74** 301–52

[112] Pearson J E 1993 Complex patterns in a simple system *Science* **261** 189–92

[113] Pego R L and Weinstein M I 1994 Asymptotic stability of solitary waves *Commun. Math. Phys.* **164** 305–49

[114] Pierce R D and Knobloch E 1992 Spiral vortices in finite cylinders *Ordered and Turbulent Patterns in Taylor–Couette Flow (NATO ASI Series B 297)* ed C D Andereck and F Hayot (New York: Plenum) pp 83–90

[115] Pierce R D and Knobloch E 1994 On the modulational stability of travelling and standing water waves *Phys. Fluids* **6** 1177–90

[116] Pierce R D and Wayne C E 1995 On the validity of mean-field amplitude equations for counterpropagating wavetrains *Nonlinearity* **8** 769–79

[117] Popp S, Stiller O, Kuznetsov E and Kramer L 1998 The cubic complex Ginzburg–Landau equation for a backward bifurcation *Physica* D **114** 81–107

[118] Predtechensky A A, McCormick W D, Swift J B, Rossberg A G and Swinney H L 1994 Traveling wave instability in sustained double-diffusive convection *Phys. Fluids* **6** 3923–35

[119] Proctor M R E and Spiegel E A 1991 Waves of solar activity *The Sun and Cool Stars: Activity, Magnetism, Dynamos (Lecture Notes in Physics 380)* ed I Tuominen, D Moss and G Rüdiger (New York: Springer) pp 117–28

[120] Proctor M R E, Tobias S M and Knobloch E 2000 Noise-sustained structures due to convective instability in finite domains *Physica* D **145** 191–206

[121] Raquepas J B and Dockery J D 1999 Dynamics of a reaction–diffusion equation with non-local inhibition *Physica* D **134** 94–110

[122] Renardy M 1999 A note on bifurcation problems in large containers *Fluid Dyn. Res.* **24** 189–99

[123] Riecke H and Kramer L 2000 The stability of standing waves with small group velocity *Physica* D **137** 124–42

[124] Ristow G H 2000 *Pattern Formation in Granular Materials* (Berlin: Springer)

[125] Roberts M, Swift J W and Wagner D H 1986 The Hopf bifurcation on a hexagonal lattice *Contemp. Math.* **56** 283–318

[126] Rucklidge A M and Silber M 1998 Bifurcations of periodic orbits with spatio-temporal symmetries *Nonlinearity* **11** 1435–55

[127] Rucklidge A M, Silber M and Fineberg J 2002 Secondary instabilities of hexagons: a bifurcation analysis of experimentally observed Faraday wave patterns *Bifurcation, Symmetry and Patterns* ed J Buescu, S Castro, A P Dias and I Labouriau (Boston, MA: Birkhäuser) to appear

[128] Sandstede B and Scheel A 2000 Absolute and convective instabilities of waves on unbounded and large bounded domains *Physica* D **145** 233–77

[129] Sandstede D and Scheel A 2000 Absolute versus convective instability of spiral waves *Phys. Rev.* E **62** 7708–14

[130] Santagiustina M, Colet P, San Miguel M and Walgraef D 1997 Noise-sustained convective structures in nonlinear optics *Phys. Rev. Lett.* **79** 3633–6

[131] Scheel A 1997 Subcritical bifurcation to infinitely many rotating waves *J. Math. Anal. Appl.* **215** 252–61
[132] Scheel A 1998 Bifurcation to spiral waves in reaction–diffusion systems *SIAM J. Math. Anal.* **29** 1399–418
[133] Schlichting H 1932 Berechnung ebener periodischer Grenzschichtströmungen *Phys. Z.* **33** 327–35
[134] Schmiegel A and Eckhardt B 1997 Fractal stability border in plane Couette flow *Phys. Rev. Lett.* **79** 5250–3
[135] Schmiegel A and Eckhardt B 2000 Persistent turbulence in annealed plane Couette flow *Europhys. Lett.* **51** 395–400
[136] Schmitz R and Zimmermann W 1996 Spatially periodic modulated Rayleigh–Bénard convection *Phys. Rev.* E **53** 5993–6011
[137] Schneider G 1994 Error estimates for the Ginzburg–Landau approximation *ZAMP* **45** 433–57
[138] Schöpf W and Kramer L 1991 Small-amplitude periodic and chaotic solutions of the complex Ginzburg–Landau equation for a subcritical bifurcation *Phys. Rev. Lett.* **66** 2316–19
[139] Silber M and Knobloch E 1991 Hopf bifurcation on a square lattice *Nonlinearity* **4** 1063–107
[140] Silber M and Proctor M R E 1998 Nonlinear competition between small and large hexagonal patterns *Phys. Rev. Lett.* **81** 2450–3
[141] Spina A, Toomre J and Knobloch E 1998 Confined states in large aspect ratio thermosolutal convection *Phys. Rev.* E **57** 524–45
[142] Strogatz S H 2000 From Kuramoto to Crawford: exploring the onset of synchronization in populations of coupled oscillators *Physica* D **143** 1–20
[143] Stuart J T and Di Prima R C 1978 The Eckhaus and Benjamin–Feir resonance mechanisms *Proc. R. Soc.* A **362** 27–41
[144] Sullivan T S and Ahlers G 1988 Nonperiodic time dependence at the onset of convection in a binary liquid mixture *Phys. Rev.* A **38** 3143–6
[145] Sushchik M M and Tsimring L S 1994 The Eckhaus instability in hexagonal patterns *Physica* D **74** 90–106
[146] Swift J W 1984 *PhD Thesis* University of California at Berkeley
[147] Swift J W 1988 Hopf bifurcation with the symmetry of the square *Nonlinearity* **1** 333–77
[148] Swinton J and Elgin J 1990 Stability of travelling pulse solutions to a laser equation *Phys. Lett.* A **145** 428–33
[149] Thual O and Fauve S 1988 Localized structures generated by subcritical instabilities *J. Physique* **49** 1829–33
[150] Tobias S M and Knobloch E 1998 Breakup of spiral waves into chemical turbulence *Phys. Rev. Lett.* **80** 4811–14
[151] Tobias S, Proctor M R E and Knobloch E 1998 Convective and absolute instabilities of fluid flows in finite geometry *Physica* D **113** 43–72
[152] Trefethen L N, Trefethen A E, Reddy S C and Driscoll T A 1993 Hydrodynamic stability without eigenvalues *Science* **261** 578–84
[153] Trefethen L N 1997 Pseudospectra of linear operators *SIAM Rev.* **39** 383–406
[154] Tsameret A and Steinberg V 1991 Noise modulated propagating patterns in a convectively unstable system *Phys. Rev. Lett.* **76** 3392–5

[155] Tse D P, Rucklidge A M, Hoyle R B and Silber M 2000 Spatial period-multiplying instabilities of hexagonal Faraday waves *Physica* D **146** 367–87

[156] Umbanhowar P, Melo F and Swinney H L 1996 Localized excitations in a vertically vibrated granular layer *Nature* **382** 793–6

[157] van Gils S and Mallet-Paret J 1986 Hopf bifurcation and symmetry: Travelling and standing waves on a circle *Proc. R. Soc.* A **104** 279–307

[158] van Harten A 1991 On the validity of the Ginzburg–Landau equation *J. Nonlin. Sci.* **1** 397–422

[159] van Hecke M 1998 Building blocks of spatio-temporal intermittency *Phys. Rev. Lett.* **80** 1896–9

[160] Vega J M, Knobloch E, and Martel C 2001 Nearly inviscid Faraday waves in annular containers of moderately large aspect ratio *Physica* D **154** 313–36

[161] Wittenberg R W and Holmes P 1997 The limited effectiveness of normal forms: A critical review and extension of local bifurcation studies of the Brusselator PDE *Physica* D **100** 1–40

[162] Worledge D, Knobloch E, Tobias S and Proctor M 1997 Dynamo waves in semi-infinite and finite domains *Proc. R. Soc.* A **453** 119–43

Chapter 7

Is chaos relevant to fluid mechanics?

Tom Mullin
The University of Manchester

My objective is to provide a resumé of experience gained by research into the role of ideas from low-dimensional dynamical systems applied to the understanding of complicated motions in fluid flows. When I began working in this area the words turbulence and chaos were used synonymously in conference titles. Now we have apparently reached the stage where low-dimensional chaos is completely 'understood' and still the problem of turbulence remains as deep and mysterious as ever. It therefore seems natural to enquire whether the important ideas encompassed within finite-dimensional dynamical systems have shed much light on disordered fluid motion.

The work that I will discuss has been carried out over a period of 20 years with a wide variety of collaborators. The person who was most influential in the early part of this research was Brooke Benjamin, and there are many other collaborators listed at the end of this chapter, among whom Andrew Cliffe and Gerd Pfister have had the greatest influence on my thinking. All my collaborators have helped me gain some understanding of the apparently simple problem that has occupied my thoughts over this period. The problem which will be central to the discussion is Taylor–Couette flow between concentric rotating cylinders. This is outlined in the next section. Insights gained from detailed investigations of this particular flow have subsequently been used to shed light on other flows and these will be commented on.

7.1 Taylor–Couette flow

Taylor–Couette flow is concerned with flow between concentric rotating cylinders. It has a long and interesting history dating from the original work of Taylor [36] in the 1920s with over 1000 papers written on the topic, many of which are reviewed by DiPrima and Swinney [8] and Tagg [34]. We will

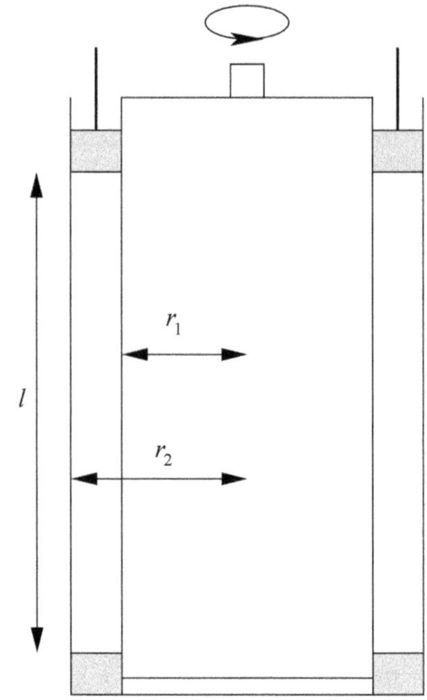

Figure 7.1. A schematic diagram of the Taylor–Couette apparatus.

concentrate on a particularly simple version of the problem where the inner cylinder rotates and the outer is held fixed. A schematic diagram of the set-up is shown in figure 7.1. The principal dynamical parameter is the *Reynolds number* Re, which is defined as

$$Re = \frac{r_1 \omega d}{\nu} \tag{7.1}$$

where r_1 is the radius of the inner cylinder, ω is the angular rotation speed of that cylinder, $d = r_2 - r_1$ is the width of the gap between the cylinders and ν is the kinematic viscosity of the fluid. There are also two geometrical parameters. The *aspect ratio* Γ is defined as

$$\Gamma = \frac{l}{d} \tag{7.2}$$

where l is the distance between the two horizontal end plates. The *radius ratio* η is defined as

$$\eta = \frac{r_1}{r_2} \tag{7.3}$$

where r_2 is the radius of the stationary outer cylinder. In any single study η is held fixed and so the problem can be treated as a two-parameter problem in practice.

Unlike many other outstanding problems in physics, the equations of motion of fluid flows, the Navier–Stokes equations, are generally accepted as being a good model which ought to be able to describe the observed behaviour. The principal difficulty arises in obtaining solutions to these equations for situations appropriate to realistic flows. A distinct advantage of the Taylor–Couette system is that modern numerical methods can be used to obtain solutions to the equations set on physically relevant boundary conditions. Quantitative comparison can then be made between the numerical predictions and observations from carefully controlled experiments. Namely, the only free parameter available is the viscosity of the fluid which can be determined independently. Hence, this approach provides an interesting challenge since the opportunity for absolute quantitative comparison between calculation on the full equations of motion and experiment is quite rare in fluid mechanics.

Taylor–Couette flow is often cited as a typical example of transition to turbulence via the Ruelle, Takens, Newhouse (RTN) route where disordered motion arises in the form of a strange attractor after a finite number of Hopf bifurcations as Re is increased [30, 31]. This is by way of contrast with Landau's ideas [19] of transition via an infinite sequence of Hopf bifurcations which arise sequentially over a finite range of Re so that a continuum of modes are excited. The idea that a few modes could be responsible for turbulence was revolutionary and created significant impact. The important work of Gollub and Swinney [12] on Taylor–Couette flow provided evidence to support this finite-dimensional idea which, in many ways, is more appealing than the continuum approach. However, I have never been able to convince myself of the merit of Taylor–Couette flow as an exemplar of the RTN transition to chaos since our experimental investigations suggested that the temporal behaviour may contain some non-generic features [25] that are discussed later. As with many things in research, it is relatively easy to be negative and express what you believe to be not so. It has taken me twenty years to try and understand what is actually going on and to make a more positive statement.

Our approach has been to treat the Taylor–Couette flow as a laboratory model of the Navier–Stokes equations. As an experimentalist it is comforting that one can always relate back to these equations and, in particular, compare them with the results of numerical calculations. What we focus on primarily in our work is seeking codimension-two points where paths of qualitatively different bifurcations meet in control parameter space [14]. These special points mark the creation of centres of global homoclinic bifurcations that result in robust chaos of the type commonly found in ODE models. In practical terms, if a complex dynamical landscape is encountered it is prudent to identify the waymarkers, and codimension-two points provide them. The objective is to seek organizing centres in the underpinning solution structure which pull together the various strands of behaviour such as period-doubling and intermittency. An attempt may then be made to draw a connection between the PDEs operating in the experiment and the ODE models via the discretized equations of motion used in the numerical

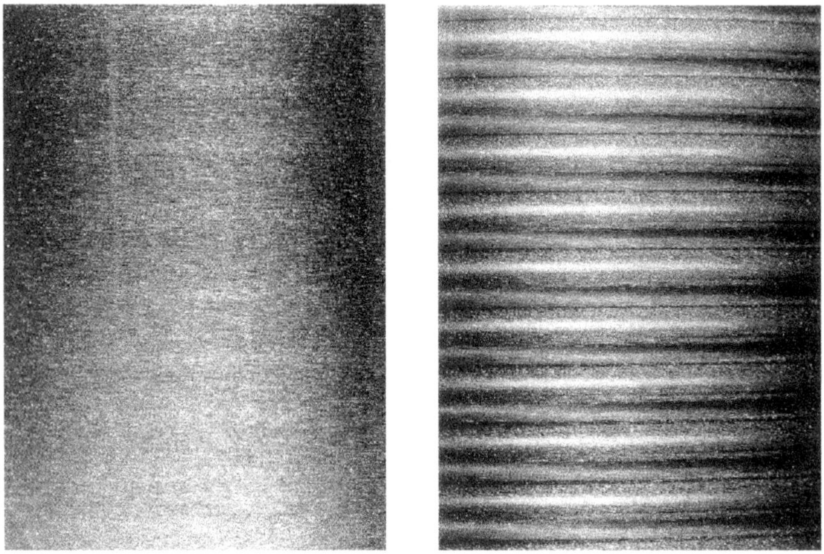

(*a*) (*b*)

Figure 7.2. Front view of Taylor–Couette flow, when *Re* is below the range at which cells appear (*a*), and visualization of Taylor cells (*b*). Reprinted by permission of Oxford University Press from T Mullin 1993 *The Nature of Chaos* (Oxford: Oxford University Press).

investigations. As long as the appropriate physically relevant symmetries are taken into account in the models, they can be used to great effect in interpreting and understanding the observations.

7.2 Preliminary observations

Demonstration of the basic features of Taylor–Couette flow is very easy to carry out in practice. In fact, some years ago Cowels [7], while still at school, produced some impressive pictures of the flow field between a pair of rotating test tubes after he read an article of mine [21] in *New Scientist*. So what is typically seen? Starting at small Reynolds number the flow is basically featureless as shown in figure 7.2(*a*). Increasing *Re* gives rise to the striking pattern shown in figure 7.2(*b*). The fluid used in the photographs contains small anisotropic platelets that reflect the incident light and thereby highlight the Taylor vortices which are wrapped around the inner cylinder. If a cross-sectional view is taken as shown in figure 7.3 secondary vortices are seen in the cylindrical gap. The main flow travels round with the inner cylinder; superposed on top of this are the secondary flows, so that a series of toroidal vortices lie stacked upon one another along the length of the cylinder. Numerical calculations can be carried

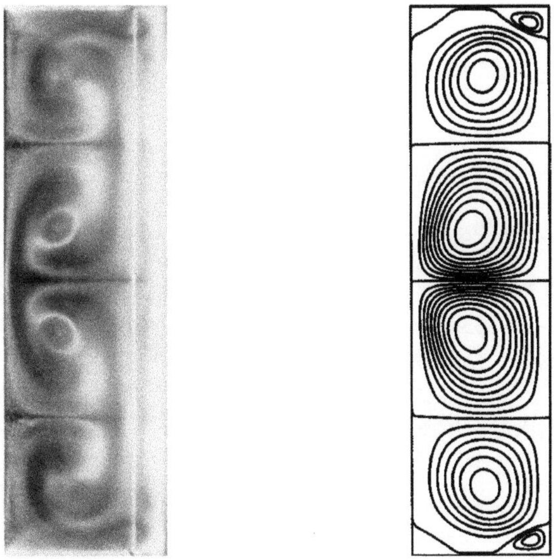

(*a*) (*b*)

Figure 7.3. A cross-sectional view of Taylor cells (*a*) shown in comparison with calculated streamlines (*b*). Reprinted by permission of Oxford University Press from T Mullin 1993 *The Nature of Chaos* (Oxford: Oxford University Press).

out on this flow, and a comparison of a typical streamline and flow visualization pattern is shown in figure 7.3. The vortices are approximately square in cross section when *Re* is increased smoothly. However, the flow exhibits multiplicity or non-uniqueness and many other stable steady states exist, each of which may be created by discontinous jumps in *Re* as discussed in [4].

When the Reynolds number is increased above a critical value, travelling waves appear as shown in figure 7.4(*a*). The waves travel at a fixed ratio of the rotation rate of the inner cylinder and, as for the steady flows, there is non-uniqueness in the number and speed ratios of the waves. Further increase in *Re* leads to the flow becoming quasi-periodic as shown in figure 7.4(*b*). Now there are two incommensurate frequencies present and the snapshot shown in figure 7.4(*b*) has been chosen to represent this state. A more complicated flow state then arises at higher *Re*, which appears to be irregular as shown in figure 7.5. In practice, the Taylor vortex structure is very evident but the small-scale disordered motion can be revealed in short time exposure photographs such as figure 7.5.

These alluring images give a vivid impression of an apparently simple sequence of the breakdown from order to disorder as the Reynolds number is increased. It may perhaps reasonably be thought that one only needs to establish whether the sequence of Hopf bifurcations which gives rise to the waves is finite or not. However, significant difficulties arise when taking Taylor–

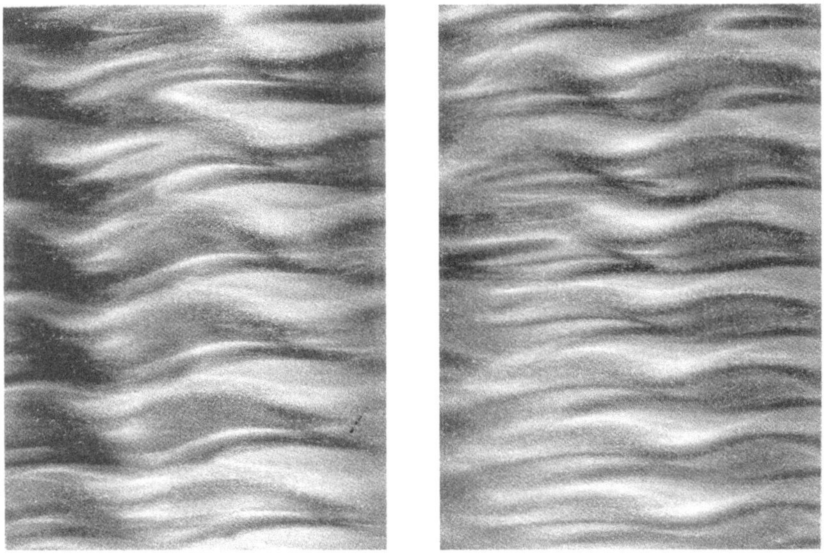

(*a*) (*b*)

Figure 7.4. Front view of Taylor–Couette flow with travelling waves (*a*), and visualization of a quasi-periodic flow (*b*). Reprinted by permission of Oxford University Press from T Mullin 1993 *The Nature of Chaos* (Oxford: Oxford University Press).

Couette flow from this demonstration level to a scientific experiment from which repeatable quantitative measurements may be made. It is an enormous technical challenge where all length scales, temperatures, speeds etc must be accurately controlled. Equally important, the rich tapestry of static and dynamic states must be appreciated since both the number of fixed points and complicated dynamical attractors increase rapidly with Re, so that exacting control of the experiment is vital.

Our observations [25] suggested that the sequence highlighted in these photographs was rather special. The waves were observed to appear and disappear over ranges of Reynolds number with the characteristics of sequences of resonant-type behaviour. This is a consequence of SO(2)-symmetry in the azimuthal direction around the annular gap. The stability of a prescribed steady cellular state may be tested by increasing Re until travelling waves appear at a defined critical value. The process can then be repeated over the range of aspect ratios for which the particular steady mode exists and the resulting stability limits are found to be typically of the form shown in figure 7.6 for four different steady states [24]. They have the appearance of a set of resonance curves and have been calculated numerically for other parameter ranges [18]. Hence we claim that the waves are special features of the geometry so that the dynamics associated with them are non-generic.

Figure 7.5. Disordered motion in Taylor–Couette flow. Reprinted by permission of Oxford University Press from T Mullin 1993 *The Nature of Chaos* (Oxford: Oxford University Press).

7.3 Symmetry considerations

In my view the most important symmetry in this flow is mirror-plane $\mathbb{Z}_2$-symmetry as illustrated in figure 7.7 where we show a pair of asymmetric four-cell states that break the mirror plane symmetry of the end boundaries. Any model that is to be used to describe an aspect of Taylor–Couette flow ought to contain this simple discrete symmetry since it is the breaking of this which appears to control the important dynamics. Even cell non-trivial cellular states break this symmetry through pitchfork bifurcations. At these bifurcation points, initially symmetric flows break the midplane symmetry when *Re* is varied to form pairs of asymmetric states. In practice, the bifurcations are disconnected by small imperfections in the experiment which is generally observed to be at the level of a few percent. There are many other examples of $\mathbb{Z}_2$-symmetry breaking pitchfork bifurcations in fluid flows, including open flows [9,32]. It has been my experience that whenever a simple symmetry is present in a fluid flow, the tendency will be for the motion to break the symmetry in the nonlinear regime, that is, asymmetric states are more commonly found than symmetric ones.

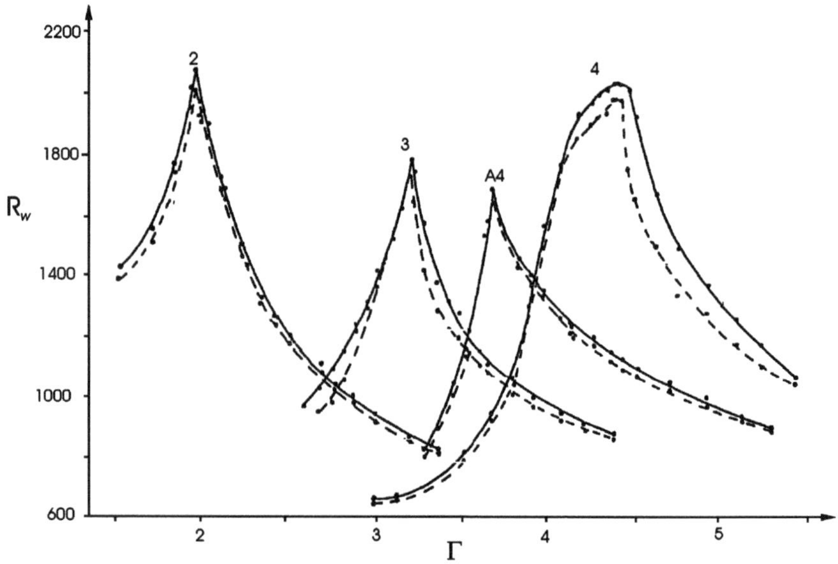

Figure 7.6. The experimentally determined critical Reynolds numbers for the onset of travelling waves plotted as a function of aspect ratio for four different steady modes. Reprinted by permission from T Mullin and T B Benjamin 1980 The transition to oscillatory motion in the Taylor experiment *Nature* **288** 567–9 ©1980 Macmillan Publishers Ltd.

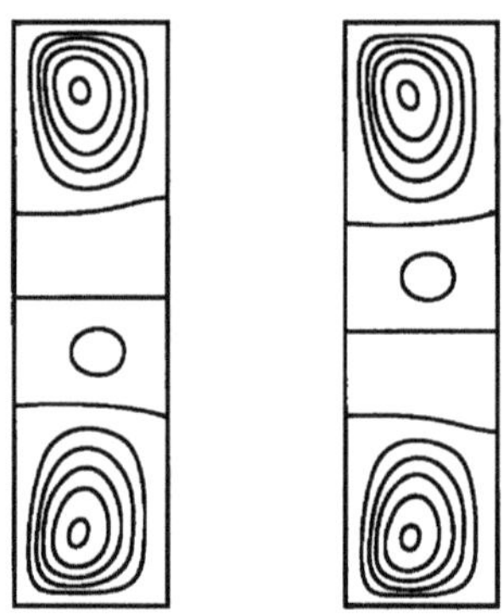

Figure 7.7. A pair of asymmetric four-cell states that break the midplane mirror symmetry.

7.4 Codimension-two bifurcations

Here we provide a brief outline of experimental and numerical evidence of codimension-two bifurcations. The central idea is to seek organizing centres [14]

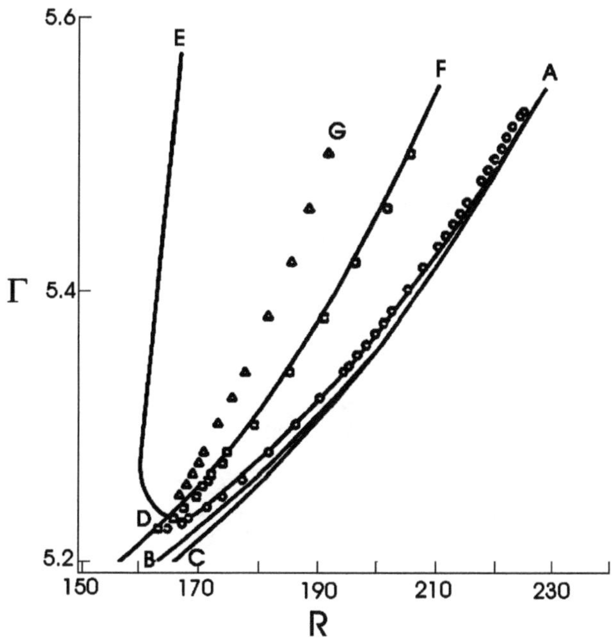

Figure 7.8. Experimental and numerical bifurcation set for the symmetry breaking saddle-node interaction. The solid curves are numerically calculated and the points are all experimental. ADE is the path of axisymmetric Hopf bifurcations, FD the path of Hopf bifurcations to wavy oscillations, AB and AC the paths of pitchfork and saddle-nodes, respectively. The further set of experimental points in the centre of the figure are torus bifurcations. Reproduced from T Mullin, K A Cliffe and G Pfister 1987 *Phys. Rev. Lett.* **58** 2212–15 ©1987 by the American Physical Society.

for the dynamics to see whether the underpinning solution structure can provide indications of the type of behaviour one might expect by appealing to relevant ODE models. Indeed, a long-term goal was to perform a centre manifold reduction on the discretized equations of motion to attempt to make a connection between the PDEs and ODE models. After all, there exists a zoo of different types of dynamics in these problems over very small parameter ranges. What hope is there for understanding such problems if, every time an experiment is performed, different sorts of behaviour are observed? A reductionist view would suggest that there may be a unifying structure and the tools of experiment, numerics and ODE models are available to help uncover it.

The two parameters which may be varied most usefully in Taylor–Couette flow are the aspect ratio Γ, and the Reynolds number *Re*. The example we have chosen is the interaction between a symmetry breaking pitchfork and saddle-node bifurcation which results in the necessary creation of a path of Hopf

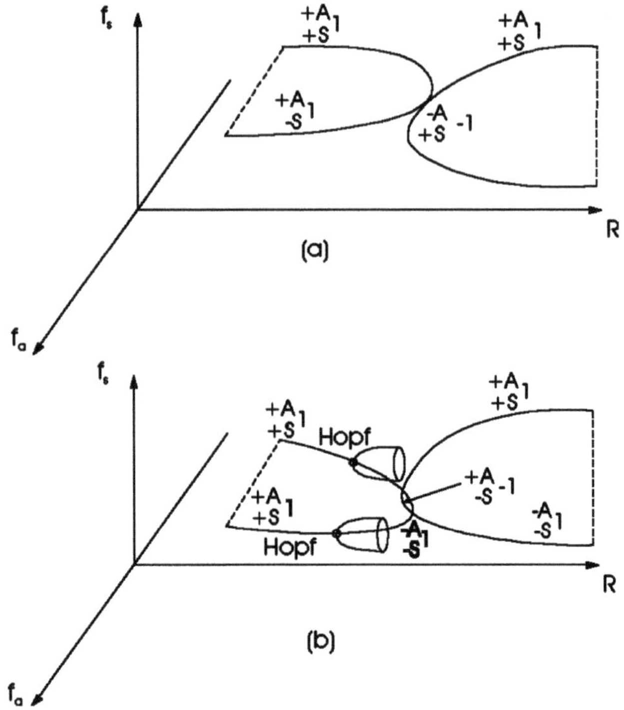

Figure 7.9. Schematic bifurcation diagram of saddle-node pitchfork interaction.

bifurcations [26]. The bifurcation set for the interaction is shown in figure 7.8 where excellent quantitative agreement between experimental and numerical results can be seen. The set consists of a curve of saddle-nodes AB and a curve of pitchforks AC, and the two curves meet at the codimension-two point near the point labelled A in figure 7.8. Variation of the control parameters moves the pitchfork from the stable to the unstable side of the saddle-node. The branches that originate at the pitchfork on the stable surface are stable to both symmetric and antisymmetric perturbations as indicated in the schematic bifurcation diagram in figure 7.9. However, when the pitchfork moves on to the unstable surface the branches will both become doubly unstable.

Following an argument from Guckenheimer [13] these branches may be restabilized through a Hopf bifurcation as shown schematically in figure 7.9. This prediction from the model is borne out by the experimental and numerical results from the Navier–Stokes equations as shown in figure 7.8, where ADE is the path of predicted Hopf bifurcations. The oscillations which arise at these Hopf bifurcations are interesting in that they are axisymmetric and their frequency goes to zero at the codimension-two point. There are also secondary Hopf bifurcations to travelling waves and they interact with the axisymmetric oscillation producing a curve of torus bifurcations in the experiment. However, the essential issue is that the mechanism based on ideas from an ODE model correctly predicts the creation of a curve of Hopf bifurcations at which novel axisymmetric oscillations originate in the experiment.

The axisymmetric oscillations discussed here turn out to be rather common in Taylor–Couette flows and their variants. The most recognizable characteristic, which makes it relatively easy to differentiate between them and the standard travelling waves, is their very low frequency. This has led to them being called a very-low-frequency or VLF-mode by Pfister *et al* [5]. These oscillations are important in that they can lead to global bifurcations that are known to play an important role in organizing the dynamics originating in the local bifurcation structure of dynamical systems [14]. Specifically we will first consider the case where a periodic orbit merges with a saddle point as a parameter of the system is varied. In the simple planar case, the amplitude of the orbit may grow as a parameter is increased and its period will tend to infinity as the distance between the orbit and the saddle point diminishes. A more complex bifurcation structure can arise in the case of three and more dimensions [33]. Depending on the eigenvalues of the saddle point a Sil'nikov wiggle involving period doubling sequences, folds and chaotic attractors can form around the homoclinic orbit [11, 16, 23]. It has been found that homoclinic orbits act as organizing centres for the low-dimensional dynamics of physical systems. Examples of these have been observed in lasers [2], chemical oscillators [3, 15] electronic circuits [16, 17] and fluid flows [28].

The particular case we will focus on here is for the interaction between an axisymmetric oscillation created by the mechanism outlined earlier which then interacts with a saddle-node. The flow configuration is Taylor–Couette flow where both ends are also made to rotate forming a rotating annulus; details may be found in [35]. A three-dimensional model that is appropriate for such interactions has been produced and studied numerically [20]. It is

$$\begin{aligned}
\dot{X} &= XZ - WY \\
\dot{Y} &= XW + YZ \\
\dot{Z} &= P + Z - \tfrac{1}{3}Z^3 - (X^2 + Y^2)(1 + QX + EZ)
\end{aligned}$$

where $W = 10$, $E = 0.5$, $Q = 0.7$ and P is the bifurcation parameter. The parameter values have been chosen with guidance from the experiments. A prediction of models of this type is that low-dimensional chaos will arise through period-doubling [20] after the torus has become locked. This sequence has indeed been observed in the vicinity of the respective codimension-two point and details can be found in [22]. We show an example of a chaotic attractor formed at the end of such a sequence in figure 7.10. In principle, it is possible to produce such a model by applying Lyapunov–Schmidt reduction to the discretized Navier–Stokes equations used in the numerical study. This has been performed by Cliffe [6] who found behaviour consistent with the experiment. Unfortunately, the parameters

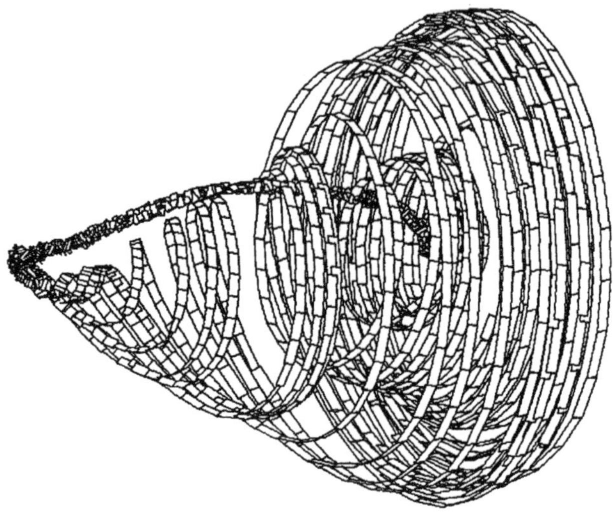

Figure 7.10. A reconstructed attractor from the rotating annulus experiment at $Re = 132.7$, $\Gamma = 4.25$ showing near-homoclinic chaos. Reproduced from T Mullin 1991 Finite dimensional dynamics in Taylor–Couette flow *IMA J. Appl. Math.* **46** 109–19 by permission of the Institute of Mathematics and its Applications.

of the model are nonlinear combinations of Re and Γ and so establishing a connection between the models, numerics and experiment remains an outstanding challenge.

The robustness of Sil'nikov dynamics is illustrated in figure 7.11 where we show reconstructed attractors from two other fluid flows: electrohydrodynamic convection in a nematic liquid crystal [29] and convection in a liquid metal with an applied magnetic field [27]. The bifurcation structure that underpins the illustrated dynamics has been studied using the combined numerical and experimental approach developed for the Taylor–Couette problem, and it is encouraging that this has led to the uncovering of robust homoclinic dynamics.

The results of an experimental investigation of the dynamical behaviour in the vicinity of a pair of codimension-two points in the rotating annulus flow is shown in figure 7.12. The paths of experimental points consist of primary and secondary Hopf bifurcations together with boundaries of chaotic behaviour. The details are unimportant here and the principal point of interest is the existence of qualitatively different dynamical behaviour in neighbouring ranges of control parameter space. These are torus-doubling, homoclinicity and intermittency which all emanate from the codimension-two points showing that the global behaviour is controlled by the local bifurcation structure.

These observations give rise to interesting questions concerning the system size. If the experiment is made very large, presumably the number of such

(*a*) (*b*)

Figure 7.11. Sil'nikov chaos in electrohydrodynamic convection in a nematic liquid crystal (*a*), and in convection in liquid gallium in the presence of a magnetic field (*b*); see also colour plate 2. Reproduced from T Mullin, A Juel and T Peacock 2001 Sil'nikov chaos in fluid flows *Intermittency in Turbulent flows* ed C Vaccilicos (Cambridge: Cambridge University Press) ©2001 by Cambridge University Press.

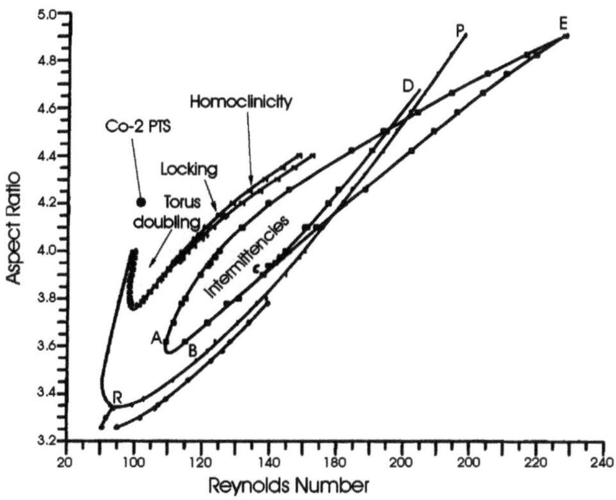

Figure 7.12. Experimental data measured in the vicinity of the Hopf/saddle-node codimension-two points. The approximate locations of three different routes to chaos (homoclinic, torus doubling and intermittency) are indicated.

codimension-two points and associated dynamical behaviour will grow rapidly. Hence, the observed dynamics which is apparently high-dimensional may result

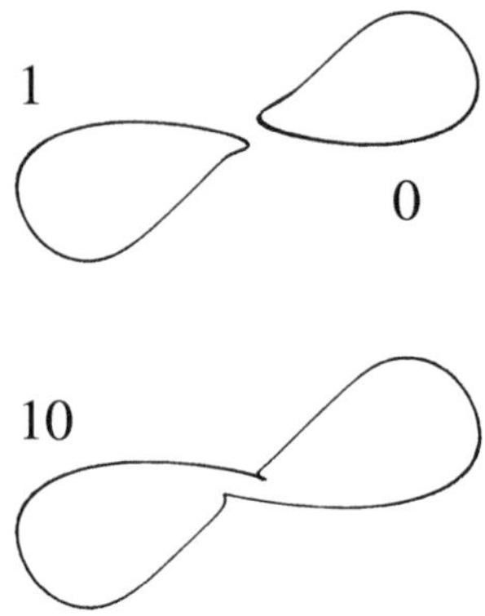

Figure 7.13. Gluing of two periodic orbits to form a single large orbit as a parameter is varied. Reproduced from J Abshagen, G Pfister and T Mullin 2001 *Phys. Rev. Lett.* **87** 224501 ©2001 by the Americal Physical Society.

from overlapping low-dimensional attractors, each of which contains qualitatively different dynamical behaviour. Alternatively, it is not clear whether all types of low-dimensional dynamics will persist in large systems. It is this aspect that we will comment on next.

7.5 Imperfect gluing bifurcations

Gluing bifurcations commonly arise in systems that undergo pitchfork symmetry-breaking bifurcations followed by pairs of Hopf bifurcations on each of the asymmetric branches. The resulting periodic orbits can grow and glue together via a homoclinic orbit to form a single large orbit when the bifurcation parameter is varied as shown in figure 7.13. This example is taken from experiments on an nonlinear electronic oscillator [10]. The period of the two individual orbits, which are labelled 0 and 1 in figure 7.13, will approach infinity at the gluing bifurcation and the period of the large-scale orbit will decrease from infinity. Simple gluing such as this is valid for a system with perfect symmetry but this ideal symmetry never exists in an experiment. We have a reasonable approximation to the ideal case in the oscillator experiment in some parameter ranges. However, imperfections are always present and the effects of imperfections on gluing bifurcations are discussed in [10].

Taylor–Couette flows have reflectional $\mathbb{Z}_2$-symmetry and are thus prime candidates for exhibiting this behaviour. The particular example we have studied is standard Taylor–Couette flow with an aspect ratio of 8. Details may be found in [1]. The flow comprises of 10 Taylor vortices and our earlier results [4] suggests that there may be up to 25 fixed points available to the system. These will not all

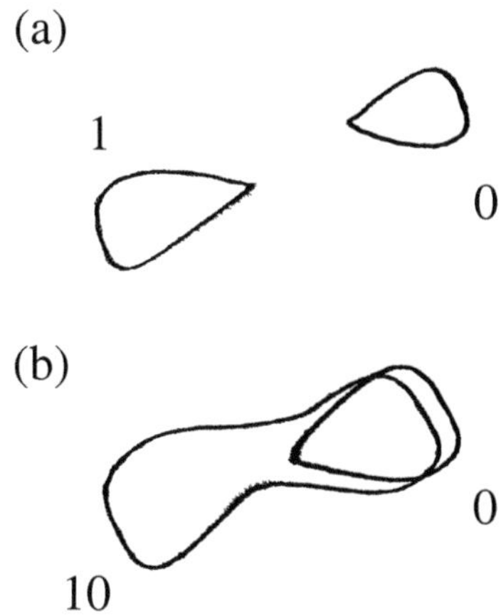

Figure 7.14. Phase portraits for the axisymmetric period orbits: separate orbits 1 and 0 at $Re = 378.3$ (*a*), and a large orbit 10 together with a separate orbit 0 at $Re = 400.8$ (*b*). Reproduced from J Abshagen, G Pfister and T Mullin 2001 *Phys. Rev. Lett.* **87** 224501 ©2001 by the Americal Physical Society.

be stable, of course, but with so many fixed points in the vicinity one might expect the flow to be rather complicated. Further complications are expected since our investigation was performed in the dynamical regime where many attractors will co-exist.

Note that the Reynolds numbers are far in excess of those used in the study of the low-dimensional dynamics. The gluing process was studied in the presence of travelling waves of the type discussed earlier. In practice, we found that they played no role in the dynamics and so we simply filtered this component out of the time series. The gluing process involves the axisymmetric oscillation alone and this appears to ride on top of the travelling wave without interacting with it. The axisymmetric oscillation has the same characteristics as those described previously which were created at the pitchfork/saddle-node interaction. Examples of the orbits 0, 1 and 10 are shown in figure 7.14.

At a slightly greater value of Γ the gluing process involves chaotic motion as shown in figure 7.15. According to the theory [10], when imperfections are present the gluing process will involve irregular orbits in parameter ranges between the individual and glued orbits. This is indeed the case and, moreover, we also observed complicated periodic orbits such as 100 as shown in figure 7.15. The overall agreement between theory and experiment is good and includes the creation of extra complicated periodic orbits in intervening parameter ranges. The fact that we can observe the details of this procedure suggests that it is robust, and it is interesting that it should be realizable in a system where many other

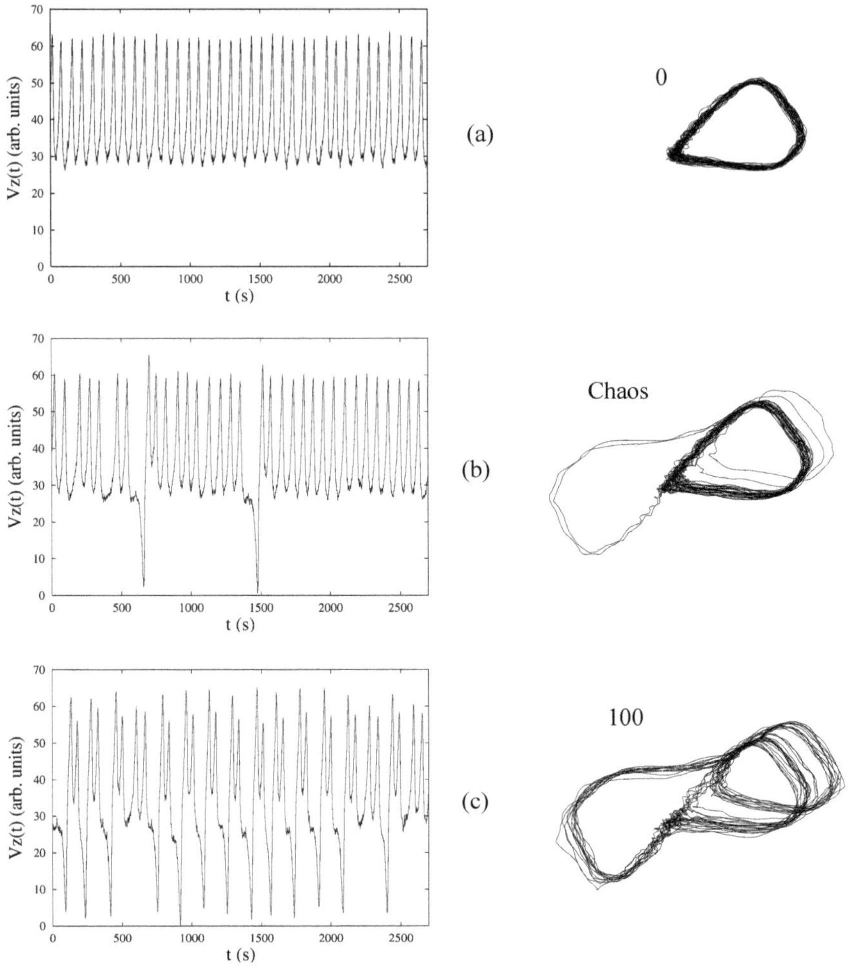

Figure 7.15. The sequence of dynamical states for $\Gamma = 8.2$: periodic orbit 0 at $Re = 459.9$ (*a*) chaos at $Re = 460.7$ (*b*), and orbit 100 at $Re = 462.4$ (*c*). Reproduced from J Abshagen, G Pfister and T Mullin 2001 *Phys. Rev. Lett.* **87** 224501 ©2001 by the Americal Physical Society.

possibilities are available. Therefore, we suggest that it may have relevance to a wider class of fluid flows.

7.6 Conclusion

The principal conclusion to be drawn is that global bifurcations are robust and apply to a variety of different flows over a range of length scales, that is, in both

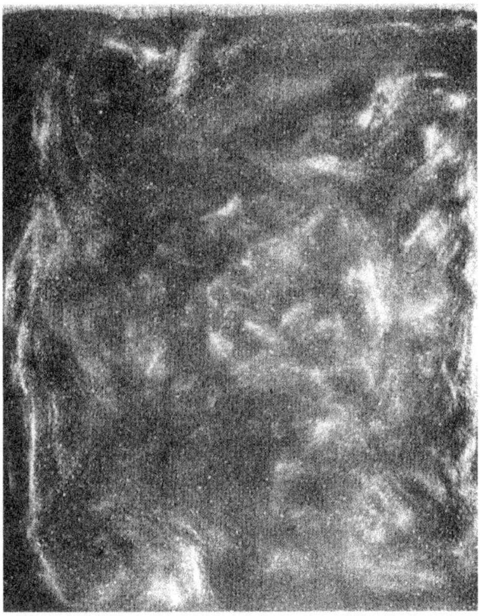

Figure 7.16. Turbulent motion in a box.

small and large boxes. Indeed, I have used many of the ideas outlined here to help industrialists with some practical flows. Notions from dynamical systems theory certainly have been helpful in gaining understanding of complicated fluid flows provided the models contain the appropriate symmetries. A physically relevant example is mirror-plane $\mathbb{Z}_2$-symmetry which many fluid flows contain. The mathematical difficulty is that the breaking of this symmetry often takes place on non-trivial flows so that it is difficult to justify the application of the theory. Despite this, progress can be made with non-trivial complicated flows.

The ideas of low-dimensional dynamics and, in particular, chaos has not shed much light on what has often been called the last unresolved problem of classical physics: turbulence. Flows such as the example shown in figure 7.16 remain largely mysterious at least from a dynamical systems perspective. This is more a measure of the depth of the problem which will probably require a combination of innovative ideas for a significant breakthrough to be made. I remain convinced that an essential ingredient of this will be based in nonlinear dynamical systems.

Acknowledgments

I was very fortunate in the early stages of my research into this topic to be led by Brooke Benjamin who instilled in me the desire to find out what was actually going on rather that what one would like to find. Andrew Cliffe has been a constant source of ideas and Gerd Pfister and Jan Abshagen set experimental standards which I aspire to. It was a treat to begin collaborations with Paul Glendinning in recent years despite his claim that he would never write a paper with an experimentalist or a dog. My ex-graduate students, Simon Tavener, Tim Price, Jonathan Kobine, Tom Peacock and Anne Juel have been, and continue to be, a source of inspiration. Pedro Reis helped prepare many of the figures.

References

[1] Abshagen J, Pfister G and Mullin T 2001 A gluing bifurcation in a dynamically complicated extended flow *Phys. Rev. Lett.* **87** 224501

[2] Allaria E, Arecchi F T, Di Garbo A and Meucci R 2001 Synchronization of homoclinic chaos *Phys. Rev. Lett.* **86** 791–4

[3] Arnéodo A, Argoul F, Elezgaray J and Richetti P 1993 Homoclinic chaos in chemical systems *Physica* D **62** 134–69

[4] Benjamin T B and Mullin T 1982 Notes on the multiplicity of flows in the Taylor–Couette experiment *J. Fluid Mech.* **121** 219–30

[5] Buzug T, von Stamm J and Pfister G 1993 Characterization of period-doubling scenarios in Taylor–Couette flow *Phys. Rev.* E **47** 1054–65

[6] Cliffe K A 2002 Private communication

[7] Cowels R 1989 Order out of chaos *New Scientist* **1693** 90

[8] DiPrima R and Swinney H L 1981 Instabilities and transition in flow between concentric rotating cylinders *Hydrodynamic Instabilities and the Transition to Turbulence* ed H L Swinney and J P Gollub (Berlin: Springer)

[9] Fearn R M, Mullin T and Cliffe K A 1990 Nonlinear flow phenomena in a symmetric sudden-expansion *J. Fluid Mech.* **211** 595–608

[10] Glendinning P, Abshagen J and Mullin T 2001 Imperfect homoclinic bifurcations *Phys. Rev.* E **64** 036208

[11] Glendinning P and Sparrow C 1984 Local and global behaviour near homoclinic orbits *J. Stat. Phys.* **43** 479–88

[12] Gollub J P and Swinney H L 1975 Onset of turbulence in a rotating fluid *Phys. Rev. Lett.* **35** 927–30

[13] Guckenheimer J 1981 *On a Codimension-2 Bifurcation (Springer Lecture Notes in Mathematics 898)* (Berlin: Springer)

[14] Guckenheimer J and Holmes P 1986 *Nonlinear Oscillations, Dynamical Systems and Bifurcations of Vector Fields (Springer Applied Mathematical Sciences 42)* (Berlin: Springer)

[15] Hauser M J B and Olsen L F 1996 Mixed-mode oscillations and homoclinic chaos in an enzyme reaction *J. Chem. Soc. Faraday Trans.* **92** 2857–63

[16] Healey J J, Broomhead D S, Cliffe K A, Jones R and Mullin T 1991 The origins of chaos in a modified Van der Pol oscillator *Physica* D **48** 322–39

[17] Herrero R, Pons R, Farjas J, Pi F and Orriols G 1996 Homoclinic dynamics in experimental Shil'nikov attractors *Phys. Rev.* E **53** 5627–36

[18] Jones C A 1982 The transition to wavy Taylor vortices *J. Fluid Mech.* **157** 135–62

[19] Landau L D and Lifshitz E M 1959 *Fluid Mechanics* (Tarrytown, NY: Pergamon) pp 102–7

[20] Langford W F 1984 Numerical studies of torus bifurcations *Int. Ser. Numer. Math.* **70** 285–93

[21] Mullin T 1989 Turbulent times for fluids *New Scientist* **1690** 52–8

[22] Mullin T 1991 Finite dimensional dynamics in Taylor–Couette flow *IMA J. Appl. Math.* **46** 109–19

[23] Mullin T 1993 *The Nature of Chaos* (Oxford: Oxford University Press)

[24] Mullin T and Benjamin T B 1982 The transition to oscillatory motion in the Taylor experiment *Nature* **288** 567–9

[25] Mullin T, Benjamin T B, Schatzel K and Pike E R 1981 New aspects of unsteady Couette flow *Phys. Lett.* A **83** 333–6

[26] Mullin T, Cliffe K A and Pfister G 1987 Unusual time-dependent phenomena in Taylor–Couette flow at moderately low Reynolds numbers *Phys. Rev. Lett.* **58** 2212–15

[27] Mullin T, Juel A and Peacock T 2001 Sil'nikov chaos in fluid flows *Intermittency in Turbulent Flows* ed C Vaccilicos (Cambridge: Cambridge University Press) pp 24–43

[28] Mullin T and Price T J 1989 An experimental observation of chaos arising from the interaction of steady and time-dependent flows *Nature* **340** 294–6

[29] Peacock T and Mullin T 2001 Homoclinic bifurcations in a liquid crystal flow *J. Fluid Mech.* **432** 369–86

[30] Ruelle D 1989 *Chaotic Evolution and Strange Attractors* (Cambridge: Cambridge University Press)

[31] Ruelle D and Takens F 1971 On the nature of turbulence *Commun. Math. Phys.* **20** 167–92 and **23** 343–4

[32] Shtern V and Hussain F 1998 Instabilities of conical flows causing steady bifurcations *J. Fluid Mech.* **366** 33–87

[33] Sil'nikov L P 1965 A case of the existence of a denumerable set of periodic motions *Sov. Math. Dokl.* **6** 163–6

[34] Tagg R 1994 The Couette–Taylor problem *Nonlin. Sci. Today* **4** 1–25

[35] Tavener S J, Mullin T and Cliffe K A 1991 Novel bifurcation phenomena in a rotating annulus *J. Fluid Mech.* **229** 438–97

[36] Taylor G I 1923 Stability of a viscous liquid contained between two rotating cylinders *Phil. Trans. R. Soc.* A **223** 289–343

Chapter 8

Time-reversed acoustics and chaos

Mathias Fink
Laboratoire Ondes et Acoustique

It is well known that the acoustic wave equation, in a non-dissipative heterogeneous medium, is invariant under a time-reversal operation. Indeed, it contains only a second-order time derivative operator. Therefore, for every burst of sound $p(\vec{r}, t)$ diverging from a source—and possibly reflected, refracted or scattered by any heterogeneous media—there exists, in theory, a set of waves $p(\vec{r}, -t)$ that precisely retrace all of these complex paths and converge in synchrony at the original source as if time were going backwards. Here, $p(\vec{r}, t) \in \mathbb{R}$ is a function of space and time. In this chapter, we take $r \in \mathbb{R}^3$, but the time-reversal symmetry is still valid if the space dimension is one or two. This gives the basic idea of time-reversal acoustics.

Taking advantage of these two properties, the concept of a time-reversal mirror (TRM) has been developed and several devices have been built to illustrate the efficiency of this concept [10–12]. In such a device, an acoustic source, located inside a lossless medium, radiates a brief transient pulse that propagates and is distorted by the medium. If the acoustic field can be measured at every point of a closed surface surrounding the medium (an acoustic retina), and retransmitted through the medium in a time-reversed chronology, then the wave will travel back to its source and recover its original shape. Note that it requires both time-reversal invariance and spatial reciprocity [13] to reconstruct the exact time-reversed wave in the whole volume by means of a two-dimensional time-reversal operation. From an experimental point of view, a closed TRM consists of a two-dimensional piezoelectric transducer array that samples the wave field over a closed surface. An array pitch of the order of $\frac{1}{2}\lambda$, where λ is the smallest wavelength of the pressure field, is needed to ensure the recording of all the information on the wave field. Each transducer is connected to its own electronic circuitry that consists of a receiving amplifier, an A/D converter, a storage memory and a programmable transmitter able to synthesize a time-reversed version of the stored signal. Unfortunately, closed TRMs are difficult to realize in practice and the TR operation is usually performed on a limited angular area, thus limiting reversal and focusing quality. A TRM typically consists of some hundred elements or time-reversal channels.

We will discuss two types of time-reversal experiments conducted with TRMs. It will be shown that the wave reversibility is improved if the wave traverses a random multiple-scattering medium before arriving at the transducer array. The multiple-scattering processes allow redirection of parts of the initial wave towards the TRM that would normally miss the transducer array. After the time-reversal operation, the whole multiple-scattering medium behaves as a coherent focusing source, with a large angular aperture for enhanced resolution. As a consequence, in multiple-scattering media, one is able to reduce the size and the complexity of the TRM. The same kind of improvement may be obtained for waves propagating in highly reverberant media such as closed cavities or waveguides. Multiple reflections along the medium boundaries significantly increase the apparent aperture of the TRM, as will be shown by a set of experiments conducted in chaotic cavities. We will see that for a reflecting cavity with chaotic boundaries, a one-channel TRM is sufficient to ensure reversibility and optimal focusing.

8.1 Time-reversal mirrors

The basic theory employs a scalar wave formulation $p(\vec{r}, t)$. Hence, this theory is strictly applicable to acoustic or ultrasonic propagation in fluid. However, the basic ingredients and conclusions apply equally well to elastic waves in solids and to electromagnetic fields.

In any propagation experiment, the solution $p(\vec{r}, t)$ in the fluid is uniquely determined by the acoustic sources and the boundary conditions. The goal, in time-reversal experiments, is to modify the initial conditions in order to generate the dual solution $p(\vec{r}, T - t)$, where T is a delay due to causality requirements. Cassereau and Fink [2] and Jackson and Dowling [13] performed a theoretical study of the conditions necessary to ensure the generation of $p(\vec{r}, T - t)$ in the entire volume of interest.

Although reversible acoustic retinas usually consist of discrete elements, it is convenient to examine the behaviour of idealized continuous retinas, defined by two-dimensional surfaces. In the case of a time-reversal cavity, we assume that the retina completely surrounds the source. The formalism of a basic time-reversal experiment can be described by the following two-step process.

In the first step, a point-like source located at $\vec{r}_0$ inside a volume V surrounded by the retina surface S, emits a pulse at $t = t_0 \geq 0$. The wave equation in a medium of density $\rho(\vec{r})$ and compressibility $\kappa(\vec{r})$ is given by

$$(L_r + L_t)p(\vec{r}, t) = -A\delta(\vec{r} - \vec{r}_0)\delta(t - t_0) \tag{8.1}$$

with

$$L_r = \vec{\nabla} \cdot \left(\frac{1}{\rho(\vec{r})} \vec{\nabla} \right) \qquad \text{and} \qquad L_t = -\kappa(\vec{r})\delta_{tt}.$$

Here, A is a dimensional constant that ensures the compatibility of physical units between the two sides of the equation; for simplicity, this constant will be omitted in the following. The solution to (8.1) reduces to Green's function $G(\vec{r}, t \mid \vec{r}_0, t_0)$. Classically, $G(\vec{r}, t \mid \vec{r}_0, t_0)$ is written as a diverging spherical wave (homogeneous and free-space case) with additional terms that describe the interaction of the field itself with the inhomogeneities (multiple scattering) and the boundaries.

We assume that we are able to measure the pressure field and its normal derivative at any point on the surface S during the interval $[0, T]$. As time-reversal experiments are based on a two-step process, the measurement step must be limited in time by a parameter T. In all that follows, we suppose that the contribution of multiple scattering decreases with time, and that T is chosen such that the information loss can be considered as negligible inside the volume V.

During the second step of the time-reversal process, the initial source at $\vec{r}_0$ is removed and we create monopole and dipole sources on the surface of the cavity that correspond to the time-reversal of those same components measured during the first step. The time-reversal operation is described by the transformation $t \to T - t$ and the secondary sources are

$$\begin{cases} p_s(\vec{r}, t) = G(\vec{r}, T - t \mid \vec{r}_0, t_0) \\ \partial_n p_s(\vec{r}, t) = \partial_n G(\vec{r}, T - t \mid \vec{r}_0, t_0). \end{cases}$$

In this equation, ∂_n is the normal derivative operator with respect to the normal direction $\vec{n}$ to S, oriented outward. Due to these secondary sources on S, a time-reversed pressure field $p_{tr}(\vec{r}_1, t_1)$ propagates inside the cavity. It can be calculated using a modified version of the Helmholtz–Kirchhoff integral

$$p_{tr}(\vec{r}_1, t_1) = \int_{-\infty}^{\infty} \mathrm{d}t \iint_S [G(\vec{r}_1, t_1 \mid \vec{r}, t)\partial_n p_s(\vec{r}, t) - p_s(\vec{r}, t)\partial_n G(\vec{r}_1, t_1 \mid \vec{r}, t)] \frac{\mathrm{d}^2\vec{r}}{\rho(\vec{r})}.$$

Spatial reciprocity and time-reversal invariance of the wave equation (8.1) yield the following expression:

$$p_{tr}(\vec{r}_1, t_1) = G(\vec{r}_1, T - t_1 \mid \vec{r}_0, t_0) - G(\vec{r}_1, t_1 \mid \vec{r}_0, T - t_0). \tag{8.2}$$

This equation can be interpreted as the superposition of incoming and outgoing spherical waves, centred at the initial source position. The incoming wave collapses at the origin and is always followed by a diverging wave. Thus the time-reversed field, observed as a function of time, from any location in the cavity, shows two wavefronts, where the second one is the exact replica of the first one, multiplied by -1.

If we assume that the retina does not perturb the propagation of the field (free-space assumption) and that the acoustic field propagates in a homogeneous fluid, the free-space Green function G reduces to a diverging spherical impulse wave that propagates with a sound speed c. Introducing its expression in (8.2) yields the following formulation of the time-reversed field:

$$p_{\mathrm{tr}}(\vec{r}_1, t_1) = K(\vec{r}_1 - \vec{r}_0, t_1 - T + t_0)$$

where $K(\vec{r}, t)$ is the kernel distribution (the propagator), given by

$$K(\vec{r}, t) = \frac{1}{4\pi|\vec{r}|}\delta\left(t + \frac{|\vec{r}|}{c}\right) - \frac{1}{4\pi|\vec{r}|}\delta\left(t - \frac{|\vec{r}|}{c}\right). \tag{8.3}$$

The kernel distribution corresponds to the difference between two spherical impulse waves. One of these converges to the origin of the spatial coordinate system, that is, the location of the initial source, and the other diverges from it. Due to this superposition, the pressure field remains finite for all time throughout the cavity, although the converging and diverging spherical waves show a singularity at the origin.

If we now consider a wide-band excitation function instead of a Dirac distribution δt, then the two wavefronts overlap near the focal point, resulting in a temporal distortion of the initial acoustic signal which leads to a temporal derivation exactly at the focal point. If one considers a monochromatic excitation of pulsation ω, then the time-reversed field has an amplitude given by the Fourier transform of (8.3) over the time variable t:

$$\tilde{K}(\vec{r}, \omega) = \frac{1}{2i\pi}\frac{\sin(\omega|\vec{r}|/c)}{|\vec{r}|} = \frac{1}{i\lambda}\frac{\sin(k|\vec{r}|)}{k|\vec{r}|}$$

where λ and k are the wavelength and wavenumber respectively. As a consequence, the time-reversal process results in a pressure field that is effectively focused on the initial source position, but with a focal spot size limited to half the wavelength.

The apparent failure of the time-reversed operation that leads to diffraction limitation can be interpreted in the following way. The second step described here is not strictly the time-reversal of the first step—during the second step of an ideal time-reversed experiment, the initial active source (that injects some energy into the system) must be replaced by a *sink* (the time-reversal of a source). An acoustic sink is a device that absorbs all arriving energy without reflecting it. Taking into account the source term in the wave equation, reversing time leads to the transformation of sources into sinks. Therefore, (8.1) which describes the wave equation with a source term, is transformed in the time-reversal operation to

$$(L_r + L_t)p(\vec{r}, -t) = -A\delta(\vec{r} - \vec{r}_0)\delta(t - t_0).$$

To achieve a perfect time reversal experimentally, the field on the surface of the cavity has to be time-reversed, and the source has to be transformed into a sink.

Only in this way may one achieve time-reversed focusing below the diffraction limit. The role of the new source term $\delta(\vec{r} - \vec{r}_0)$ is to transmit a diverging wave that exactly cancels the usual outgoing spherical wave.

Taking into account the evanescent waves concept, one can explain the necessity of replacing a source by a sink in the complete time-reversed operation as follows. In the first step a point-like source of size quite smaller than the transmitted wavelengths radiates a field whose angular spectrum contains both propagating and evanescent waves. The evanescent wave components are lost after propagation over some wavelengths. Hence, the time-reversed field that is retransmitted by the surface of the cavity in the time-reversed step, does not contain evanescent components. The role of the sink is to be a source modulated by $\delta(-t - t_0)$ that exactly radiates, with the appropriate timing, the evanescent waves that have been lost during the first step. This way the resulting field contains the evanescent part that is needed to focus below diffraction limits. With De Rosny, we recently built such a sink in our laboratory and we observed focal spot sizes well below diffraction limits (typically of the order $\lambda/20$) [17].

This theoretical model of the closed time-reversal cavity (TRC) is interesting, because it offers an understanding of the basic limitations of the time-reversed self-focusing process. However, it has several limitations, particularly in comparison with the classical experimental set-up which usually works without an acoustic sink:

- It can be proven that it is not necessary to measure and time-reverse both the scalar field (acoustic pressure) and its normal derivative on the cavity surface: measuring the pressure field and re-emitting the time-reversed field in the backward direction yield the same results, provided the evanescent parts of the acoustic fields have vanished (propagation along several wavelengths) [3]. This comes from the fact that each transducer element of the cavity records the incoming field from the forward direction, and retransmits it (after the time-reversal operation) in the *backward* direction (and not in the forward direction). The change between the forward and the backward direction replaces the measurement and the time-reversal of the normal derivative.
- From an experimental point of view, it is not possible to measure and re-emit the pressure field at any point of a two-dimensional surface: experiments are carried out with transducer arrays that spatially sample the receiving and emitting surface. The spatial sampling of the TRC by a set of transducers may introduce grating lobes. These lobes can be avoided by using an array pitch smaller than $\frac{1}{2}\lambda_{\min}$, where $\lambda_{\min}$ is the smallest wavelength of the transient pressure field. If this is the case, then each transducer senses all the wavevectors of the incident field.
- The temporal sampling of the data recorded and transmitted by the TRC has to be at least of the order of $\frac{1}{8}T_{\min}$, with $T_{\min}$ the minimal period, to avoid secondary lobes [14].

- In general, it is difficult to use acoustic arrays that completely surround the area of interest and the closed TRM is usually replaced by a finite angular aperture TRM. As a result, the size of the focal spot, which is usually related to the mirror angular aperture observed from the source, increases substantially.

8.2 Time-reversal experiments

8.2.1 Time reversal through random media

Derode *et al* [4] carried out the first experimental demonstration of the reversibility of an acoustic wave propagating through a random collection of scatterers with strong multiple-scattering contributions. In an experiment such as the one depicted in figure 8.1, a multiple-scattering sample is placed between the source and an array made of 128 elements. The whole set-up is in a water tank. The scattering medium consists of a set of 2000 randomly distributed parallel steel rods (diameter 0.8 mm). The sample thickness is $L = 40$ mm, and the average distance between rods is 2.3 mm. The source is 30 cm away from the TRM and transmits a short (1 μs) ultrasonic pulse (3 cycles of 3.5 MHz).

Figure 8.2(*a*) shows the signal received by one of the TRM elements when the source radiates a 1 μs signal through water. Figure 8.2(*b*) shows the new waveform received by the element when a scattering medium is introduced between the source and the TRM. The signal spreads over more than 200 ms, i.e. roughly 200 times the initial pulse duration. After the arrival of a first wavefront corresponding to the ballistic wave, a long incoherent wave is observed, which results from the multiply scattered contribution. In the second step of the experiment, all 128 signals are time-reversed and transmitted. A hydrophone measures the time-reversed wave around the source location. Two different aspects of this problem have been studied: the property of the signal recreated at the source location (time compression) and the spatial property of the time-reversed wave around the source location (spatial focusing).

The time-reversed wave traverses the rods back to the source, and the signal received at the source is represented in figure 8.2(*c*). An impressive compression is observed, because the received signal lasts about 1 μs, as opposed to over 200 μs for the scattered signals. The pressure field is also measured around the source in order to get the directivity pattern of the beam emerging from the rods after time-reversal; see figure 8.3. Surprisingly, multiple scattering has not degraded the resolution of the system. Indeed, the resolution is found to be six times finer (bold line) than the classical diffraction limit (thin line)! This effect does not contradict the laws of diffraction, though. The intersection of the incoming wavefront with the sample has a typical size D. After time reversal, the waves travel back along the same scattering paths and focus at the source as if they were passing through a converging lens with size D. The angular aperture of this pseudo-lens is much wider than that of the array alone, which implies an

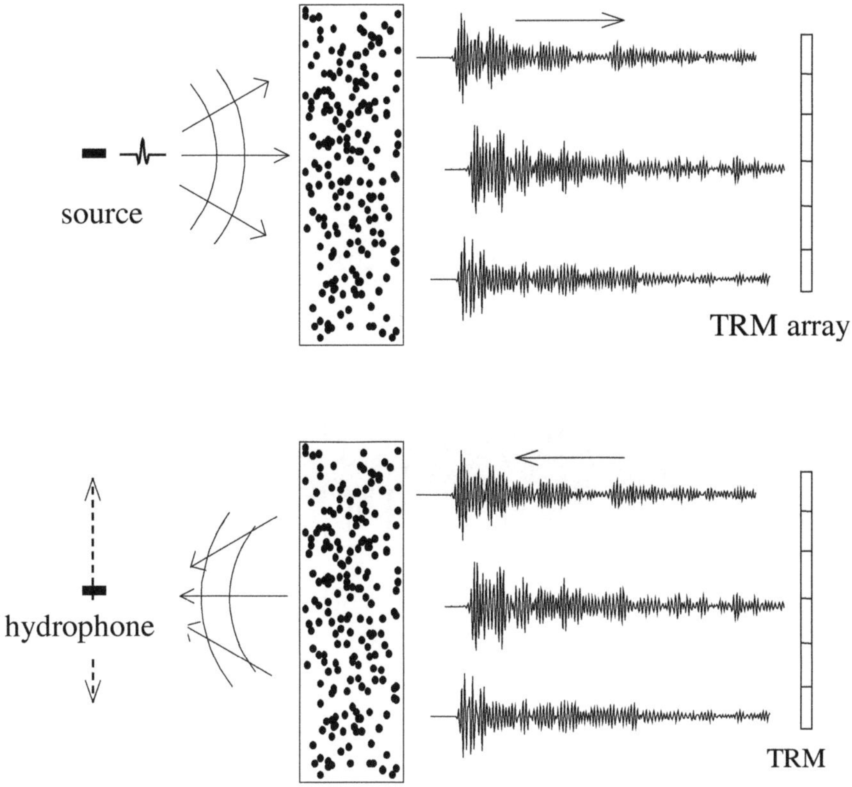

Figure 8.1. Sketch of the multiple-scattering experiment.

improvement in resolution. In other words, because of the scattering sample, the array is able to detect higher spatial frequencies than in a purely homogeneous medium. High spatial frequencies that would have been lost otherwise are redirected towards the array, due to the presence of the scatterers in a large area.

This experiment also shows that the acoustic time-reversal experiments are surprisingly stable. The recorded signals were sampled with 8-bit analog-to-digital converters, which introduce discretization errors, but the focusing process still works. One can compare this with time-reversal experiments involving particles moving like balls on an elastic billiard of the same geometry. Computation of the direct and reversed particle trajectories moving in a plane among a fixed array of some thousand convex obstacles (a Lorentz gas) shows that the complete trajectory is irreversible. Indeed, such a Lorentz gas is a well-known example of a chaotic system that is highly sensitive to initial conditions. The finite precision of the computer leads to an error in the trajectory of the time-reversed particle that grows exponentially with the number of scattering encounters.

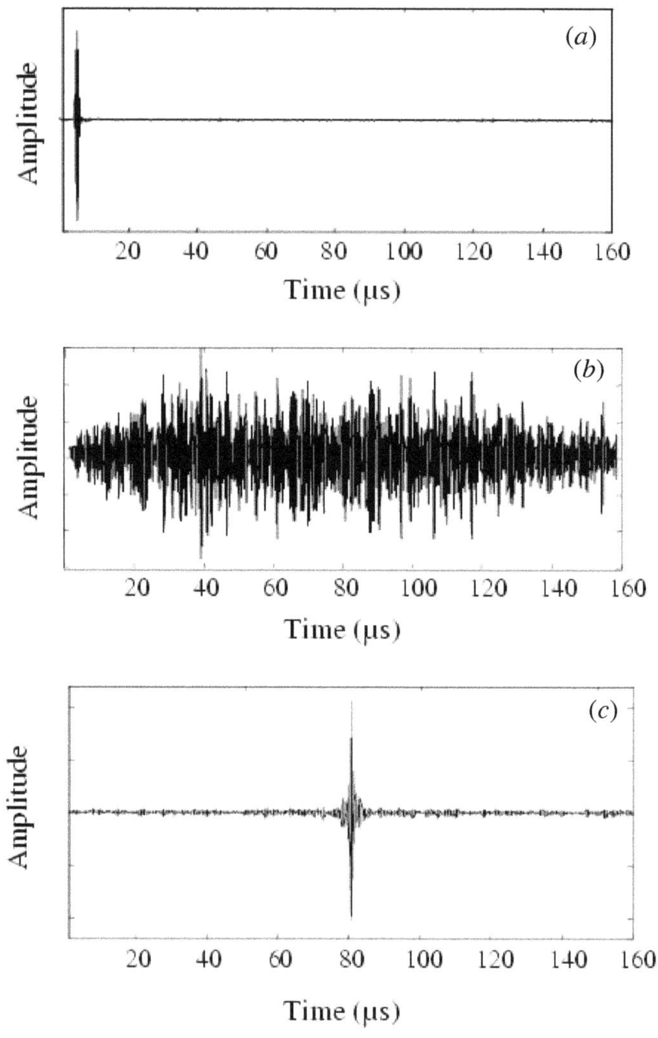

Figure 8.2. Signal transmitted in water and received at transducer 64 (*a*). Signal transmitted through the multiple-scattering sample and received at transducer 64 (*b*). Time-reversed signal at the source (*c*).

Snieder and Scales [19] performed numerical simulations to point out the fundamental difference between waves and particles in the presence of multiple scattering by random scatterers. In fact, they used time reversal as a diagnostic of wave and particle chaos: in a time-reversal experiment, a complete focusing on the source will only take place if the velocity and positions are known exactly. The degree δ to which errors in these quantities destroy the quality of focusing

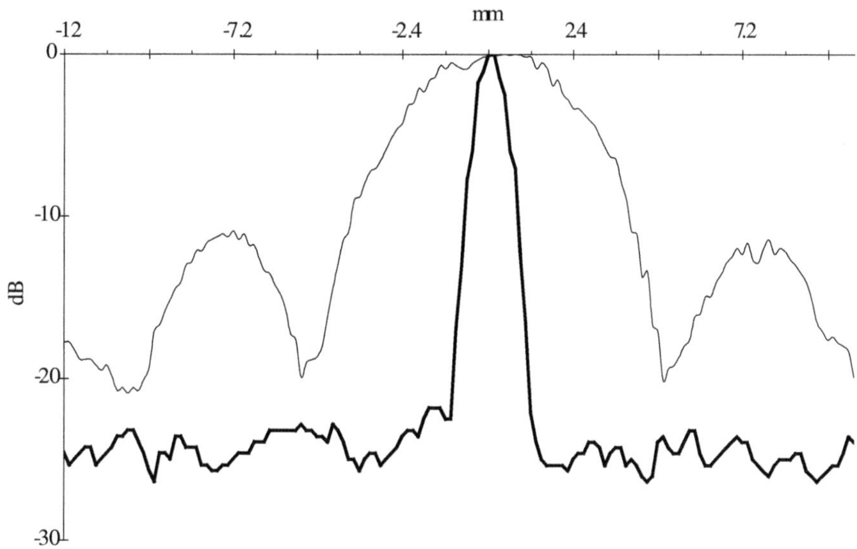

Figure 8.3. Directivity patters of the time-reversed field.

is a diagnostic of the stability of the wave or particle propagation. Intuitively, the consequences of a slight deviation δ in the trajectory of a billiard ball will become more and more pronounced as time goes on, and as the ball undergoes more and more collisions. Waves are much less sensitive to initial conditions than particles. Specifically, in a multiple-scattering situation, the critical length scale δ that causes a signification deviation at a particular time t in the future, decreases exponentially with time in the case of particles, but it only decreases as the square-root of time for waves in the same situation.

Waves and particles react in fundamentally different ways to perturbations of the initial conditions. The physical reason for this is that each particle follows a well-defined trajectory whereas waves travel along all possible trajectories, visiting all the scatterers in all possible combinations. While a small error in the initial velocity or position makes the particle miss one obstacle and completely change its future trajectory, the wave amplitude is much more stable because it results from the interference of all the possible trajectories and small errors in the transducer operations will sum up in a linear way resulting in small perturbation.

8.2.2 Time-reversal as a matched-filter or time correlator

As with any linear and time-invariant process, wave propagation through a multiple-scattering medium may be described as a linear system with different impulse responses. If a source, located at $\vec{r}_0$ sends a Dirac pulse $\delta(t)$, then the jth transducer of the TRM will record the impulse response $h_j(t)$ that corresponds, for a point transducer, to the Green function $G(\vec{r}, t \mid \vec{r}_0, t_0)$. Moreover, due to reciprocity, $h_j(t)$ is also the impulse response describing the propagation of a pulse from the jth transducer to the source. Thus, neglecting the causal time delay T, the time-reversed signal at the source is equal to the convolution product $h_j(t) * h_j(-t)$. This convolution product, in terms of signal analysis, is typical of a *matched filter*. Given a signal as input, a matched filter is a linear filter whose output is optimal in some sense. Whatever the impulse response $h_j(t)$, the convolution $h_j(t) * h_j(-t)$ is maximal at time $t = 0$. This maximum is always positive and equal to $\int h_j^2(t)\,\mathrm{d}t$, that is, the energy of the signal $h_j(t)$. This has an important consequence. Indeed, with an N-element array, the time-reversed signal recreated at the source can be written as a sum

$$p_{\mathrm{tr}}(\vec{r}_0, t) = \sum_{j=1}^{j=N} h_j(t) * h_j(-t).$$

Even if the $h_j(t)$s are completely random and apparently uncorrelated signals, each term in this sum reaches its maximum at time $t = 0$. Hence, all contributions add constructively near $t = 0$, while at earlier or later times uncorrelated contributions tend to destroy one another. Hence, the recreation of a sharp peak after time-reversal on an N-element array can be viewed as an interference process between the N outputs of N matched filters.

The robustness of the TRM can also be accounted for through the matched filter approach. If, for some reason, the TRM does not exactly retransmit $h_j(-t)$, but rather $h_j(-t) + n_j(t)$, where $n_j(t)$ is an additional noise on channel j, then the recreated signal becomes

$$\sum_{j=1}^{j=N} h_j(t) * h_j(-t) + \sum_{j=1}^{j=N} h_j(t) * n_j(t).$$

The time-reversed signals $h_j(-t)$ are tailored to exactly match the medium impulse response, which results in a sharp peak. Additional small noise is not matched to the medium and, given the extremely long duration involved, it will generate a low-level long-lasting background noise rather than a sharp peak.

8.2.3 Time reversal as a spatial correlator

Another way to consider the focusing properties of the time-reversed wave is to follow the impulse response approach and treat the time-reversal process as a spatial correlator. If we denote by $h'_j(t)$ the propagation impulse response from the jth element of the array to an observation point $(\vec{r}, r_1)$ different from the source location $(\vec{r}, r_0)$, then the signal recreated at $(\vec{r}, r_1)$ at time $t_1 = 0$ is

$$p_{\mathrm{tr}}(\vec{r}_1, 0) = \int h_j(t) h'_j(t)\,\mathrm{d}t. \qquad (8.4)$$

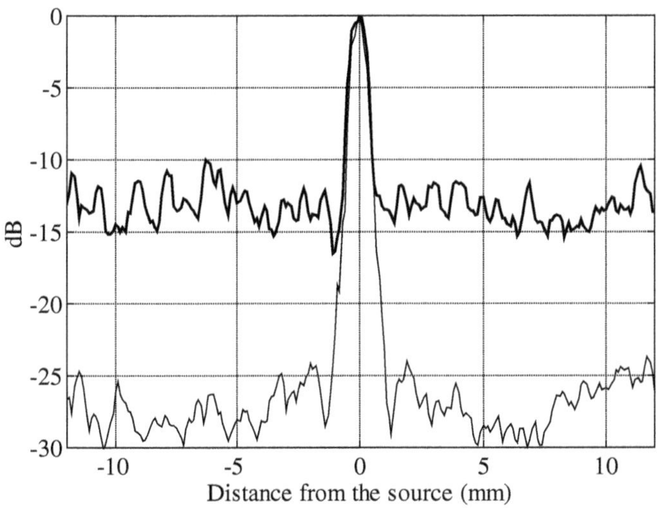

Figure 8.4. Directivity patterns with $N = 122$ transducer (thin line) and $N = 1$ transducer (bold line).

Note that this expression can be used as a way to define the directivity pattern of the time-reversed waves around the source. Now, due to reciprocity, the source and the receiver can be exchanged, that is, $h'_j(t)$ is also the signal that would be received at $\vec{r}_1$ if the source were the jth element of the array. Therefore, we can imagine this array element as being the source and the transmitted field as being observed at the two points $\vec{r}_1$ and $\vec{r}_0$. The spatial correlation function of this wave field would be the average $\langle h_j(t)h'_j(t)\rangle$ of the product of the impulse responses over different realizations of the disorder. Therefore, equation (8.4) can be viewed as an estimation of this spatial correlation function. Note that in one time-reversal experiment we only have access to a single realization of the disorder. However, the ensemble average can be replaced by a time average, by a frequency average or by a spatial average on a set of transducers. As is quoted in [6]:

> In that sense, the spatial resolution of the time-reversal mirror (i.e. the -6 dB width of the directivity pattern) is simply an estimation of the correlation length of the scattered wave field.

This has an important consequence. Indeed, if the resolution of the system essentially depends on the correlation properties of the scattered wave field, it should become independent of the array's aperture. This is confirmed by the experimental results. Figure 8.4 presents the directivity patterns obtained through a 40 mm thick multiple-scattering sample, using either one array element or all 122 array elements as a TRM. In both cases, the spatial resolution at -6 dB is the same, namely approximately 0.85 mm. *In total contradiction with what happens in a homogeneous medium*, enlarging the aperture of the array does not change the -6 dB spatial resolution. However, even though the number N of active array elements does not influence the typical width of the focal spot, it does have a strong impact on the background level of the directivity pattern (approximately -12 dB for $N = 1$ and approximately -28 dB for $N = 122$); see also figure 8.4.

Finally, the fundamental properties of time reversal in a random medium come from the fact that it is both a space and a time correlator. The time-reversed waves can be viewed as an estimate of the space and time autocorrelation functions of the waves scattered by a random medium. The estimate becomes better as the number of transducers in the mirror increases. Moreover, since the addition of small noise to the scattered signals (e.g. by digitizing on a reduced number of bits) may alter the noise level, but does not drastically change the correlation time or the correlation length of the scattered waves, the system is not sensitive to small perturbations. Even in the extreme case where the scattered signals are digitized *on a single bit*, Derode has shown recently that the time and space resolution of the TRM remain practically unchanged [5], which is striking evidence for the robustness of wave time reversal in a random medium.

8.3 Time reversal in chaotic cavities

In the time-reversal cavity approach, the transducer array samples a closed surface surrounding the acoustic source. In the previous section, we saw how the multiple-scattering processes in a large sample widen the effective TRM aperture. The same kind of improvement may be obtained for waves propagating in a waveguide or in a cavity. Multiple reflections along the medium boundaries significantly increase the apparent aperture of the TRM. The basic idea is to replace one part of the TR cavity transducers by reflecting boundaries that redirect one part of the incident wave towards the TRM aperture. Thus spatial information is converted into the time domain and the reversal quality depends crucially on the duration of the time-reversal window, i.e. the length of the recording to be reversed.

Experiments conducted by Roux in rectangular ultrasonic waveguides show the effectiveness of the time-reversal processing to compensate for multi-path effects [18]. Impressive time compression has been observed that compensated for reverberation and dispersion. Furthermore, similar to the multiple-scattering experiment, the time reversal is focused on a spot that is much thinner than the one observed in free water. This can be interpreted by the theory of images in a medium bounded by two mirrors: for an observer located at the source point, the TRM seems to be escorted by an infinite set of virtual images related to multi-path propagation. This has lead to an effective aperture that was 10 times larger than the real aperture.

Acoustic waveguides are currently also found in underwater acoustics, especially in shallow water, and TRMs can compensate for the multi-path propagation in oceans that limits the capacity of underwater communication systems. The problem arises because acoustic transmissions in shallow water bounce off the ocean surface and floor, so that a transmitted pulse gives rise to multiple copies of it that arrive at the receiver. Recently, underwater acoustic experiments have been conducted by Kuperman and his group from San Diego University in a 120 m-deep sea water channel, with a 24-element TRM working at 500 Hz and 3.5 kHz. They observed focusing and multi-path compensation at distances up to 30 km [15].

In this section, we are interested in another aspect of multiply-reflected waves; namely, waves confined in closed reflecting cavities such as elastic waves propagating in a silicon wafer. With such boundary conditions, no information can escape from the system and a reverberant acoustic field is created. If, moreover, the cavity satisfies ergodic and mixing properties, and absorption is negligible, one may hope to collect all information at only one point.

Draeger and Fink [7–9] have shown experimentally and theoretically that, in this particular case, a time reversal can be obtained *using only one time-reversal channel* operating in a closed cavity. The field is measured at one point over a long period of time and the time-reversed signal is re-emittted at the same position. The experiment is two-dimensional and has been carried out using elastic surface waves propagating along a monocrystalline silicon wafer whose shape is a D-shape stadium. This geometry is chosen to avoid quasi-periodic orbits. Silicon was selected for its weak absorption properties. The elastic waves that propagate in such a plate are *Lamb waves*. An aluminium cone coupled to a longitudinal transducer generates these waves at one point of the cavity. A second transducer is used as a receiver. The central frequency of the transducers is 1 MHz and its bandwidth is 100%. At this frequency, only three Lamb modes are possible (one flexural and two extensional). The source is isotropic and considered point-like because the cone tip is much smaller than the central wavelength. A heterodyne laser interferometer measures the displacement field as a function of time at different points in the cavity. Assuming that there are hardly any mode conversions between the flexural mode and other modes at the boundaries, we only need to deal with one field, namely the flexural-scalar field.

The experiment is again a two-step process as described above; see also figure 8.5. In the first step, one of the transducers, located at point A, transmits a short omni-directional signal of duration 0.5 μs into the wafer. Another transducer, located at point B, observes a very long chaotic signal that results from multiple reflections of the incident pulse along the edges of the cavity, and which continues for more than 30 ms corresponding to some hundred reflections along the boundaries. Then, a 2 ms portion of the signal is selected, time-reversed and re-emitted at point B. Since the time-reversed wave is a flexural wave that induces vertical displacements of the silicon surface, it can be observed using the optical interferometer that scans the surface near point A; see figure 8.5.

One observes both an impressive time compression at point A and a refocusing of the time-reversed wave around the origin, with a focal spot whose radial dimension is equal to half the wavelength of the flexural wave; see

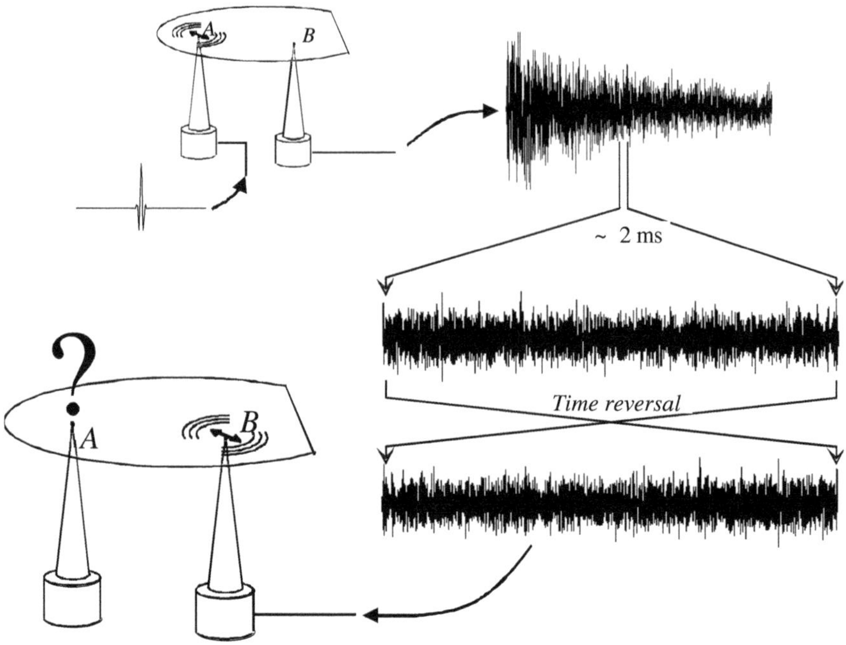

Figure 8.5. Time-reversal experiment conducted in a chaotic cavity with flexural waves. As a first step, a point transducer located at point A transmits a 1 μs long signal. The signal is recorded at point B by the second transducer. The signal spreads over more than 30 ms due to reverberation. In the second step of the experiment, a 2 ms portion of the recorded signal is time-reversed and retransmitted back into the cavity.

figures 8.6(a) and (b). Using reflections at the boundaries, the time-reversed wave field converges towards the origin from all directions and gives a circular spot, like the one that could be obtained with a closed time-reversal cavity covered with transducers. The 2 ms time-reversed waveform is the time sequence needed to focus exactly on point A.

The success of this time-reversal experiment is particularly interesting for the following two reasons. First, it proves again the feasibility of time reversal in wave systems with chaotic ray dynamics. Paradoxically, in the case of one-channel time reversal, chaotic dynamics is not just harmless, but it is even useful, because it guarantees ergodicity and mixing. Second, using a source of vanishing aperture, we obtain an almost perfect focusing quality. The procedure approaches the performance of a closed TRM, which has an aperture of 360°. Hence, a one-point time reversal in a chaotic cavity produces better results than a TRM in an open system. Using reflections at the edge, focusing quality is not limited to the aperture and, in addition, the time-reversed collapsing wavefront approaches the focal spot from all directions.

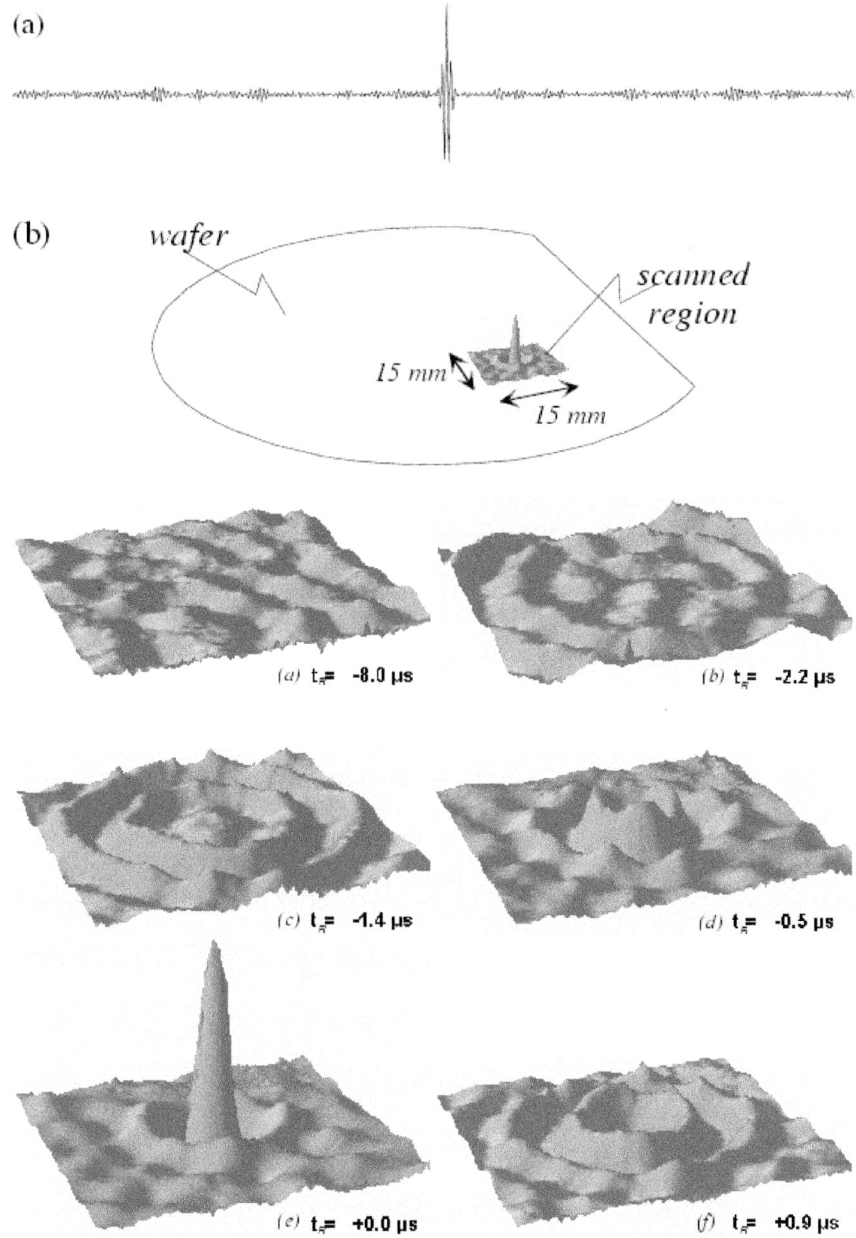

Figure 8.6. Time-reversed signal observed at point A. The observed signal is 210 μs long (a). Time-reversed wave field observed at different times around point A on a square of 15 mm $\times$ 15 mm (b).

Although one obtains excellent focusing, a one-channel time reversal is not perfect, because a weak noise level can be observed throughout the system. Residual temporal and spatial side lobes persist even for time-reversal windows of infinite size. These are due to multiple reflections passing through the locations of the time-reversal transducers and have been expressed in closed form by Draeger [8]. Using an eigenmode analysis of the wave field, he shows that, for long time-reversal windows, there is a saturation regime that limits the signal-to-noise ratio. The reason is the following: once the saturation regime is reached, all the eigenmodes are well-resolved. However, A and B are always located at the nodes of some eigenmodes and these eigenfrequencies cannot be transmitted or received in the cavity.

8.3.1 Time reversal as a temporal correlator

If we neglect the acousto-electric responses of the transducers and take into account the modal decomposition of the impulse response $h_{AB}(t)$ on the eigenmodes $\Psi_n(\vec{x})$ of the cavity with eigenfrequency ω_n, then we get

$$h_{AB}(t) = \sum_n \Psi_n(A)\Psi_n(B)\frac{\sin(\omega_n t)}{\omega_n} \qquad (t > 0).$$

This signal is recorded at B and a part $\Delta T = [t_1, t_2]$ is time-reversed and re-emitted as

$$h_{AB}^{\Delta T}(-t) = \begin{cases} h(-t) & t \in [-t_1, -t_2], \\ 0 & \text{otherwise.} \end{cases}$$

Hence, the time-reversed signal at A reads:

$$\begin{aligned} p_{\mathrm{tr}}^{\Delta T}(A, t) &= \int_{t_1}^{t_2} h_{AB}(t_R + \tau) h_{AB}(\tau)\, \mathrm{d}\tau \\ &= \left[\sum_n \frac{1}{\omega_n}\Psi_n(A)\Psi_n(B)\right]\left[\sum_m \frac{1}{\omega_m}\Psi_m(A)\Psi_m(B)\right] I_{mn}, \end{aligned}$$

with I_{mn} equal to

$$\begin{aligned} I_{mn} &= \int_{t_1}^{t_2} [\sin(\omega_n \tau) \sin(\omega_n[\tau + t])\, \mathrm{d}\tau \\ &= \tfrac{1}{2}\sin(\omega_n t) \int_{t_1}^{t_2} \sin([\omega_m - \omega_n]\tau) + \sin([\omega_m + \omega_n]\tau)\, \mathrm{d}\tau \\ &\quad + \tfrac{1}{2}\cos(\omega_n t) \int_{t_1}^{t_2} \cos([\omega_m - \omega_n]\tau) + \cos([\omega_m + \omega_n]\tau)\, \mathrm{d}\tau. \end{aligned}$$

The second term of each integral gives a contribution of order $(\omega_m + \omega_n)^{-1} \ll \Delta T$, which can be neglected. Thus we obtain the approximation

$$I_{mn} \approx \begin{cases} \dfrac{\sin(\omega_n t)}{2(\omega_m - \omega_n)}[-\cos((\omega_m - \omega_n)t_2) + \cos((\omega_m - \omega_n)t_1)] & \\ \quad + \dfrac{\cos(\omega_n t)}{2(\omega_m - \omega_n)}[\sin((\omega_m - \omega_n)t_2) & \\ \quad - \sin((\omega_m - \omega_n)t_1)] & \text{if } \omega_m \neq \omega_n \\ \frac{1}{2}\Delta T \cos(\omega_n t_R) & \text{if } \omega_m = \omega_n. \end{cases}$$

If we assume that the eigenmodes are not degenerate, $\omega_m = \omega_n \Leftrightarrow m = n$ and the second term represents the diagonal elements of the sum over n and m. The first term is only important if the difference $\omega_m - \omega_n$ is small, i.e. for neighbouring eigenfrequencies. In the case of a chaotic cavity, nearest neighbours tend to repel each other. If the characteristic distance $\Delta\omega$ is sufficiently large such that $\Delta T \gg 1/\Delta\omega$, then the non-diagonal terms of I_{mn} will be negligible compared to the diagonal contributions and one obtains

$$I_{mn} = \frac{1}{2}\delta_{mn}\Delta T \cos(\omega_n t) + O\left(\frac{1}{\Delta\omega}\right).$$

Hence, in the limit $\Delta T \to \infty$, the time-reversed signal observed at A by a reversal in B is given by

$$p_{\text{tr}}^{\Delta T}(A, t) = \frac{1}{2}\Delta T \sum_n \frac{1}{\omega_n^2}\Psi_n^2(A)\Psi_n^2(B)\cos(\omega_n t).$$

This expression gives a simple interpretation of the residual temporal lobes, which are observed experimentally in figure 8.6. The time-reversed signal observed at the origin cannot simply be reduced to a Dirac distribution $\delta(t)$. Even for $\Delta T \gg 1/\Delta\omega$, it is equal to another cross-correlation product, namely

$$\begin{aligned} C_{\Delta T}(t) &= h_{AA}(-t) \overset{*}{_{(\Delta T)}} h_{BB}(t) \\ &= \int_{t_1}^{t_2} h_{BB}(t + \tau)h_{AA}(\tau)\,\mathrm{d}\tau \\ &= \left[\sum_n \frac{1}{\omega_n}\Psi_n^2(B)\right]\left[\sum_m \frac{1}{\omega_m}\Psi_m^2(A)\right] I_{mn} \end{aligned}$$

where the impulse responses h_{AA} and h_{BB} describe the backscattering properties of A and B due to the boundaries of the cavity. Each impulse response is composed of a first peak at $t = 0$, followed by multiple reflections that pass over the source point even after the excitation has ended. Hence, the signal observed in A, after a Dirac excitation, can be described by $h_{AA}(t)$. Therefore, a perfect time-reversal operation (i.e. we simply count time backwards) would give at A the signal $h_{AA}(-t)$, that is, some multiple reflections with a final peak at $t = 0$. For the same reason, the reversed point B cannot exactly transmit a particular waveform in the cavity. Due to the boundaries, a Dirac excitation at B will also give rise to a transmitted signal $h_{BB}(t)$. Hence, in the limit as we increase the length of the time-reversal windows, we get for a one-channel time-reversal experiment the cavity formula deduced by Draeger [8]:

$$h_{AB}(-t) * h_{BA}(t) = h_{AA}(-t) * h_{BB}(t).$$

8.3.2 Time reversal as a spatial correlator

Similar to what was done for the multiple-scattering medium, the focusing properties of the time-reversed wave can be calculated using the spatial correlator approach. If we let $h_{A'B}(t)$ denote the propagation impulse response from point B to an observation point A' (with coordinates $\vec{r}_1$), different from the source location A, then the time-reversed signal recreated at A' at time $t_1 = 0$ is

$$p_{\mathrm{tr}}(\vec{r}_1, t) = \int h_{AB}(t) h_{A'B}(t)\, \mathrm{d}t.$$

Here we let the time interval ΔT tend to infinity. Hence, the directivity pattern of the time-reversed wave field is given by the cross-correlation of the Green functions that can be constructed from the eigenmodes of the cavity

$$p_{\mathrm{tr}}(\vec{r}_1, 0) = \sum_n \frac{1}{\omega_n^2} \Psi_n(A) \Psi_n(\vec{r}_1) \Psi_n^2(B). \tag{8.5}$$

Note that in a real experiment one has to take into account the limited bandwidth of the transducers. This means that a spectral function $F(\omega)$ centred at frequency ω_c and with bandwidth $\Delta\omega$ must be introduced. Then, we can write equation (8.5) in the form

$$p_{\mathrm{tr}}(\vec{r}_1, 0) = \sum_n \frac{1}{\omega_n^2} \Psi_n(A) \Psi_n(\vec{r}_1) \Psi_n^2(B) F(\omega_n). \tag{8.6}$$

Thus the summation is limited to a finite number of modes, which in our experiment is typically of the order of some hundreds. Since we do not know the exact eigenmode distribution of each chaotic cavity, we cannot evaluate this expression directly. However, one may use a statistical approach and consider the average over different realizations, which consists of summing over different cavity realizations. So we replace the product of the eigenmodes in equation (8.6) with their expected values $\langle \ldots \rangle$. We also use a qualitative argument proposed by Berry [1, 16, 20] to characterize irregular modes in a chaotic system. If chaotic rays support an irregular mode, they can be considered as a superposition of a large number of planar waves with random direction and phase. This implies that the amplitude of an eigenmode has a Gaussian distribution with $\langle \Psi^2 \rangle = \sigma^2$ and a short-range isotropic correlation function given by a Bessel function of the form

$$\langle \Psi_n(A)\Psi_n(\vec{r}_1)\rangle = J_0\left(\frac{2\pi|\vec{r}_1 - \vec{r}_0|}{\lambda_n}\right) \tag{8.7}$$

where λ_n is the wavelength corresponding to ω_n. If A and A' are sufficiently far away from B such that they are not correlated with B, then

$$\langle \Psi_n(A)\Psi_n(\vec{r}_1)\Psi_n^2(B)\rangle = \langle \Psi_n(A)\Psi_n(\vec{r}_1)\rangle\langle \Psi_n^2(B)\rangle.$$

Then, one finally obtains

$$\langle p_{\mathrm{tr}}(\vec{r}_1, 0)\rangle = \sum_n \frac{1}{\omega_n^2} J_0\left(\frac{2\pi|\vec{r}_1 - \vec{r}_0|}{\lambda_n}\right)\sigma^2 F(\omega_n). \tag{8.8}$$

The experimental results obtained in figure 8.6(*b*) agree with this prediction and show that in a chaotic cavity the spatial resolution is independent from the time-reversal mirror aperture. Indeed, with a one-channel time-reversal mirror, the directivity patterns at $t = 0$ closely resemble the Bessel function $J_0(\frac{2\pi}{\lambda_c}|\vec{r}_1 - \vec{r}_0|)$ that corresponds to the central frequency of the transducers.

In figure 8.6 one can also observe that a very good estimate of the eigenmode correlation function is experimentally obtained with only one realization. A one-channel omni-directional transducer is able to refocus a wave in a chaotic cavity, and we did not average the data over different cavities or over different positions of the transducer B.

8.3.3 Phase conjugation versus time reversal: self-averaging in the time domain

This interesting result emphasizes the great interest of time-reversal experiments, compared to phase-conjugated experiments. In phase conjugation, one only works with monochromatic waves and not with broadband pulses. For example, if one only works at frequency ω_n, so that there is only one term in equation (8.7), then one cannot refocus a wave on point A. An omni-directional transducer, located at an arbitrary point B, that works in monochromatic mode, sends a diverging wave into the cavity that has no reason to refocus on point A. The refocusing process only works with broadband pulses, with a large number of eigenmodes in the transducer bandwidth. Here, the averaging process that gives a good estimate of the spatial correlation function is not obtained by summing over different realizations of the cavity, like in equation (8.8), but by a sum over 'pseudo-realizations' that correspond to different modes in the same cavity. This comes from the fact that in a chaotic cavity we may assume a statistical de-correlation of the different eigenmodes. As the number of eigenmodes available in the transducer bandwidth increases, the refocusing quality becomes better and the focal spot pattern comes closer to the ideal Bessel function. Hence, the signal-to-noise level should increase as the square-root of the number of modes in the transducer bandwidth.

A similar result has also been observed in the time-reversal experiment conducted in a multiple-scattering medium. A clear refocusing was obtained with only a single-array element; see figure 8.4. The focusing process works with broadband pulses (the transducer centre frequency is 3.5 MHz with a 50% bandwidth at -6 dB). For each individual frequency there is no focusing and the estimate of the spatial correlation is very noisy. However, for a large bandwidth, if we have statistical de-correlation of the wave fields for different frequencies, the time-reversed field is self-averaging.

8.4 Conclusion

In this paper, we have shown how chaotic ray dynamics in random media and in cavities enhances resolution in time-reversal experiments. Multiple reverberation makes the effective size of the TRM much larger than its physical size. This study proves the feasibility of acoustic time reversal in media with chaotic ray dynamics. Paradoxically, chaotic dynamics is useful, because it reduces the number of channels needed to ensure an accurate time-reversal experiment. This result can easily be extended to electromagnetic propagation through random and chaotic media. Perhaps one of the most spectacular future applications of time-reversed technologies will be wireless communication, where the multi-pathing may enhance the amount of information that can be transmitted between antenna arrays in complex scattering environments like cities. From quantum chaos to wireless communications, time-reversal symmetry plays an important role and, contrary to long-held beliefs, chaotic scattering of acoustic or microwave signals may strongly enhance the amount of independent information that can be transmitted from an antenna array to a volume of arbitrary shape.

Acknowledgments

The material in this chapter has appeared in the invited article (M Fink and J de Rosny 2002 Time-reversed acoustics in random media and in chaotic cavities, *Nonlinearity* **15** R1–18) and is reproduced here with permission from Institute of Physics Publishing and the London Mathematical Society ©2002.

I would like to thank J de Rosny for his contribution to the acoustic sink experiment.

References

[1] Berry M V 1983 Chaotic behaviour of deterministic systems *Proc. Ecole d'été de Physique Théorique 36, Les Houches 1981* ed G Iooss, R H G Helleman and R Stora (Amsterdam: North-Holland) pp 171–271

[2] Cassereau D and Fink M 1992 Time reversal of ultrasonic fields—Part III: Theory of the closed time-reversal cavity *IEEE Trans. Ultrason. Ferroelectr. Freq. Contr.* **39** 579–92

[3] Cassereau D and Fink M 1994 Time reversal focusing through a plane interface separating two fluids *J. Acoust. Soc. Am.* **96** 3145–54

[4] Derode A, Roux P and Fink M 1995 Robust Acoustic Time reversal with high order multiple scattering *Phys. Rev. Lett.* **75** 4206–9

[5] Derode A, Tourin A and Fink M 1999 Ultrasonic pulse compression with one-bit time reversal through multiple scattering *J. Appl. Phys.* **85** 6343–52

[6] Derode A, Tourin A and Fink M 2001 Random multiple scattering of ultrasound. I—Coherent and ballistic waves *Phys. Rev.* E **64** 036605

[7] Draeger C and Fink M 1997 One channel time-reversal of elastic waves in a chaotic 2D-silicon cavity *Phys. Rev. Lett.* **79** 407–10

[8] Draeger C and Fink M 1999 One-channel time-reversal in chaotic cavities: experimental results *J. Acoust. Soc. Am.* **105** 611–17

[9] Draeger C and Fink M 1999 One-channel time-reversal in chaotic cavities: theoretical limits *J. Acoust. Soc. Am.* **105** 618–25

[10] Fink M 1992 Time reversal of ultrasonic fields—Part I: Basic principles *IEEE Trans. Ultrason. Ferroelectr. Freq. Contr.* **39** 555–66

[11] Fink M 1997 Time reversed acoustics *Phys. Today* **50** 34–40

[12] Fink M, Prada C, Wu F and Cassereau D 1989 Self focusing in inhomogeneous media with time reversal acoustic mirrors *IEEE Ultrasonics Symp. Proc., Montreal* vol 1, ed B R McAvoy (New York: IEEE) pp 681–6

[13] Jackson D R and Dowling D R 1991 Phase conjugation in underwater acoustics *J. Acoust. Soc. Am.* **89** 171–86

[14] Kino G S 1987 *Acoustics Waves (Signal Processing Series)* (Englewood Cliffs, NJ: Prentice-Hall)

[15] Kupperman W A, Hodgkiss W S, Song H C, Akal T, Ferla T and Jackson D 1998 Phase conjugation in the ocean: experimental demonstration of an acoustic time reversal mirror *J. Acoust. Soc. Am.* **103** 25–40

[16] McDonald S W and Kaufman A N 1988 Wave chaos in the stadium; statistical properties of short-wave solutions of the Helmholtz equation *Phys. Rev.* A **37** 3067–86

[17] De Rosny J and Fink M 2002 Overcoming the diffraction limit in wave physics using a time-reversal mirror and a novel acoustic sink *Phys. Rev. Lett.* to appear

[18] Roux P, Roman B and Fink M 1997 Time-reversal in an ultrasonic waveguide *Appl. Phys. Lett.* **70** 1811–13

[19] Snieder R and Scales J 1998 Time-reversed imaging as a diagnostic of wave and particle chaos *Phys. Rev.* E **58** 5668–75

[20] Weaver R and Burkhardt J 1994 Weak Anderson localization and enhanced backscatter in reverberation rooms and quantum dots *J. Acoust. Soc. Am.* **96** 3186–90

Chapter 9

Reduction methods applied to non-locally coupled oscillator systems

Yoshiki Kuramoto
Kyoto University

When we try to understand the qualitative dynamics of spatially extended nonlinear systems through mathematical analysis, we require theoretical tools to reduce the governing dynamical laws to simpler forms. Thus it is natural to ask under what general conditions such model reduction can be achieved. A possible answer is that the reduction is possible when the system involves a few special degrees of freedom whose dynamics is distinctively slower, i.e. whose characteristic time scales are far longer than those of the remaining degrees of freedom: then those degrees of freedom with short time scales may either be eliminated *adiabatically* or cancelled as a result of time-averaging of their rapid changes. The existence of slow variables is thus crucial to the reduction. Here 'slow' implies that the stability is almost neutral, so that we may call such slow degrees of freedom *neutral modes*.

A little consideration will tell us that there are at least three classes of neutral modes in nature. They are:

(i) critical eigenmodes at bifurcation points,
(ii) phase modes, and
(iii) conserved quantities.

We now make some remarks concerning classes (ii) and (iii) of neutral modes.

First, in physics terminology, phase modes are referred to as Goldstone modes and they exist when the current state breaks a continuous symmetry possessed by the system itself. For example, the spatial oscillation of a Turing pattern in an infinite homogeneous medium breaks the continuous symmetry of spatial translation. Similarly, a limit-cycle oscillation in an autonomous dynamical system breaks the temporal translational symmetry. In each case, there is a special degree of freedom called the *phase*.

Second, conserved quantities form a further important class of neutral modes. They are neutral by definition; perturbations to conserved quantities are preserved and they neither decay nor grow.

Reduction theories have been developed for each of the three classes of neutral modes. For class (i), we have the so-called centre-manifold reduction [11, 17, 25]. For class (ii), we have the so-called phase reduction [17] or phase dynamics. However, it is uncertain whether there is any established name for the reduction theories making use of neutral modes of class (iii). However, there are important examples of such reduction, e.g., the beautiful and monumental work in the statistical mechanics of irreversible processes, which is the Chapman–Enskog theory by which the Boltzmann gas kinetic equation is reduced to the Navier–Stokes equation [4]. Incidentally, the author was interested to note some time ago that there is a parallel between Chapman–Enskog theory and phase dynamics [18].

Now, it may be that neutral modes from different classes coexist, in which case the corresponding reduction theories may be combined. Such combined forms of reduction have turned out to be instrumental to the understanding of complex dynamics [5]. Similar situations may arise even within one of the three neutral mode classes, as exemplified by the multiple bifurcation theories [10] which are applied to the study of complex dynamics generated by an interplay of different types of bifurcation modes.

Conserved quantities, in combination with other types of neutral mode, may well have been important in, for example, fluid dynamics [33]. However, in this chapter we will concentrate on the first two types of reduction schemes and demonstrate how powerful these methods are in developing our understanding of the dynamics of a particular class of systems, namely spatially extended fields and large populations of coupled limit-cycle oscillators, on which the author has worked over the past few decades [16, 17, 19, 22]. Actually, the power of the centre-manifold and phase reductions applied to coupled oscillator systems has been fully demonstrated in the past through numerous examples [1, 6, 7, 9, 12–14, 32].

However, past studies have been concerned exclusively with two extreme coupling types, namely *locally* and *globally* coupled systems. The natural question is thus how we should deal with the intermediate case which we refer to as *non-locally coupled systems*. In fact, non-local coupling generates a wide range of unexpected behaviour. Some recent studies on an array of non-locally coupled oscillators revealed the occurrence of a new type of turbulent behaviour characterized by multi-scaling properties [23, 24]. In this chapter, I will discuss two different classes of collective behaviour:

(i) rotating sprial waves without phase singularity and
(ii) coexisting patterns of coherent and incoherent oscillators.

In fact, these classes are not separate, and the first class may be regarded as a special, but particularly important, subclass of the second. Some recent results on these topics obtained by the author will be presented here. We now begin with preliminary considerations of non-local coupling itself from a rather general point of view.

9.1 When coupling non-locality becomes crucial

When we speak of non-locally coupled systems, one point that we should bear in mind is that locality or non-locality is a relative concept in the sense that it depends on the level of description. This may be illustrated by the following example:

$$\dot{\boldsymbol{A}} = \boldsymbol{F}(\boldsymbol{A}) + \boldsymbol{K} \cdot B \tag{9.1}$$

$$\tau \dot{B} = -B + D\nabla^2 B + \boldsymbol{K}' \cdot \boldsymbol{A}. \tag{9.2}$$

This model constitutes a reaction–diffusion system that appears a little unconventional. There are two subsystems, represented by the variables $\boldsymbol{A}$ and B, between which linear cross-couplings $\boldsymbol{K} \cdot B$ and $\boldsymbol{K}' \cdot A$ are assumed. The second subsystem has diffusion, whereas the first does not.

Now, suppose that the dynamical system $\dot{\boldsymbol{A}} = \boldsymbol{F}(\boldsymbol{A})$ represents a limit-cycle oscillator. A possible physical interpretation of (9.1), (9.2) is as follows. The first subsystem represents a large assembly of densely distributed oscillators. There is no direct coupling among them, but each discrete oscillatory element is locally coupled with a diffusive field represented by B. That is, the activity of the discrete elements is influenced by the local concentration B of the diffusive chemical, which obeys a simple linear law; it is diffusing, decaying at a constant rate and has characteristic time scale τ. The coupling of B with $\boldsymbol{A}$ may arise if the chemical is produced by the discrete elements, with the rate of production depending on the elements' state.

The system (9.1), (9.2) may look similar to a biological population of oscillatory cells where direct mutual contact is absent, but where a chemical plays the role of a coupling agent through its diffusive nature. Since the first subsystem must involve two or more components, the system of (9.1) and (9.2) represents a reaction–diffusion system involving at least three components. A model similar to this was proposed for demonstrating the occurrence of an unconventional type of spatio-temporal chaos with scaling properties [20], and also studied independently by Nicola *et al* [28] in connection with propagating domains of striped patterns.

It is commonly stated that diffusive coupling is a typical local coupling. However, if we eliminate B by exploiting the linearity of (9.2), the effective coupling of the oscillating field $\boldsymbol{A}$ becomes non-local and given by the general form

$$\dot{\boldsymbol{A}} = \boldsymbol{F}(\boldsymbol{A}) + \int \mathrm{d}t' \int \mathrm{d}\boldsymbol{r}' \, \hat{G}(\boldsymbol{r} - \boldsymbol{r}', t - t')\boldsymbol{A}(\boldsymbol{r}', t'). \tag{9.3}$$

Note here that the non-locality is both spatial and temporal.

We now consider the conditions under which such effective non-locality can or cannot be approximated by a local coupling. For this purpose, we consider the more specific reaction–diffusion system

$$\dot{X} = f(X, Y) + K \cdot (B - X) \tag{9.4}$$

$$\dot{Y} = g(X, Y) \tag{9.5}$$

$$\tau \dot{B} = -B + D\partial_x^2 B + X \tag{9.6}$$

Here the individual oscillatory dynamics is given by a two-component vector field.

To make the point of discussion as clear as possible, we simplify and consider a one-dimensional case only, with the system length L tending to infinity, and we also assume that the time scale of B is infinitesimally short by letting τ tend to zero. Consequently, B is adiabatically eliminated and (9.4) becomes

$$\dot{X} = f(X, Y) + \frac{K}{2\sqrt{D}} \int_{-\infty}^{\infty} \mathrm{d}x' \exp\left(-\frac{|x - x'|}{\sqrt{D}}\right) [X(x') - X(x)]. \tag{9.7}$$

Thus, the coupling is no longer non-local in time, and in space it is a simple exponential with a coupling range given by $\sqrt{D}$. This equation combined with (9.5) constitutes our system. Observe that if the spatial pattern has a sufficiently long wavelength for its variation within the coupling range to be small, then the coupling reduces to a diffusion coupling or

$$\dot{X} = f(X, Y) + \tilde{D}\frac{\partial^2 X}{\partial x^2} \tag{9.8}$$

where the effective diffusion constant is given by

$$\tilde{D} = KD. \tag{9.9}$$

In this way, we recover an ordinary two-component oscillatory reaction–diffusion system such as the FitzHugh–Nagumo nerve conduction equation [27], Brusselator [29], Gray–Scott model [31] or Keener–Tyson model [15], each of which includes diffusion. Since there is a large literature on such two-component reaction–diffusion models, there is not much new to say here on this subject.

The consistency of the diffusion-coupling approximation or local-coupling approximation may be checked by estimating the characteristic wavelength of the pattern generated by the reduced reaction–diffusion equation thus obtained, and comparing it with the effective coupling range. It is clear that the characteristic wavelength l_p scales like $l_\mathrm{p} \sim \sqrt{\tilde{D}} = \sqrt{KD}$, while the effective coupling range is given by $\sqrt{D}$.

Thus, if the coupling K is sufficiently strong then l_p becomes larger than the effective coupling range, which is consistent with the local-coupling approximation. In contrast, if K is sufficiently small, the pattern will exhibit considerable spatial variation within the coupling range, which leads to a contradiction with the local-coupling approximation. In the next section, we show an example of the peculiar dynamics which occurs in the latter situation.

9.2 Rotating spiral waves without phase singularity

Unusual types of spiral waves can appear in the reaction–diffusion system:

$$\dot{X} = f(X, Y) + K \cdot (B - X) \tag{9.10}$$
$$\dot{Y} = g(X, Y) \tag{9.11}$$
$$\tau \dot{B} = -B + D\nabla^2 B + X \tag{9.12}$$

where

$$f(X, Y) = \epsilon^{-1}(X - X^3 - Y) \tag{9.13}$$
$$g(X, Y) = aX + b - cY \tag{9.14}$$

and ∇^2 is the two-dimensional Laplacian. These belong to the same class of reaction–diffusion systems discussed in the preceding section except that the spatial dimension is now two. The local oscillatory dynamics is given by a slightly generalized form of the FitzHugh–Nagumo model called the Bonhoeffer–Van der Pol model.

Before explaining the behaviour of this system, we recall a basic fact concerning traditional spiral waves in oscillatory and excitable reaction–diffusion media [39]. Suppose we have a two-dimensional rotating spiral pattern which may be rotating rigidly and steadily, and which may or may not have a meandering core [3, 34]. Figure 9.1(*b*) shows a schematic blow-up of the central core of the spiral covered with an imaginary grid. At a given time, each grid point finds itself to be in a certain state in the one-oscillator phase space. Thus, we have a mapping between the two sets of points in the physical space of figure 9.1(*b*) and the phase space of figure 9.1(*c*). The resulting grid in the phase space is nothing but a smooth deformation of the regular grid in the physical space without any topological change. If the grid points in the physical space are infinitely dense, the corresponding object in the phase space gives a continuous membrane-like object which is simply connected. If the phase-space dimension is higher than two, one may imagine this two-dimensional object being suspended in that larger space.

Now, our reaction–diffusion system (9.10)–(9.14) can also sustain spiral waves whose overall pattern appears similar to the usual spiral waves discussed earlier. However, if we look more closely at the core structure, we observe a qualitative difference. In figure 9.2, we depict the mappings between the physical and phase space for the reaction–diffusion system (9.10)–(9.14) under oscillatory conditions, where the mapped grid points are projected onto the XY-plane and, for reasons which shall become clear, all grid lines are deleted. Observe that when the coupling is strong, we obtain the usual spiral wave picture, but as the coupling becomes weaker, the grid points become more sparsely distributed in the inner region of the phase-space picture.

Let us imagine a continuum limit. It is not clear from such a rough numerical analysis, with relatively few grid points, whether or not there is a critical value in

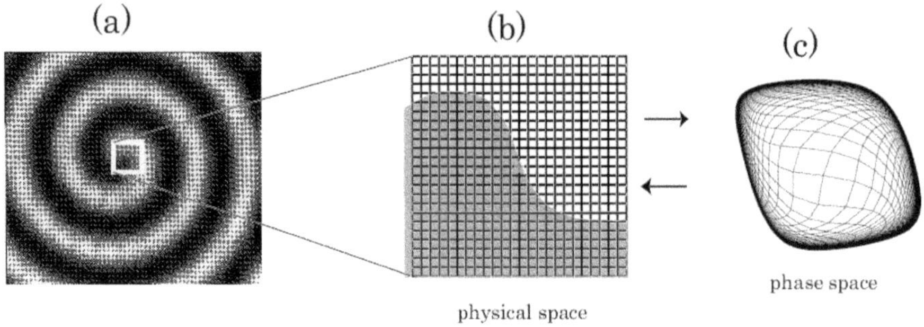

Figure 9.1. A small square area near the core of a two-dimensional spiral pattern, schematically shown in (*a*), is covered with an imaginary grid (*b*). This imaginary grid is then mapped onto the phase space giving a deformed grid as shown in (*c*). If the grid points in (*b*) are infinitely dense, the corresponding deformed grid in the phase space becomes a two-dimensional membrane which is simply connected unless the spatial continuity of the pattern is lost.

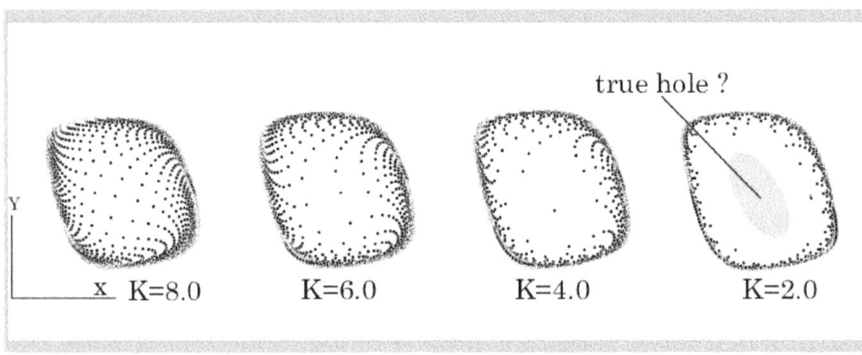

Figure 9.2. Distribution of points in the (X, Y) phase plane mapped from grid points in the core region of a spiral pattern obtained from the reaction–diffusion model (9.10)–(9.14). For sufficiently small coupling constant K, a hole appears in the central part of the phase pictures, implying the occurrence of spatial discontinuity. The values of the other parameters are fixed at $a = 1.0$, $b = 0.2$, $c = 0$, $\tau = 0.01$.

the coupling strength below which a hole is created in the continuous membrane. However, as we describe later, a theory based on centre-manifold reduction predicts the creation of a hole. The existence of a hole implies that two points which are infinitesimally close in physical space may no longer be close in the phase space. In other words, the spatial continuity of the pattern has been lost. It seems that loss of spatial continuity is something which can occur quite easily in non-locally coupled systems. In the previous discussion, we have assumed that the local dynamics is self-oscillatory. However, the appearance of a hole in the phase space can also be shown under non-oscillatory but excitable conditions.

We now apply centre-manifold reduction to reaction–diffusion system (9.10)–(9.14), in order to provide a clearer perspective to the problem. The standard technique is applied to the situation where the individual oscillators are close to Hopf bifurcation, and their coupling with the diffusive field is weak, the weakness here being proportional to the smallness of the bifurcation parameter. Under these conditions, the reduced amplitude equation takes the form

$$\dot{A}(\boldsymbol{r}) = (1 + \mathrm{i}\omega_0)A - (1 + \mathrm{i}b)|A|^2 A + K \cdot (1 + \mathrm{i}a)(Z(\boldsymbol{r}) - A(\boldsymbol{r})) \tag{9.15}$$

where

$$Z(\boldsymbol{r}) = \int \mathrm{d}\boldsymbol{r}'\, G(\boldsymbol{r} - \boldsymbol{r}')A(\boldsymbol{r}'). \tag{9.16}$$

This equation has the form of a complex Ginzburg–Landau (CGL) equation with non-local coupling. The coupling function G is assumed to be normalized, and its long-distance behaviour is given by exponential decay. In some cases, it is useful to regard the complex quantity Z as a given driving force acting commonly on the individual oscillators. Of course, this forcing should ultimately be determined by the dynamics of the individual A in a self-consistent manner, and we will soon return to this issue. Note that the coupling constant K in (9.15) is not the original K but is scaled with the bifurcation parameter, so that an $O(1)$ value of the new K corresponds to a very small value of the original K.

As for the reaction–diffusion system before reduction, we find that when K is sufficiently large, the spatial scale of the pattern is longer than the coupling range, making the diffusion-coupling approximation valid, so that we recover the standard CGL. The corresponding spiral waves are thus standard. However, we will now see how the core of the standard spiral wave collapses when K is decreased.

Figure 9.3 is similar to figure 9.2 and illustrates the collapse of the spiral core with a series of pictures, each depicting the distribution of the grid points in the complex amplitude plane obtained by a mapping from the core region, in the same manner as for the original reaction–diffusion system. For sufficiently large K, the distribution appears to be standard, while for smaller K the central part becomes void. The critical value of K appears to exist somewhere between figure 9.3(*b*) and figure 9.3(*c*), and theoretically it is predicted to be one. The reason is as follows. Suppose first that the coupling is sufficiently strong for the spiral to be standard. Under suitable parameter values, the spiral rotates rigidly with some frequency ω. Then, the radial profile of the modulus S of the complex amplitude A across the centre of rotation has a profile as shown in figure 9.4(*a*). S vanishes strictly at the centre. The internal forcing field (which will sometimes be referred to as the *mean field*) Z has the same symmetry as A, so that its modulus R shows a profile similar to S (see figure 9.4(*b*)), and is also vanishing at the centre of rotation.

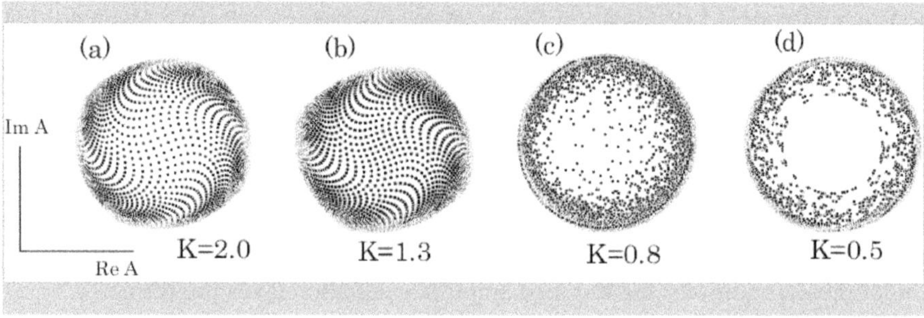

Figure 9.3. Pictures similar to those in figure 9.2, but the model equation used here is the non-locally coupled Ginzburg–Landau equation (9.15), (9.16). Parameter values are $a = 0$, $b = 1.0$. The coupling function $G(\boldsymbol{r})$ is given by the kernel of the integral representation of the solution B of the equation $-B + D\nabla^2 B + X(\boldsymbol{r}) = 0$ with two-dimensional no-flux boundary conditions.

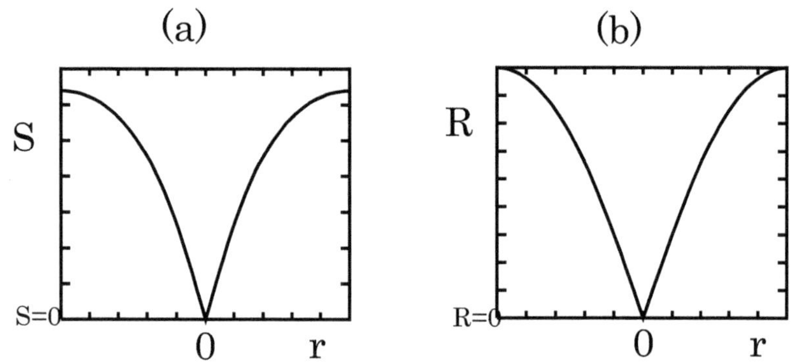

Figure 9.4. (*a*) Radial profile of the modulus S of the complex amplitude A, and (*b*) that of the internal field Z, each across the centre of rotation of a steadily rotating spiral pattern genrerated by the non-locally coupled Ginzburg–Landau equation. Here the coupling constant K is large enough for the spatial continuity of the pattern to be preserved.

If we interpret the situation in terms of the forced-oscillator picture mentioned earlier, we may say that each local oscillator is completely synchronized with the internal forcing. Our question now is under what condition this synchrony is violated.

Actually, the loss of synchrony is initiated by the central oscillator. Since the internal field is vanishing at the very centre, the oscillator there is free from forcing and simply obeys

$$\dot{A} = (1 - K - \mathrm{i}\omega_0')A - (1 + \mathrm{i}b)|A|^2 A \tag{9.17}$$

where $\omega_0' = \omega_0 - aK$. It is clear that if K is larger than 1, the vanishing amplitude

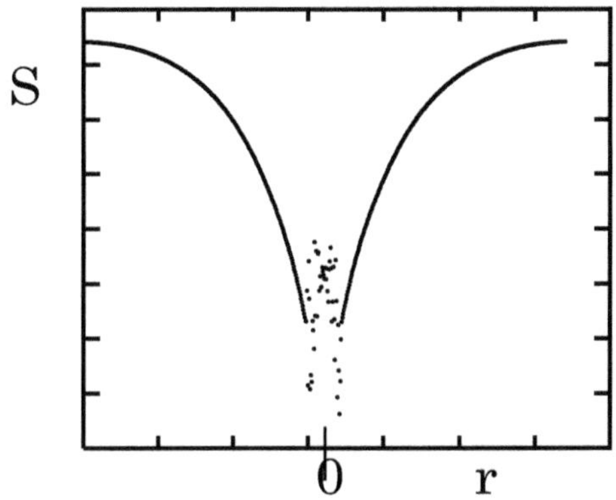

Figure 9.5. Instantaneous radial profile of the modulus S of A across the approximate centre of rotation of a spiral wave for K slightly less than K_c (a schematic diagram). A small group of oscillators near the central core fails to be entrained by the oscillation of the internal field Z, so that it forms a group of quasi-periodic oscillators with distributed frequencies.

is stable, while if K is less than 1, the oscillator becomes Hopf unstable and starts to oscillate with a frequency which is usually different from that of the forcing or spiral rotation.

The oscillators which are near the centre are not entirely free of the internal forcing: a small forcing term will appear additively on the right-hand side of (9.17). On decreasing K further, these oscillators will successively become Hopf unstable. Each such instability will produce a new frequency.

Therefore, for a given value of K less than one, there is a critical radius r_c such that inside the circle this radius the oscillators are Hopf unstable with a second frequency Ω independent of the forcing frequency, while outside the circle the oscillators remain synchronized with the forcing. Since the forcing amplitude is space-dependent, the new frequencies thus produced in the inner region are also space-dependent. This implies that the assembly of these quasi-periodic oscillators will be phase-randomized. The corresponding situation is illustrated in figure 9.5 which gives a schematic image of a snapshot of the amplitudes of the individual oscillators for K slightly smaller than its critical value.

In the argument developed here, there is the implicit assumption that the internal forcing remains strictly time-periodic even for K less than K_c ($= 1$), even though the central group of oscillators would be random in phase. This time-periodicity assumption may be valid near K_c where the incoherent region is small. However, the assumption appears contradictory if we recall the self-consistency relation required between forcing and the motion of the individual oscillators.

9.3 Simpler case: coexistence of coherent and incoherent domains

We now develop the discussion which concluded the previous section, concerning the consistency of a periodic internal forcing and an incoherent assembly of oscillators. To this end, it is appropriate to examine the problem from a wider viewpoint. Suppose we are working with a class of non-locally coupled oscillator systems governed by

$$\dot{\boldsymbol{A}}(\boldsymbol{r},t) = \boldsymbol{R}(\boldsymbol{A}) + \boldsymbol{Z}(\boldsymbol{r},t) \tag{9.18}$$

$$\boldsymbol{Z}(\boldsymbol{r},t) = \int \mathrm{d}\boldsymbol{r}'\, \hat{G}(\boldsymbol{r}-\boldsymbol{r}')\boldsymbol{A}(\boldsymbol{r}',t) \tag{9.19}$$

which includes our CGL-type dynamics as a special case. The coupling term may be looked upon as a common forcing, as noted earlier. It may happen that the forcing field $\boldsymbol{Z}$ is essentially time-periodic but has a certain spatial structure. Then, in some parts of the system the forcing amplitude may be large enough to entrain the individual oscillators, whilst in the remaining parts it is too weak for entrainment. In this way, the whole space will be divided into domains each of which is composed of all coherent or all incoherent oscillators. The spiral wave case is just one such example, but it seems natural to look for simpler problems which are more suited for detailed analysis.

In this vein, we consider the same CGL-type equation as (9.15) but with a spatial dimension equal to one, so that the coupling function is given by a simple exponential or

$$Z(x,t) = \frac{1}{2}\int \mathrm{d}x'\, \exp(-\kappa|x-x'|)A(x',t). \tag{9.20}$$

Under periodic boundary conditions and with a suitable initial condition, the system separates into coherent and incoherent domains. Figure 9.6 shows a distribution of the phases of the complex amplitudes of the oscillators at a given time. The basic structure of the distribution stays stable in a statistical sense. In contrast to the spiral wave case, the incoherent domain appears dominant, and therefore this example is a severe test of the hypothesis that an incoherent subpopulation of oscillators can generate periodic internal forcing.

In fact, we may further simplify the situation if we apply phase reduction to the amplitude equation. This can be done when the coupling is weak enough for the modulus of the complex amplitude A to take almost its full value. The result of reduction is a phase oscillator model with non-local coupling, and with the phase-coupling function given by a simple sine function [17], i.e. we have

$$\dot{\phi}(x) = \omega - \int \mathrm{d}x'\, G(x-x')\sin(\phi(x)-\phi(x')+\alpha) \tag{9.21}$$

where $G(x) = \exp(-\kappa x)$. This simple non-local model gives rise to patterns similar to figure 9.6, as is shown in figure 9.7 [2]. We now try to develop a theory of this model.

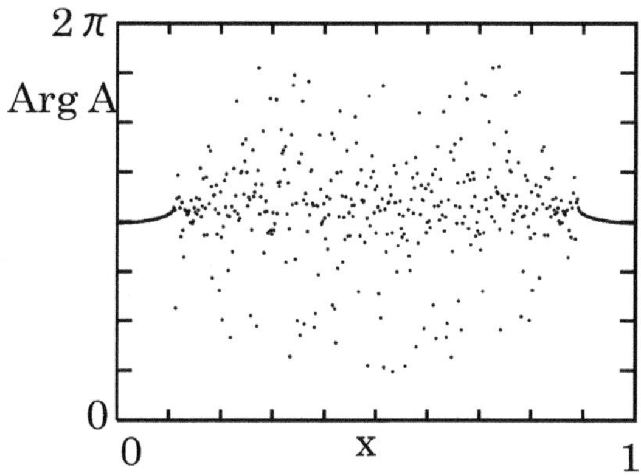

Figure 9.6. Instantaneous spatial distribution of the phase of the complex amplitude obtained from the non-locally coupled one-dimensional Ginzburg–Landau (9.15), (9.20). The system length is one, and periodic boundary conditons are assumed. Parameter values are $a = -1.00$, $b = 0.88$, $\kappa = 4.0$. Under these conditions, spatially uniform oscillation is linearly stable. However, if we start with a suitable single-humped initial phase pattern, the majority of the oscillators becomes strongly incoherent, although a small group of coherent oscillators is left near the system boundaries.

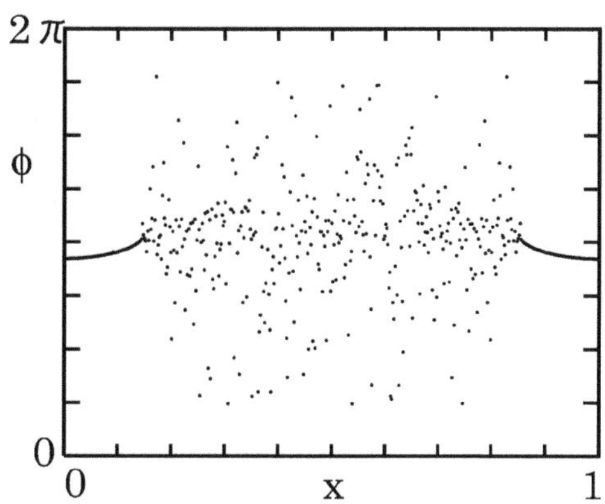

Figure 9.7. The non-local phase equation (9.21) produces a quite similar pattern to that of figure 9.6. Parameter values are $\alpha = 1.457$, $\kappa = 4.0$. The system length is one.

In the phase-oscillator limit, the definition of the internal field, given by (9.20), reduces to

$$Z(x,t) = \int \mathrm{d}x'\, G(x-x')\exp(\mathrm{i}\phi(x'))$$
$$\equiv R(x,t)\exp(\mathrm{i}(\Omega t + \Theta(x,t))). \tag{9.22}$$

By separating the phase-drift term $\mathrm{i}\Omega t$ from the definition of the phase of the internal field, we are anticipating that the dominant motion of the internal forcing is periodic with some frequency Ω, so that Θ gives a residual phase. By using R and Θ, the non-local phase equation can be formally expressed in terms of forced one-oscillator dynamics:

$$\dot{\psi}(x,t) = \omega - \Omega - R(x,t)\sin(\psi(x,t) + \alpha - \Theta(x,t)) \tag{9.23}$$

where we have also used the residual phase $\psi \equiv \phi - \Omega t$ for the individual oscillators, rather than the original phase ϕ. At the same time, using ψ, our definition of the forcing field simplifies a little:

$$R(x,t)\exp(\mathrm{i}\Theta(x,t)) = \int \mathrm{d}x'\, G(x-x')\exp(\mathrm{i}\psi(x')). \tag{9.24}$$

From a combination of (9.23) and (9.24), it is possible to develop a self-consistent theory for finding the forcing field pattern. However, this can only be done if R and Θ are time-independent, and numerical experiments indicate that this is a good approximation. Partial evidence is presented in figure 9.8, which shows the temporal evolution of the real part of the forcing field measured at the midpoint of the system, where the effect of phase disorder would be maximal. Actually, the fluctuations in the oscillator amplitude and phase look rather small.

However, R and Θ have pronounced spatial structure. For instance, the (long-time average) spatial structure of R is depicted in figure 9.9. It can be seen that near the system boundary where the forcing amplitude is large, the oscillators are entrained to its periodic motion so that their frequencies, shown in figure 9.10, have a flat profile there which is close to the forcing frequency Ω. However in the inner region, where the forcing amplitude is relatively small, the oscillators are not entrained and their frequencies are distributed.

Because R and Θ turned out to be approximately time-independent, we are ready to develop a self-consistent theory. Note that there is a strong parallel between the theory given here and that of collective synchronization of globally coupled phase oscillators [16], which was developed by the author more than a quarter century ago and inspired by the pioneering work of Art Winfree in 1967 [38].

First, we reiterate that although R and Θ are assumed to be time-independent, they have spatial structure. We thus try to find the solution of (9.22) for each local oscillator, as a function of the local values of R and Θ. However, there are two possible types of solution depending on the local value of R. One is

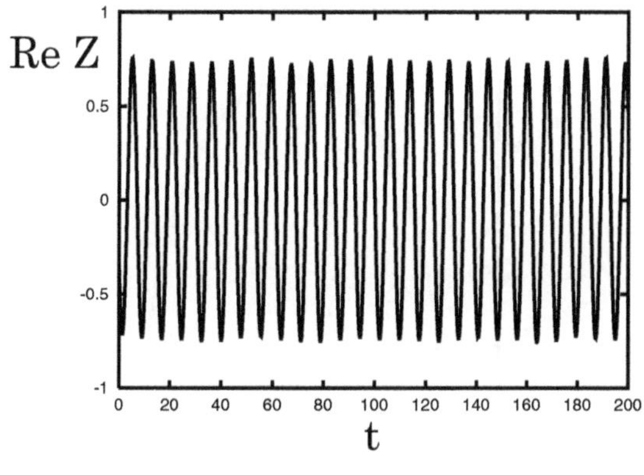

Figure 9.8. Temporal evolution of the real part of the forcing field $Z(t)$ measured at the midpoint of the system ($x = \frac{1}{2}$). Although there are small fluctuations, the oscillation is still quite regular, in contrast to the random phase distribution of the oscillators as illustrated in figure 9.7.

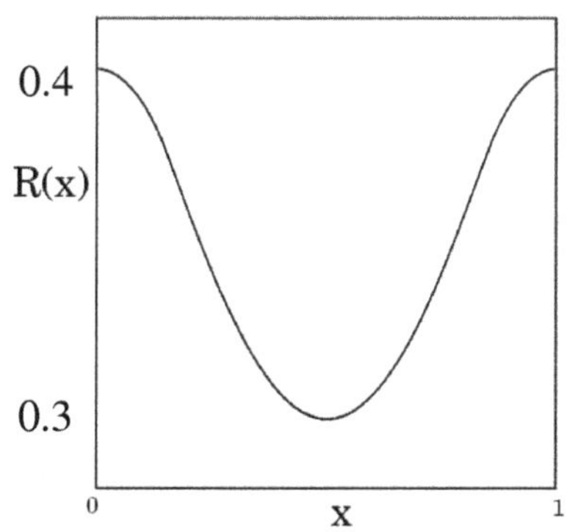

Figure 9.9. Spatial structure of the modulus R of the internal forcing Z, obtained from long-time averages.

the stationary or phase-locked solution which occurs for larger R and the other is the drifting solution which occurs for smaller R.

If the local solution is of the phase-locked type, the corresponding value of ψ may be inserted directly into (9.23). If the solution is of the drifting type, there is no fixed value of ψ to be inserted into (9.23), but instead we have a definite invariant measure for ψ which is inversely proportional to the ψ-dependent drift velocity, i.e. the right-hand side of (9.22). Further, one may evaluate the factor $\exp(i\psi)$ in (9.22) by replacing it with its statistical average calculated using this invariant measure. In this way, (9.23) gives R and Θ as a function of R and Θ

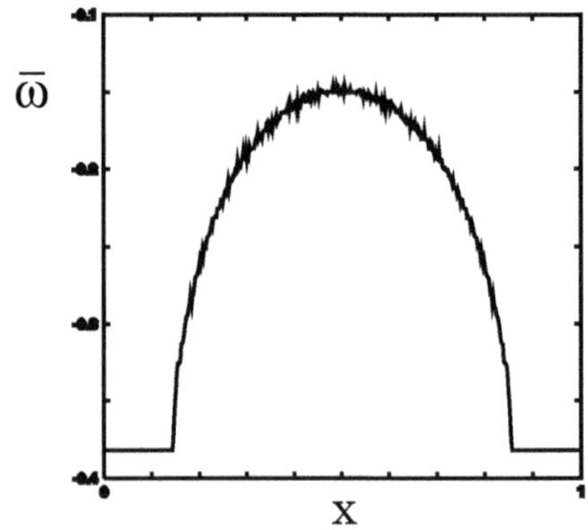

Figure 9.10. Frequency distribution of the oscillators obtained from a long-time average of the instantaneous individual frequencies. A plateau is seen near each system boundary, implying synchronization of the oscillators there with the internal forcing. However, frequencies are distributed in the inner region, implying de-synchronization.

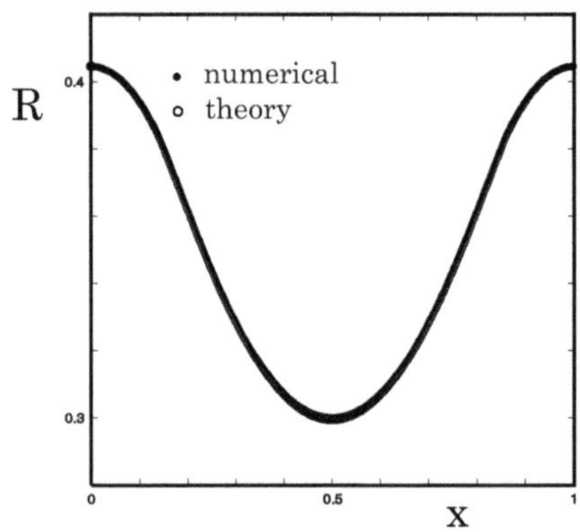

Figure 9.11. Comparison of the theoretical curve for R, which is calculated from a self-consistency equation for the internal field, with the simulation curve of figure 9.9. The agreement is so good that the curves are indistinguishable.

themselves, i.e. a self-consistent equation. This represents a nonlinear eigenvalue problem because the oscillation frequency Ω is still to be determined. In any case, it is at least possible to solve this self-consistent equation numerically, and this calculation was performed in [21].

Figure 9.11 shows a comparison between the theory and numerical simulations for the spatial distributions of R and the actual oscillator frequencies. They agree almost perfectly, comfirming the validity of the theory, and also the power of the phase reduction. The distribution of the local frequencies can

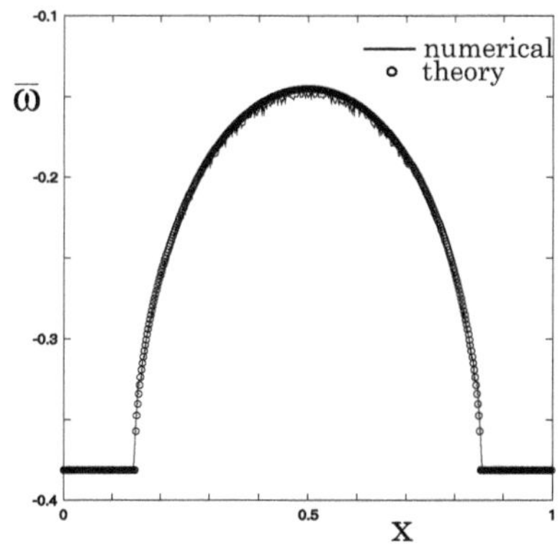

Figure 9.12. Comparison of the theoretical curve for the individual frequencies, which is calculated from (9.23) using the solutions R and Θ of a self-consistency equation for the internal field, with the simulation curve of figure 9.10.

also be calculated theoretically, and this also shows an excellent agreement with numerical results: see figure 9.12.

An unsatisfactory aspect of this argument is that the phenomenon discussed is peculiar to a finite-size system, and therefore may be of little interest from a physical point of view. There is a slightly more interesting case which occurs when the phase-coupling function is generalized by including the second harmonic

$$\dot{\phi}(x) = \omega + \int \mathrm{d}x'\, G(x - x')\Gamma(\phi(x) - \phi(x')) \tag{9.25}$$

where

$$\Gamma(y) = -\sin(y + \alpha) + r\sin(2y). \tag{9.26}$$

Otherwise, the model is the same as before. As a result of this modification, a different type of coexistence of coherent and incoherent domains can appear, and this is depicted in figure 9.13. The coexistent state here seems to remain stable in arbitrarily large systems. For the most part, the system is coherent and has a constant gradient phase, and the incoherent part is localized. Since for this model incoherence produces higher frequency due to the decrease in the local amplitude of the internal forcing, this localized assembly of oscillators behaves collectively like a pacemaker.

Different possibilities have been proposed for the appearance of autonomous or self-organized pacemakers in self-oscillatory media [8, 26, 30]. Our results described here suggest a new mechanism arising from the non-locality in coupling. However, beware that the non-locality could arise from the elimination of a rapidly diffusing variable from a locally coupled system. In a manner similar to the phase-coupling case without the second harmonic, a self-consistent theory can be formulated, although this time we need two complex amplitudes for the

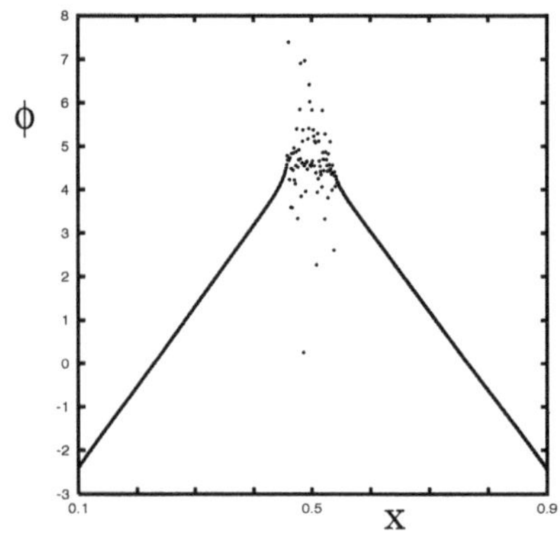

Figure 9.13. Instantaneous spatial distribution of the phases obtained for the phase model (9.25), (9.26), where $G(x) = \exp(-\kappa x)$. Parameter values are $\alpha = 1.05$, $r = 0.25$ and $\kappa = 32.0$.

internal field, corresponding to the first and second harmonics in the phase-coupling function.

9.4 Concluding remarks

In this chapter, I have tried to demonstrate the usefulness of reduction approaches in nonlinear dynamics, by expounding a case study in a class of coupled oscillator systems. The potential of the reduction concept for uncovering new behaviours in coupled systems is still very great.

I would like to stress again the importance of non-locally coupled systems. Non-local coupling is not merely an intermediate case between local and global coupling, but in fact can give rise to its own new dynamics. In terms of the characteristic length scales involved, this is because non-locally coupled systems have a distinct asympototic regime clearly separated from those of systems with purely local or purely global coupling. The key length scales are:

(i) the coupling radius r_c,
(ii) the linear dimension of the system L (which was assumed to be infinite in the previous discussion) and
(iii) the characteristic length scale l_p of the generated pattern.

Global coupling implies $r_c \gg L$, while local coupling implies $l_p \gg r_c$. The third asymptotic regime corresponding to non-local coupling is given by $L \gg r_c \gg l_p$, although near the boundary between the last two asymptotic regimes, i.e. $l_p \simeq r_c$, types of behaviour peculiar to non-local coupling may also emerge. There may be an additional length scale r_0 which characterizes the 'interatomic distance' or 'lattice spacing', which is the typical distance of a given dynamical element from its nearest neighbours.

In this chapter we have worked with a continuum of limit-cycle oscillators by letting r_0 tend to zero. This idealization is allowed if $r_c \gg r_0$ and under such a condition, the non-local coupling may be called 'extremely long-ranged'. In effect, we supposed that each dynamical element interacts with a large number of other dynamical elements.

The statistical mechanics of spin systems, as well as other cooperative systems in thermal equilibrium, and the theory of Vlasov plasma tell us that under long-range coupling the mean-field picture gives a precise description of the dynamics. An analogy between globally coupled oscillators and Vlasov plasma was successfully used by Strogatz *et al* [35] to explain the order-parameter damping in oscillator systems analogous to Landau damping. Also, for non-locally coupled systems, we have taken full advantage of the mean-field picture, and the corresponding concept of *internal forcing* has been very useful in the analysis of our system and the interpretation of the results obtained.

Although non-locally coupled systems have been sidestepped in the past, presumably because of anticipated mathematical difficulties as compared to local and global coupling systems, such difficulties may, to a large extent, be resolved through the application of mean-field dynamics which is valid for quasi-continua of coupled dynamical elements.

Finally, we should mention two areas for future work in the field of non-locally coupled oscillatory systems. First, it would be desirable to develop experimental systems with similar structures to those in our models: for example, see the recent work of Vanak and Epstein [36, 37], which concerns the Belousov–Zhabotinski (BZ) reaction with water-in-oil aerosol OT micro-emulsion. This experimental system mimics a large population of micro-organisms with the use of BZ oscillators in the water droplets, and these oscillators interact via diffusive chemical components in the oil medium. By careful choice of the coupling agent, such experiments could form an ideal test bed for the study of the pattern dynamics peculiar to non-locally coupled systems.

Second, we should emphasize the application of non-local coupling to neural field dynamics, where non-local coupling has a *structural* origin, in contrast to the present study where *effective* non-locality arose from the elimination of a rapidly changing diffusive component. In the near future, the rich variety of spatio-temporal patterns observed in non-locally coupled oscillatory systems may provide new interpretations of the dynamics of cortical neurons which have been observed in physiological experiments.

References

[1] Battogtokh D and Mikhailov A 1996 Controlling turbulence in the complex Ginzburg–Landau equation *Physica* D **90** 84–95

[2] Battogtokh D 2001 Personal communication

[3] Barkley D 1992 Linear stability analysis of rotating spiral waves *Phys. Rev. Lett.* **68** 2090–3

[4] Chapman S and Cowling T G 1970 *The Mathematical Theory of Nonuniform Gases (Lectures in Statistical Mechanics)* ed G E Uhlenbeck and G W Ford (Cambridge: Cambridge University Press) ch IV

[5] Coullet P and Iooss G 1990 Instabilities of one-dimensional cellular patterns *Phys. Rev. Lett.* **64** 866–9

[6] Crawford J D and Davies K T R 1999 Synchronization of globally coupled phase oscillators *Physica* D **125** 1–46

[7] Cross M C and Hohenberg P C 1993 Pattern formation outside of equilibrium *Rev. Mod. Phys.* **65** 851–1112

[8] Deissler R J and Brand H R 1994 Periodic, quasi-periodic, and localized solutions of the quintic complex Ginzburg–Landau equation *Phys. Rev. Lett.* **72** 478–81

[9] Ermentrout G B and Kopell N 1984 Frequency plateaus in a chain of weakly coupled oscillators *SIAM J. Math. Anal.* **15** 215–37

[10] Golubitsky M and Schaeffer D G 1985 *Singularities and Groups in Bifurcation Theory* (Berlin: Springer)

[11] Guckenheimer J and Holmes P 1983 *Nonlinear Oscillations, Dynamical Systems, and Bifurcations of Vector Fields* (Berlin: Springer)

[12] Hakim V and Rappel W J 1992 Dynamics of globally coupled Ginzburg–Landau equation *Phys. Rev.* A **46** R7347–50

[13] Hansel D, Mato G and Meunier C 1993 Phase dynamics for weakly coupled Hodgkin–Huxley neurons *Europhys. Lett.* **23** 367–72

[14] Hoppensteadt F and Izhikevich E M 1997 *Weakly Connected Neural Networks* (Berlin: Springer)

[15] Keener J P and Tyson J J 1988 Singular perturbation theory of travelling waves in excitable media *Physica* D **32** 327–61

[16] Kuramoto Y 1975 Self-entrainment of a population of coupled nonlinear oscillators *International Symposium on Mathematical Problems in Theoretical Physics (Lecture Notes in Physics 39)* ed H Araki (Berlin: Springer) pp 420–2

[17] Kuramoto Y 1984 *Chemical Oscillations, Waves, and Turbulence* (Berlin: Springer)

[18] Kuramoto Y 1989 On the reduction of evolution equations in extended systems *Prog. Theor. Phys.* **99** 244–62

[19] Kuramoto Y 1991 Collective synchronization of pulse-coupled oscillators and excitable units *Physica* D **50** 15–30

[20] Kuramoto Y 1995 Scaling behaviour of turbulent oscillators with non-local interaction *Prog. Theor. Phys.* **94** 321–30

[21] Kuramoto Y in preparation

[22] Kuramoto Y and Nishikawa I 1987 Statistical macrodynamics of large dynamical systems *J. Stat. Phys.* **49** 569–605

[23] Kuramoto Y and Nakao H 1996 Origin of power-law correlations in distributed oscillators and maps with non-local coupling *Phys. Rev. Lett.* **76** 4352–5

[24] Kuramoto Y, Nakao H and Battogtokh D 2000 Multi-scaled turbulence in large populations of oscillators in a diffusive medium *Physica* A **288** 244–64

[25] Marsden J E and McCracken M 1976 *The Hopf Bifurcation and Its Applications* (Berlin: Springer)

[26] Mikhailov A 1992 Stable autonomous pacemakers in the enlarged Ginzburg–Landau model *Physica* D **55** 99–112

[27] Nagumo J S, Arimoto S and Yoshizawa S 1962 An active pulse transmission line simulating nerve axon *Proc. IRE* **50** 2061–71

[28] Nicola E M, Or-Guil M, Wolf W and Baer M Drifting pattern domains in a reaction–diffusion system with non-local coupling *Preprint* Server nlin.PS/0005043

[29] Nicolis G and Prigogine I 1977 *Self-Organization in Nonequilibrium Systems* (New York: Wiley)

[30] Ohta T, Hayase Y and Kobayashi R 1996 Spontaneous formation of concentric waves in a two-component reaction–diffusion system *Phys. Rev.* E **54** 6074–83

[31] Pearson J E 1993 Complex patterns in a simple system *Science* **261** 189–92

[32] Sakaguchi H 1993 Phase dynamics and localized solutions to the Ginzburg–Landau type amplitude equations *Prog. Theor. Phys.* **89** 1123–46

[33] Siggia E D and Zipperius A 1981 Pattern selection in Rayleigh-Bénard convection near threshold *Phys. Rev. Lett.* **47** 835–838

[34] Skinner G S and Swinney H 1991 Periodic to quasi-periodic transition of chemical spiral rotation *Physica* D **48** 1–16

[35] Strogatz S H, Mirollo R and Matthews P C 1992 Coupled nonlinear oscillators below the synchronized threshold *Phys. Rev. Lett.* **68** 2730–3

[36] Vanak V K and Epstein I R 2001 Inwardly rotating spiral waves in a reaction–diffusion systems *Science* **294** 835–7

[37] Vanak V K and Epstein I R 2001 Pattern formation in a tunable medium: The Belousov–Zhabotinsky reaction in an aerosol OT microemulsion *Phys. Rev. Lett.* **87** 1228301

[38] Winfree A T 1967 Biological rhythms and the behaviour of populations of coupled oscillators *J. Theor. Biol.* **16** 15–42

[39] Winfree A T 1974 Rotating chemical reactions *Sci. Am.* **230** 82–95

Chapter 10

A prime number of prime questions about vortex dynamics in nonlinear media

Art Winfree
University of Arizona

People have been fascinated by fluid flows and turbulence at least since such time as some neolithic ancestor, with a noisy wind in her ears may have marvelled before the spectacle of immense ripples rolling across a field of grass. I experienced just the same at Celtic Field in Clifton Down one afternoon during the meeting reported in this book. On both scales of time and space—acoustic noise and eddies of landscape proportions—an invisible continuum conserves momentum and energy under simple rules, but the outcome seems lawless. Fluid flow was also my own favorite mystery as an undergraduate engineer, in part due to its resemblance to physiological processes: turbulent flows seem almost 'alive'. After a century of scientific study and half that duration of mathematical and computational modelling, I think it is still safe to guess that students of such spatio-temporal dynamics have most of their historic adventures yet before them. This chapter relates one such adventure as a good prospect for the discovery of unforeseen principles, in this case again in a transparent three-dimensional continuum, but with rotating chemical and electrical patterns replacing momentum eddies, and with a unique pattern time scale rather than one that cascades from seconds to milliseconds.

The organizers of this symposium on 'Nonlinear Dynamics and Chaos: Where do we go from here?' invited examples of topics that have not yet benefited much from theory of nonlinear dynamics, with the special proviso that if applied mathematicians and physicists would have a look at them, they would probably make lots of discoveries. The topic I picked is the dynamics of phase singularities in motionless uniform continua that support limit-cycle oscillations or shock wave propagation. What 'moves' in such media is a pattern of chemical concentrations or voltages, not the medium itself (quite unlike fluids). Such media are simpler than fluids in that they constitute scalar rather than vector fields. They also have some qualities that prove indispensable for their unique behaviours but have no analogy in fluids. Several years ago numerical exploration of this kind of field equation turned up a dozen seemingly stable distinctive patterns that seem parametrically robust. So it seems reasonable to expect the like in a wide variety of excitable physiological and chemical media. To name just a few with characteristic scale around 1 cm, there are the flickering glows of oxidizing phosphorus vapor [43, p 301], oxidation waves in certain chemical gels [43, ch 13], and electrical depolarization waves in thick muscle of the pregnant uterus and in the ventricular muscle of the heart [43, ch 17] (figure 10.1). In the book you now read, chapter 11 by Bressloff and Cowan and chapter 12 by Ermentrout bring other biological examples involving the oscillations and waves in brain tissue. Computational examples range from the complex Ginzburg–Landau field through a wide variety of excitable media.

One might not expect such diverse media to have much in common. But in 2D they all do support a rotating spiral wave organized around a phase singularity. (Note that 'singularity' always means 'of the phase description': there is no discontinuity of any physical quantity.) In 3D this spiral becomes a scroll and the phase singularity becomes a singular filament. The filament typically closes in a ring. Rings typically link and knot. Thus we come to configurations of linked and knotted rings, called 'organizing centres'. This chapter addresses questions about their stability or persistence on various time scales.

I am only going to talk about numerical experiments in this chapter, experiments about a profoundly idealized excitable medium in the form of a simple partial differential equation. Just this modest beginning presented more mysteries than foreseen, almost all of which remain unsolved.

10.1 Stable organizing centres

By way of preview, figure 10.2 presents six of the ostensibly stable 3D configurations of phase singularities in one numerical excitable medium that supports propagation of shock-like waves. They are all wiggly vortex rings with invariant topological characteristics. You see them reproduced here on a uniform scale: the arrow shows the wavelength, λ_0, of the 2D vortex, the product of its rotation period and the shock propagation speed. The tubes bound regions of field rotation around the singular filament, out of which a shock radiates in the form of a spiral in 2D or a scroll in 3D. The tube drawn around each thin singular filament may be thought of as a rough level contour for the amplitude of fluctuations, which is of course zero right at the filament, the rotation centre. A contour could be chosen with arbitrarily small girth around this pivot point; here it is chosen as λ_0, at which radius the amplitude is almost up to that of the propagating shock fronts outside. The purpose of dressing the singular filament in such tubes is to show how closely such vortex tubes fit together: they get no closer than $0.4\lambda_0$, a little more than the tube diameter λ_0/π. These topologically distinct ways that any

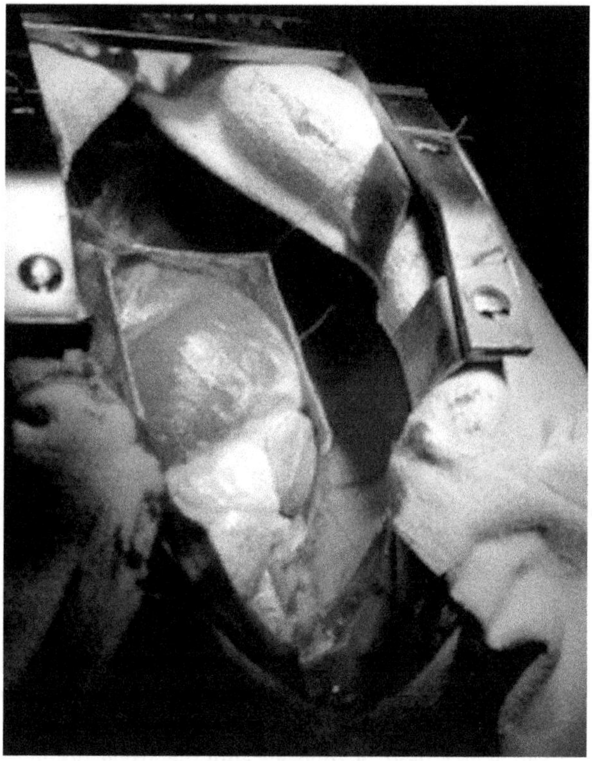

Figure 10.1. In April 1997 Frank Witkowski and I did the experiments presented in *Nature* [49, 50]. I took this snapshot from near the dog's head just after we opened its pericardium and just before removing the heart to Langendorff plumbing for observation. The ventricles are at the ostensible top here, and white fat covers the left atrium. The heart is 12 cm long, top-bottom. Its wall is only 1–2 cm thick, too thin to support such fanciful objects as discussed in this chapter. Contemplation of excitability in heart muscle merely initiated thinking about vortex rings; there is no reason to think closed rings are relevant to cardiac electrophysiology.

given uniform excitable medium can stay persistently self-organized in space and time are called 'organizing centres' because without them the medium becomes uniformly quiescent, but if one of these things is created anywhere, it persists as a source of roughly periodic shock waves that organize the reaction sequences throughout the whole volume. These persistent vortex rings are alternative modes of the same partial differential equation (PDE). Each such particle-like source of periodic waves has its own distinctive self-consistent geometry. Very little is yet understood about their dynamics.

Now let us back up and see where these pictures come from. Let us start with the idealized medium, the FitzHugh–Nagumo (FHN) model for electrical

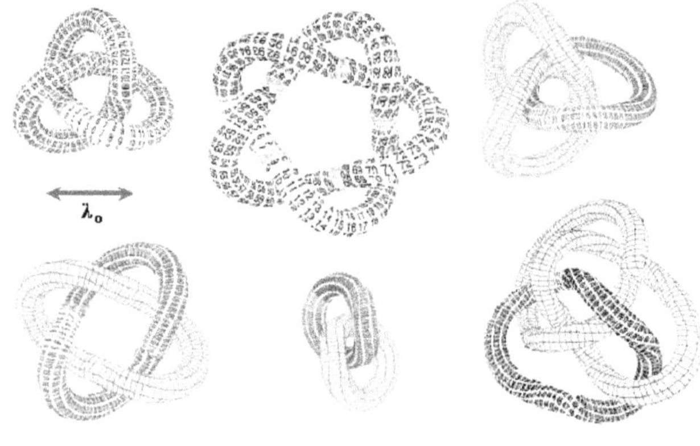

Figure 10.2. All these objects are better described later in this chapter, one by one. Here and in figures 10.8–10.13 the vortex tubes are all drawn with the same arbitrary diameter $= \lambda_0/\pi$. The drawings are adapted from the computations of Henze [11] using FitzHugh–Nagumo kinetics at a particular parameter setting.

dynamics in nerve or heart cell membranes (figure 10.3). What ideas lead to this sort of equation? A first idea is that the excitable tissue is an electrically resistive continuum, like salt water, with a time-varying electrical potential, V, at each point. So $\frac{\partial V}{\partial t} = D\nabla^2 V$, where the potential diffusion coefficient D (cm^2 s^{-1}) is the inverse product of intrinsic bulk resistivity (Ω cm) by capacitance per unit area of cell membrane occupying the volume (farad cm^{-3}). A second idea is that V corresponds to a potential gradient as large as 100 000 V cm^{-1} across the thin cell membrane, strong enough to deform the electrically charged parts of protein-built ionic channels embedded in that membrane. According to the deformation, they leak sodium and potassium ions into or out from the cell at various rates, adding to the time derivative of V as shown. A third idea is the cubic rate function. It comes from V not only changing sodium channel resistivity in a roughly quadratic way, but also providing the driving force for ions through that channel. Finally, a fourth idea: p here represents another ionic process, also involving protein pores in the membrane, which transpires very much more slowly than the shock of V depolarization. It entails potassium ions crossing the membrane in the opposite direction to sodium, so they have an effect on $\frac{\partial V}{\partial t}$ opposite to that of V, eventually effecting a recovery of excitability.

The literature presents many enormously elaborate dynamical equations of this general form, meticulously tailored to the many ionic channels of each particular kind of cell and to more realistic visions of how cells are electrically connected together. I invoke here the simplest generic representative called the

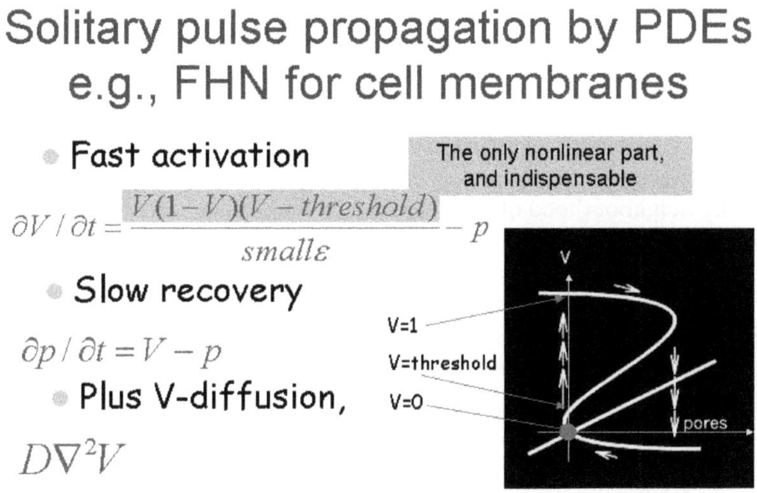

Figure 10.3. The reaction–diffusion equation (a scalar field equation) of a typical excitable medium. The thumbnail sketch illustrates nullclines: loci of horizontal flow $\frac{\partial V}{\partial t} = 0$ and of vertical flow $\frac{\partial p}{\partial t} = 0$ of the ODE part, the local kinetics without Laplacian coupling to neighbours. The key feature is that all vectors lead to a unique steady state, but from beyond a threshold distance away in the V direction (V for voltage) they first very rapidly lead farther away (this part makes the shock front propagate), then slowly return from the p direction (p for recovery process involving potassium protein pores). $D\nabla^2$ can be written as $\partial^2/\partial(x/\sqrt{D})^2$ etc, so we see that changing D changes nothing but the spatial scale of the solutions, supposing initial conditions rescaled by $\sqrt{D}$ and boundary conditions likewise or remote.

FHN model [9]:

$$\frac{\partial V}{\partial t} = \frac{1}{\epsilon} V(1 - V)(V - V_0) - p \tag{10.1}$$

$$\frac{\partial p}{\partial t} = V - gp \tag{10.2}$$

where $\epsilon \ll 1$, V_0 is the threshold $\ll 1$ mentioned in figure 10.3, and g (ignored in the figure) is $O(1)$. The equation is written in this form for transparency, but is actually computed from an equivalent version with more explicit parameters given later (see equations (10.4), (10.5)). Note the cubic nonlinearity; everything else is linear. The phase portrait of such an ordinary differential equation (ODE) pair is represented in the thumbnail within figure 10.3. The cubic (V for voltage) and linear (p for potassium protein pores) nullclines intersect in a globally attracting steady state (large dot) at the cell's 'resting potential'. A stimulus that displaces it far enough vertically ('depolarizing' the membrane) puts the system in a region of fast V increase (note the small ϵ in the rate equation) until it is caught below the cubic's upper nullcline branch. Then p evolves slowly to the right until a catastrophic drop to the lower branch, just above which the membrane leisurely recovers its excitability toward the unique steady state, and patiently awaits another stimulus large enough to pull it out of the immediate attracting neighbourhood of that large dot. Such stimuli can come from adjacent excited cells, because cells are resistively coupled by 'gap junctions'. Making a continuum average over microscopic discrete cells, this coupling amounts to adding a Laplacian operator to the V evolution equation (10.1); now we have the PDE

$$\frac{\partial V}{\partial t} = D\nabla^2 V + \frac{1}{\epsilon} V(1 - V)(V - V_0) \qquad (10.3)$$

and if the parameters are in the physiological range, it propagates the pulse-like excitation-recovery process as a shock at a characteristic speed on the order of $\sqrt{(D/\epsilon)}$. With D on the order of 1 $\text{cm}^2\ \text{s}^{-1}$ and ϵ on the order of $(1000\ \text{s})^{-1}$, this is roughly 30 cm s^{-1}. Such an impulse may perpetually circulate on a long enough ring of connected cells, which is to say the cells may circulate along the indicated arrows on the phase portrait in figure 10.3 like circus elephants linked trunk to tail. They do not just all pour, one after the other, into the attracting steady state, because they are linked in a ring, so a cell already caught up in the fast upward arrows will pull a neighbour up out of its immediate range of attraction, whereupon it too streaks away vertically, and pulls up its next neighbour. Thus such media are not just excitable, but also *re*-excitable after a sufficient interval. This need not be as long as the time needed to follow the arrows back to the neighbourhood of the attracting steady state: if ϵ is small so that the vertical motion is swift then re-excitation can occur from the lower branch far to the right of the steady state, and the drop down from the upper branch then tunnels through that nullcline's turn-around in the upper right with comparable prematurity. In this way arises the possibility of *rotating* shocks and, in two- or three-dimensional simply-connected continua, phase singularities (pivot points or curves) inside them. Figure 10.4 illustrates one such situation in 2D. (See the PowerPoint lecture from [51].) As illustrated, this pattern rotates rigidly, throwing out a spiral shock like the magnetic field around a spinning neutron star, in the form of an involute spiral with radial spacing equal to the circle's perimeter. Inside this evolute circle, called a 'rotor', ∇V is necessarily transverse to ∇p, and iso-V contours (contours of fixed V in this stopped-time snapshot) somewhere cross perpendicularly through iso-p contours. But outside this little disk around the pivot the V and p contours are both parallel to propagating wavefronts.

Another way to say this is that almost any chemical composition (V, p) found in such a medium is found everywhere along a curve (or in 3D, a 2D surface) behind the shock front that propagates forward at nearly uniform speed in the form of an involute spiral (or corresponding scroll-like surface in 3D). The

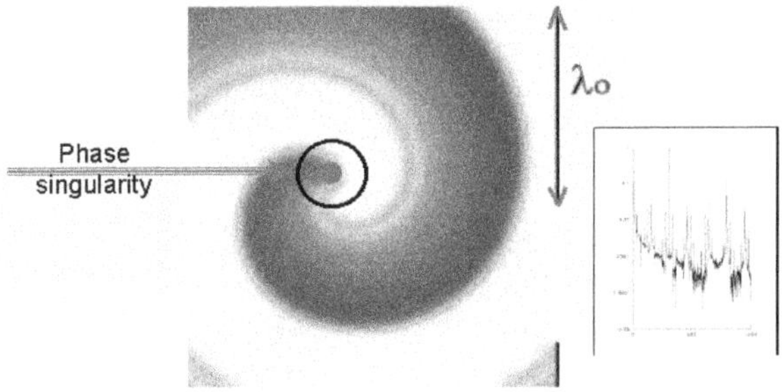

Figure 10.4. One frame from a FHN computation in 2D, from initial conditions arranged to evolve a phase singularity inside the black circle. Its diameter is λ_0/π of the involute spiral shock you see radiating/rotating away. Rotor motion is typically more complex than illustrated here: the zone circled typically 'meanders' quasi-periodically with at least one additional discrete frequency. Quite a few others are shown on a vertical logarithmic amplitude scale in the thumbnail (referring to a different medium, but quite like FHN.) This is not yet well understood.

set of such compositions makes a closed ring in the (V, p) state space. But compositions found well inside that locus in the state space instead correspond to barely moving points (or 1D space curves in 3D). They are all bundled together in the vortex core. In 3D this 1D space curve, being the intersection of an iso-V surface with an independent iso-p surface, generically closes in a ring.

The simple equations of (10.2), (10.3) determine the rotor's structure in figure 10.4 and, at such parameters as those used, impel its asymptotically rigid rotation. But it should be noted that at more typical parameters the grey rotor area around the pivot not only rotates but also translates along a path whose curvature is modulated with an additional frequency or collection of discrete frequencies. The thumbnail insert in figure 10.4 depicts one such spectrum (from a different excitable medium (see [43, p 463]) with richly coincidental relations between its discrete spectral peaks. These remain today, after many years of study, still unaccounted for.

This presents the first question of the ten invited by the organizers of the Colston Research Symposium:

Question 10.1. *How are we to understand the richness of these typical spectra?*

Chapter 9 by Kuramoto, which concerns rotors without phase singularities in non-excitable (oscillating) media, may provide a new approach to understanding this richness. If I understood his presentation correctly, numerical grid points near the centre of rotation become entrained to a newly bifurcated frequency, typically higher than that of the pre-existing oscillations. Oscillators near the centre become entrained to this new frequency because that is where its amplitude

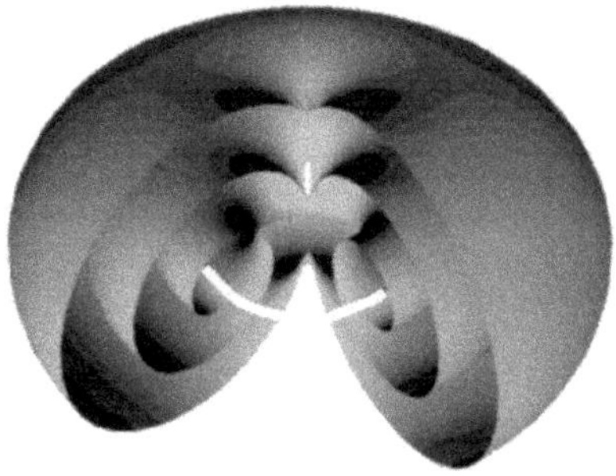

Figure 10.5. Here we see the scroll-like shock front sectored open to reveal its spiral cross section. This is half of the iso-V surface, the other half (not pictured, in the wake of the shock, where V is decreasing again) being a time-delayed version of the same. In 3D, the 2D centre of rotation or phase singularity necessarily becomes a vortex line or singular filament. It is the intersection of a certain iso-V surface with a certain iso-p surface, thus generically a closed ring, around each point of which the (V, p) field spins at (almost) the characteristic frequency, ω_0, of 2D rotors in figure 10.4. It will be seen naked again in figures 10.14–10.17 but figures 10.2, 10.8–10.13, 10.20, and 10.21 clothe it in a tubular sheath corresponding to the black circle at the source of the spiral in figure 10.4. This encloses a disk of rotating transverse V and p gradients, drawn at girth equal to the outer spiral wave spacing, λ_0.

is greatest, while the basic limit cycle's amplitude has a zero there. In *excitable* media the local dynamics has no limit cycle and does not entrain, but it does bifurcate a new frequency ('meander'), whose amplitude is high only within a couple rotor diameters from centre. Are these scenarios relatable? In chapter 9 by Kuramoto, if the disorderly scattering of dots near the centre of rotation persists in the continuum limit and somehow corresponds to a multi-frequency (hyper-) meander, might it provide a way to begin understanding the peculiar spectra observed in meanders?

The FHN equations without diffusion can be tuned to exhibit a limit cycle, but that is not done in this chapter. Our local dynamics is excitable but has no limit cycle. Moreover with diffusion added, the PDE supports rotors at a characteristic frequency, and with a number of additional frequencies if parameters are pushed across certain bifurcations. But that is not done, up to figure 10.14: all numerical experiments represent the narrow parameter-space domain of single-frequency rigid rotation. Only in figure 10.14 and beyond will we cross a supercritical Hopf bifurcation locus in the medium's parameter space to bring one additional frequency of the 2D rotor up to large amplitude as a sort of 'perturbation'. The intent is that this should provide something like thermal agitation to slightly expand the vortex rings and to prevent their hibernation into a comfortably compact exact symmetry.

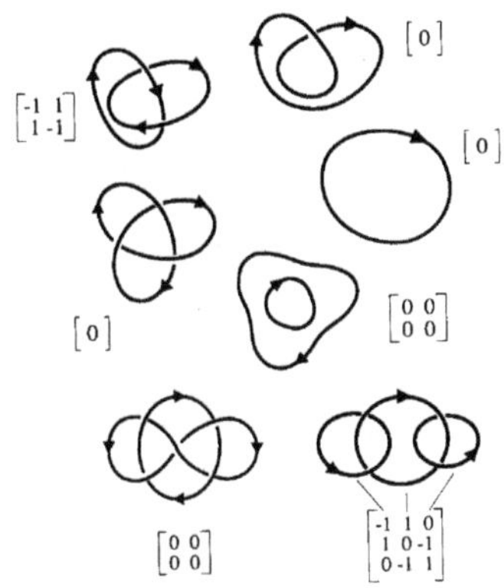

Figure 10.6. Vortex rings may link and knot without violating any known constraint of geometry, physical chemistry, etc. Each filament is given an orientation so that the direction of wave rotation around the filament is given by the right-hand rule. Matrices of linkage integers (positive and negative) are shown. Reprinted by permission from A T Winfree 1990 Stable particle-like solutions to the nonlinear wave equations of three-dimensional excitable media *SIAM Rev.* **32** 1–53.

Figure 10.5 extends figure 10.4 with cylindrical symmetry into 3D. The phase singularity, while a vortex centre point in 2D, becomes a circular filament in 3D. See the PowerPoint lecture from [51] for animation of these scroll waves.

Circular singular filaments have been known to exist in real chemical excitable media since 1973 [33], and to shrink to nothing in time like (initial radius)$^2/2D$. It happens that only the fast variable, V, is diffusing in most computations shown here; it makes little difference in practice, but shrinkage can be derived analytically in the limit of huge radius (compared to λ_0) if the slow variable, p, diffuses at the same rate as the fast variable, V. Specifically, the spontaneous rate of shrinkage of a *sufficiently large* ring would be close to $\frac{\mathrm{d}r}{\mathrm{d}t} = -D/r$, i.e. the area enclosed would shrink at rate $\frac{\mathrm{d}A}{\mathrm{d}t} = -2\pi D$ [24]. Another way to express this is to imagine the filament has a uniform, length-independent tension proportional to D, which drives each curved arc segment through an equivalent viscous medium at fixed speed D/r.

In practice even some *small* solitary rings do this very nearly in the laboratory and in the computer unless the numerical medium be only marginally excitable. But this law is not quite right for other cases. If the slow variable, p, does not diffuse, as in computations shown here, the coefficient (of the other variable) is not exactly D. And more importantly, if vortex rings are linked or knotted (figure 10.6), then shrinkage typically stops when (inevitably) transverse filaments get too close together (like the diameter of the nominal vortex tubes in figure 10.2). This gives us the second question, which will recur in several forms:

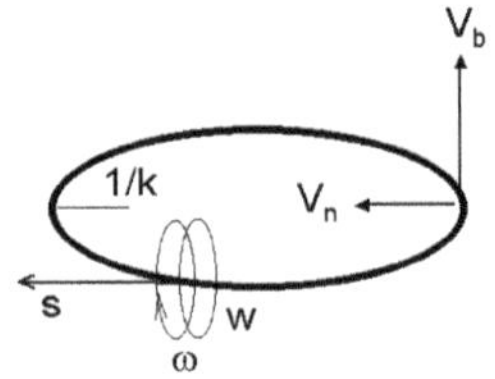

Figure 10.7. Labels for the essential quantities of a description of local filament motion in terms of local curvature, k, and twist, w, supposing the rotating concentration field looks the same, up to rotations, in every local normal plane near an elliptical filament. The two fine circles are supposed to suggest rotation at angular velocity ω with phase graded at w radians per unit distance, s.

Question 10.2. *What stops the shrinkage of filaments?*

This question leads to an interest in the 'laws of motion' of such singular filaments. A natural guess is that local motion would be determined by the local geometry of (V, p) distributed in normal planes everywhere along the filament as in figures 10.4 and 10.5. The simplest description is then in terms of $k(s)$, the curvature of the filament as a function of arc length along it, together with the phase of the concentration field's rotation, in a snapshot at fixed time. In figure 10.5 both are constant along the filament but they need not be so (figure 10.7). Averaging over a rotation period, the absolute phase proves unimportant, but its arc-length derivative, called its 'twist', w, remains important. It has no analogy in hydrodynamic vortices; in reaction–diffusion media such as those governed by equations (10.2), (10.3) it is real-valued and limited only by the requirement that a circuit of the ring must restore the starting phase. Then local filament motion is conveniently described in the filament's Frenet frame oriented by its tangent and its osculating circle. Resolving this motion into a normal component of velocity, V_n, toward the local centre of curvature, and a binormal component, V_b, we ignore any component along the tangent as being unobservable in principle. (A caution: these subscripted Vs mean velocity components, not the local intensive variable, Voltage.) Then question 10.2 as formulated in 1987 became: 'How do local V_n and V_b depend on local k and w?' [45]. From symmetry considerations it would seem that twist could enter this act only as an even function, and, as previously noted, argument from the form of reaction–diffusion PDEs [24] said $V_n = Dk$ when k is small, $w = 0$, and all diffusion coefficients are equal (i.e. if equation (10.2) has an additional term $D\nabla^2 p$ on its right-hand side, as it indeed does when we model excitability in chemical solutions rather than in cell membranes). Question 10.2 then becomes: 'Is that all?'

A surprising answer came from J P Keener [15] in a form abstract enough to apply not only to such excitable media as figure 10.2, but even to fields of coupled limit cycle oscillators or harmonic oscillators such as the complex Ginzburg–Landau PDE. Keener showed that k and w also affect the local frequency of rotation, that $\frac{dw}{ds}$ is an important causal factor and that in the limit of small k and w this results in a dynamic that consists of three linear combinations of these three quantities, with nine undetermined coefficients that are believed to depend only on the local ODE and the diffusion matrix:

$$\begin{aligned} V_n &= \underline{b}_2 k + \bar{c}_2 \frac{dw}{ds} - \bar{a}_2 w^2 \\ V_b &= c_3 k + \bar{c}_4 \frac{dw}{ds} - \bar{a}_3 w^2 \\ \Delta\omega &= \bar{c}_1 k + \underline{b}_1 \frac{dw}{ds} - a_1 w^2 \end{aligned} \tag{10.4}$$

My efforts in 1988–91 to estimate these coefficients numerically gave consistent results only for the three coefficients of k in the movement equations and the coefficient of w^2 in the spin-rate equation. The rest kept coming out helter-skelter on both sides of zero, suggesting maybe they are zero in this approximation and some other influences affect filament motion when twist and curvature are larger in relation to spiral wavelength. Further pursuing Keener's ansatz analytically Biktashev *et al* [5] argued that five of the nine coefficients must be zero in every case (with small k and w as usual), specifically the overlined coefficients in (10.4), $a_{2,3}$ and $c_{1,2,4}$. Both Keener and Biktashev *et al* agree that if all reactants (e.g. V and p in figure 10.3) diffuse equally, then coefficients $c_{1,3}$ must be zero, and two others, the underlined coefficients $b_{1,2}$, are equal to D. A surviving innovation in (10.4) is the idea that b_1 is not zero. Note that in equations (10.1), as amended in (10.3) and (10.2), and their numerical solutions in figures 10.8–10.17, only V diffuses. But if both V and p diffuse equally, the scheme (10.4) simplifies back to $V_n = D/R$ (toward the centre of curvature R space units distant), $V_b = 0$ (no drift perpendicular to the plane of the curve), and $\Delta\omega = (D\frac{dw}{ds} - a_1 w^2)$, where $-a_1$ is $O(D)$, with w expressed in radians per space unit. So as far as filament *motion* is concerned, we are back to the original formulation of Panfilov and Pertsov [24], in which twist was neglected or implicitly supposed to affect only spin rate, at least for a solitary isolated filament. Note also that in this equal diffusion case the law of motion plainly precludes persistence of rings, because every arc always moves directly toward its local centre of curvature, faster and faster as the ring shrinks and its curvature increases. Presumably the meander frequency also depends on local geometry—in a similar way?

Question 10.3. *For the isolated barely curved and barely twisted filament presupposed in (10.4), how to evaluate the single undetermined coefficient, a_1 (and all the others, in case of general D matrix, e.g. with only the fast exciting variable diffusing as in figure 10.2), short of empirical measurement?*

Some algebraic estimates of a_1 are attempted in [41]. One is $-QD/8\pi$, where Q is the dimensionless 'quality factor' of any rotor, $\lambda_0^2\omega_0/(2\pi D)$. Here subscript 0 denotes the unique 2D rotor and the spiral wave it radiates; in 3D curvature and twist somewhat alter the corresponding values. And D means the diffusion coefficient of the species with fast kinetics, which propagates the excitation. For excitable media in which *only* the fast species is free to diffuse, this Q has theoretical minimum value 6π and in the diverse laboratory examples known (including many with multiple species diffusing) stays $O(10^2)$ while physical parameters range over a dozen orders of magnitude [43, p 278] (for 3D graphs see PowerPoint lectures from [51]). Exception: *marginally* excitable media, in which Q goes to infinity as parameters are adjusted to make rotors become inviable just before propagation also fails. This estimate of a_1 is only for small w, and only for media in which frequency times propagation speed is constant, as is expected for small ϵ; the coefficient doubles for dispersion-free media, as expected in marginal excitability. In a range of small ϵ called the *Fife regime* Margerit and Barkley [20, 21] provide analytically exact answers for arbitrary w at $k = 0$. Augment their coefficient by about half again and you have the result determined in [41] using a numerical model of the Belousov–Zhabotinsky medium; augment that by about half again and you have my $-QD/8\pi$. The only estimate from laboratory experiments in the Belousov–Zhabotinsky medium is about half the Margerit result, but it is uncertain by at least that factor [41].

I usually prefer to rewrite the dynamics (10.4) in space units equal to λ_0, in time units equal to $2\pi/\omega_0$, and in angle units of 2π rather than radians, and to make the coefficients (here capitalized) likewise dimensionless by providing appropriate multipliers that depend only on dimensionless Q:

$$
\begin{aligned}
QV_n &= B_2k + \frac{2\pi}{\sqrt{Q}}C_2\frac{\mathrm{d}w}{\mathrm{d}s} - \frac{4\pi^2}{\sqrt{Q}}A_2w^2 \\
QV_b &= C_3k + \frac{2\pi}{\sqrt{Q}}C_4\frac{\mathrm{d}w}{\mathrm{d}s} - \frac{4\pi^2}{\sqrt{Q}}A_3w^2 \\
Q\Delta\omega &= \frac{\sqrt{Q}}{2\pi}C_1k + B_1\frac{\mathrm{d}w}{\mathrm{d}s} - 2\pi A_1w^2.
\end{aligned}
\tag{10.5}
$$

It is to be understood that the six variables here differ from those used in (10.4) under the same names in that the units (and so the numerical values) are different. As before, five of the nine coefficients (C_1, C_2, C_4, A_2 and A_3) are zero according to Biktashev, and in the case of equal diffusion coefficients so is C_3 while $B_1 = B_2 = 1$. In any case my approximate value for a_1 becomes here $A_1 = 1/8\pi$. The original lower case dimensional coefficients are obtained from these dimensionless uppercase ones by multiplying C_1 by $\sqrt{(}D\omega_0/2\pi)$, B_1, B_2, C_3 and A_1 by D, and C_2, C_4, A_2 and A_3 by $\sqrt{(}2\pi D^3/\omega_0)$. Although no new physics has entered here, the advantage of this mere translation is that (to whatever extent such equations pertain to organizing centres) one can now think about filament dynamics as though organizing centres were all about the same size, with waves propagating at unit speed, and rotors turning in unit time. I imagine that the new upper case coefficients are also less sensitive to parameter variations (they are completely independent of multipliers of diffusion, and of course of the arbitrary choice of physical units), and that Q has become the principal parameter, but this guess has not been tested.

Note that the computations reviewed here followed figure 10.3 in suppressing diffusion of p, so, following Keener [15], the coefficient c_1 of k of (10.4) is *not* near zero. Biktashev *et al* [5] find it to be zero anyway, but this is only for small k, w, not for the present case. It was in fact measured in the particular medium of figures 10.8–10.13 [6] as -0.28 rad space-unit/time-unit, or in natural units (turns, wavelengths, periods)[1], close to *shockspeed*/Q, thus substantially *de*creasing the spin rate.

Since 1986 numerical experiments on linked and knotted rings with 'realistic' $k(s)$, $w(s)$ and D matrix had been showing something unforeseen in the small-(k, w) limit: they typically approach large stable $k(s)$ and $w(s)$ in a compact arrangement of no-longer-shrinking rings. Something has arrested the initial shrinkage expected and observed in arbitrarily large rings. We are back to the original form of question 10.2: 'What stops them?' I think the culprit is a violation of the assumption underlying (10.4), that the neighbourhood of the filament in every normal plane to the filament looks like the middle part of figure 10.4, so that 3D anatomy differs locally from 2D anatomy only in the factors depicted in figure 10.7, namely the slight tilt, k, and rotational shear, w, that transform one normal plane to the next. Rather, I think what is happening (but it is hard to be entirely sure, as the old computations on record did not highlight this feature) is that a distant segment of filament invariably intervenes in this tidy geometry in a non-local way, as we will see later under topic 'Dynamic B'.

Let us look at some of the stopped configurations. They are easily started by seeding 3D computations of (10.1) plus Laplacian as in (10.3), and (10.2), or similar reaction–diffusion PDEs, with appropriate initial conditions. Computation for [6] and figures 10.8–10.13, used a slightly rearranged version of the FHN model:

$$\frac{\partial V}{\partial t} = \nabla^2 V + \frac{3V - V^3 - 3p}{3\mu_1} \tag{10.6}$$

$$\frac{\mathrm{d}p}{\mathrm{d}t} = \frac{\mu_1}{2}(2V - p + 2\mu_2) \tag{10.7}$$

with parameters $\mu_1 = 0.3$ and $\mu_2 = 0.7$. Translated in terms of the original format (10.2), (10.3), at these parameters $\epsilon = 0.01$, $V_0 = 0.022$ and $g = 1.53$. This is a peculiar choice: μ_1 (proportional to $\sqrt{\epsilon}$) is so big that the medium is only marginally excitable; the intersection of nullclines is so close to the knee

[1] All the relations presented here benefit from such conversion but I leave them in basic units to facilitate checking for dimensional consistency.

of the cubic that a limit cycle is about to arise by Hopf bifurcation; the 2D rotor repels from no-flux boundaries to a distance of about a half wavelength; there are two possible kinds of stable rotor with quite different frequency and wavelength; both rotate rigidly; and the link-free, knot-free 3D vortex ring shrinks only to a certain size rather than shrinking faster and faster toward collapse as previously indicated [6]. In figures 10.14–10.17, parameter μ_1 is changed to 0.2 and μ_2 is changed to 0.9 to sample more generic behaviour [39]. The computations are all done by explicit Euler integration in cubic arrays at least $3\lambda_0$ wide so the Neumann boundary conditions demonstrably have no effect on rings shrinking toward the middle. The volume is finely enough sampled by the numerical grid so that shock waves propagate isotropically at close to the speed expected analytically or from computation on finer 2D grids. The initial conditions (ICs) are described analytically in terms of complex-valued functions with the right symmetries (for details see [36, appendix A] and [43, ch 16] or papers cited there.) The complex number then points to (V, p) values stored on the plane of a pre-computed spiral wave [12]. It should be noted that these functions typically have a number of perfect symmetries (though this is not necessary: slightly distorted ICs still evolve toward similar asymptotic solutions), and so do the solutions evolved from them unless the symmetry is unstable and some slight perturbation is provided to nucleate the breaking of symmetry. This normally comes from the coarseness of the numerical grid itself, but it is not provided in the special case that the symmetry of the ICs coincides with that of the cubic grid.

In any case, symmetry seems stable in the next six figures. These are the six panels of figure 10.2. They do not show the naked filament but rather surround it by an octagon in the normal plane at each of 100 equispaced stations along the filament where we constructed Frenet frames and quantified both local geometry and local filament motions. The octagon may be thought of as roughly a level contour of amplitude for V or p fluctuations near the zero-amplitude phase singularity, the pivot for the locally rotating patterns. The station numbers are indicated to show the orientation of the filament: forward means the rotation ahead looks clockwise. The octagon's perimeter is λ_0; it is oriented so one corner and the next are on the normal and binormal axes so the surface texture indicates torsion of the Frenet frame. (Other publications have shown instead an iso-V or iso-p level surface as in figure 10.5; for example, see the PowerPoint lecture from [51] for a movie of the scroll waves surrounding a trefoil-knotted singular filament. But the figures printed here merely show a nominal boundary of the vortex core inside which the transverse concentration fields rotate and outside which behaviour is better described as propagation at fixed speed normal to the iso-V surface.)

Figures 10.8–10.11 show four configurations of compactly interacting vortex rings that eventually (about 20 vortex rotation periods from our symmetric but otherwise imperfect initial conditions) move like a rigid body with inertia. All are centred in cubes with Neumann boundary conditions $3.1\lambda_0$ apart, wide enough that we could neglect influences from the no-flux side walls. The floor and ceiling

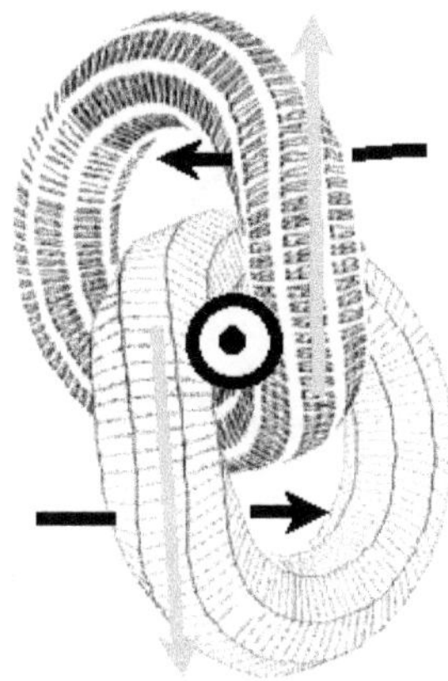

Figure 10.8. A pair of identical linked vortex tubes surround interior rings oriented (light arrows) so the right hand rule indicates shock circulation direction (darker arrows). It happens that isolated rings move two orders of magnitude slower than shock propagation through the interior of the ring. So these two would soon collide, except that filaments somehow repel and avoid one another, resulting in rigid rotation of the pair anticlockwise while they comparably slowly approach the observer as indicated by the face-on arrowhead superposed in the centre. After evolving through scores of rotation periods from initially larger diameters, each ring stopped shrinking while still nearly circular, with perimeter close to $\pi\lambda_0$. The central hole is about twice the diameter needed to accommodate the linking vortex tube. Both rings being linked once, each necessarily bears one cycle of left twist [35, 46]. With either filament orientation reversed they would both be right-twisted and the object would be the mirror-image of this one. With *both* reversed we get back again the same object as viewed from the other side; there is no left-twist/right-twist pair [46]. Note that in mirror-imaged fields all the anatomy behaves as expected, and so do shock propagation vectors, but filament orientation, being the cross-product of two gradients, does not. Note also the stable ring's diameter, λ_0. It happens in this unique medium that even the twist-free isolated circular ring with cylindrical symmetry stabilizes at a small diameter, just under $0.5\lambda_0$. But all rings shown in this chapter are bigger than that, not just because they must accommodate at least one threading tube and so must exceed diameter $(2/\pi)\lambda_0$; they actually exceed it by more because ring perimeter rather than area seems proportional to the number in linking filaments (see later). So their mechanism of resisting further shrinkage is different from that of the link-free, knot-free cylindrically symmetric ring that, in this peculiar medium alone, stabilizes before shrinking to extinction. I imagine that this link- or knot- dependent mechanism involves Dynamics A, B and C speculatively presented later.

implement periodic boundary conditions and the observed glide of the organizing centre is oriented by choice of initial conditions to exactly that direction.

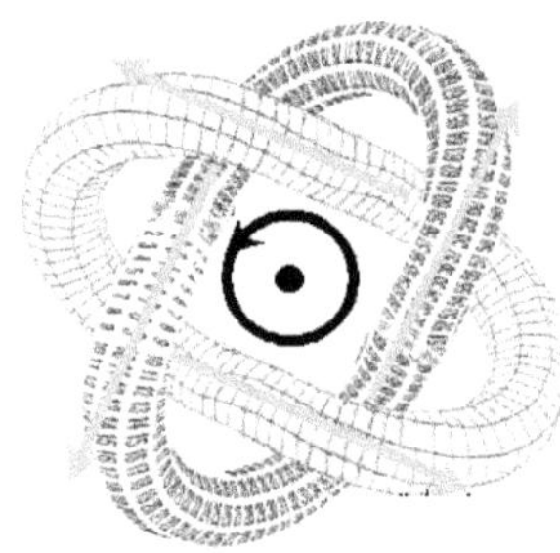

Figure 10.9. Like figure 10.8 but each ring is *twice* linked by the other, accordingly bears *two* cycles of left twist, and is about twice as long (5.76 wavelengths of the 2D spiral, i.e. $5.76\lambda_0$), so it even more generously encloses the two vortex tubes. Its maximum local twist, $w(s)$, stays just below a certain threshold discussed later under 'Dynamic A'. These rings are conspicuously non-circular and non-planar, but have almost no Wr (as in all linked rings in figures 10.8–10.11.) As in figure 10.8, there is also a mirror-image version with opposite linkage, twist and writhing. There can also be distinct versions in which one or the other ring's orientation is reversed (these are mirror-images), thus reversing linkages but leaving twist and writhing free to adjust in other ways. These have never been computed.

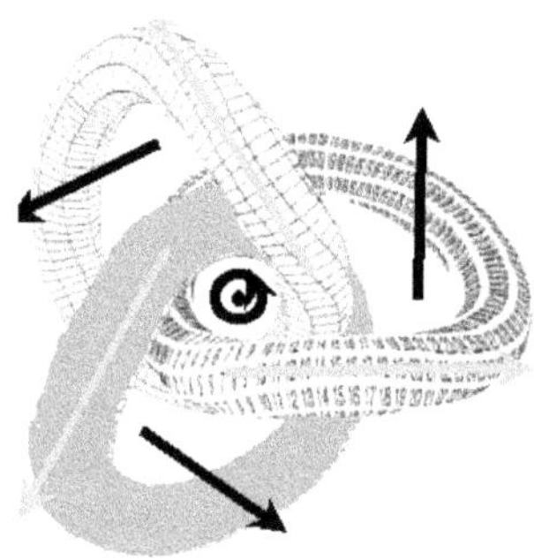

Figure 10.10. Like figure 10.8 but each of three rings is twice linked by the other, accordingly bears *two* cycles of left twist as in figure 10.9, and is only a little shorter ($4.95\lambda_0$) than in figure 10.9 ($5.76\lambda_0$). The maximum $w(s)$ in this and all such solutions stays a bit under the sproing threshold. Three-fold symmetry persists despite its incompatibility with the cubic numerical grid. These rings glide/precess about the z-axis. Because they are tilted and well off centre, about one-third of each ring is moving opposite to the positive V_b observed in isolated rings. As in figure 10.8, there is also a mirror-image version and there is a distinct version in which two rings are linked by the others in opposite directions, and so bear no total net twist, while the third ring bears two twists (and a mirror-image of this). These also have never been computed.

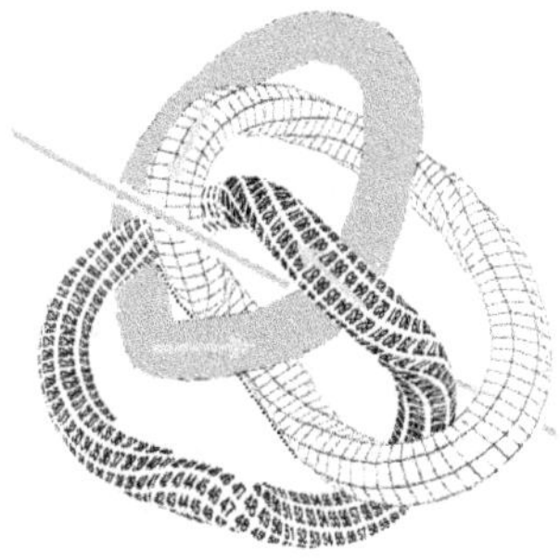

Figure 10.11. Like figure 10.10 but these three rings evolved from three flat 2:1 ellipses configured as Borromean rings [52], so named from the crest of the Italian Borromeo family. They are $6.7\lambda_0$ long and, while inextricably entangled, are not pairwise linked. (In imagination remove the shaded one: now you can pull the other two free from one another). Integration around any ring accordingly accumulates no *total* twist, Tw, in one circuit (however w^2 locally waxes and wanes even more extravagantly than in figures 10.8–10.10). This object slowly glides/precesses as a rigid body along the long arrow. As in figures 10.8–10.10, all three rings have evolved identically from their identical initial conditions, to this asymptotically stable configuration. There is also a mirror-image version. As in prior figures, these rings evolve from initially planar configuration without increase of Wr, even though they do develop torsion out of the plane and are neither mirror-symmetric nor inscribed on the surface of a sphere. No set of nine coefficients in (10.4) describes the motion of figure 10.11, nor do coefficients fitted to figures 10.8–10.14 agree with one another.

Question 10.4. *Is it really true that these solutions asymptotically approach a unique shape that transforms in time by mere rotation and translation, at least in media that support rigid rotation of the vortex in 2D?*

Experience is limited. Computations have gone only up to about 100 vortex rotation periods. While this generously exceeds collapse time for link-free knot-free rings in the same medium, it falls far short of reaction–diffusion characteristic time L^2/D if L is taken to mean the perimeter of the knot. A knot's perimeter seems to be settling asymptotically in early computations [12, figure 6] but who would dare predict the future in the absence of sound theory? And a skeptic could reasonably argue that adding even a little persistent random noise to such calculations might expose an instability. The possibility is also not excluded that the final shape is, to some extent, determined by initial conditions or that it deforms plastically under such perturbations as arise from transient encounter with a wall or another travelling ring. Rebound to a preferred shape should be tested in the medium of figures 10.8–10.13 by modestly shearing the numerical grid then resuming calculations. Preliminary trials demonstrated rebound only for the cylindrically symmetric twist-free solitary ring [6], which is unique in that its stability does not involve twist or the putative mechanisms mainly discussed in this chapter. The only trial on a knotted filament (a decade ago, unpublished) induced a lasting change of shape (!). This needs re-doing at longer duration.

Suppose (10.4) were as true of highly curved and twisted filaments as of the cosmic-scale rings for which it was derived. Then any stable structure must have $\Delta\omega$ independent of s, with spiral shock waves outgoing at identical frequency from every arc. They must then collide at internal collision interfaces that nowhere touch the filament. These resemble a fusion of the tubes drawn in figures 10.8–10.13. With all frequencies equal, is their position uniquely determined? I suspect not. Moreover this required uniformity of $\Delta\omega$ constrains the linear combination of k, w, and $\frac{\mathrm{d}w}{\mathrm{d}s}$ to two degrees of freedom. What combinations are compatible, through the V_n and V_b equations (10.4), with the observed asymptotically rigid glide/spin of a closed curve whose shape is prescribed by $k(s)$ and torsion $\tau(s)$?

Note that torsion plays no role in (10.4), and in the ring integral of $w(s)$ there is an additional constraint involving torsion, namely that this integral must be the linkage integer minus the writhing. Writhing, Wr, is a real-valued global measure on a space curve, related to the arc-length integral of torsion or to the area swept out on a unit sphere by the curve's tangent vector [47]. $Wr = 0$ for closed curves that are mirror-symmetric or confined to the surface of a sphere or plane. If (10.4) had simplified to the 'local induction approximation' of fluid mechanics, $V_b = Dk$, rather than to $V_n = Dk$ as in equal-diffusion excitable media, then Wr would have had more transparent importance as a quantity conserved under that motion [18].

In figures 10.8–10.13 the organizing centre rigidly glides, two orders of magnitude slower than waves propagate in this medium, along a straight path. Meanwhile it precesses comparably slower than the rotor spins. The glide is along the direction of periodic boundary conditions, toward the viewer, except in Figure 10.11, where the spin/glide axis of helical progress is instead along the cube's body diagonal and the viewpoint is almost but not quite down that axis. The arrangement of the pattern of figure 10.11 was initialized as perpendicular ellipses, whereupon they started shrinking while slowly drifting in the direction of shock propagation through the inside of each ring (apply the right-hand rule to filament orientations indicated by the pale arrows). In the present view, this is somewhat toward the viewer and somewhat anticlockwise. Then each encounters a second ring, cross-braced by a third, bends it, and is bent by the cross-brace. Shrinkage (V_n) stops and parts of the ring move with $V_b > 0$ while other parts move with $V_b < 0$. These motions completely violate (10.4); there is no combination of nine coefficients that comes close to connecting the quantitatively observed rigid-body motions to the quantified local k, w^2 and $\mathrm{d}w/\mathrm{d}s$ [11]. This brings us to the next question:

Question 10.5. *What to do if twist* cannot even be consistently defined*? This comes up because it turned out [12] that the rotation of concentration gradient*

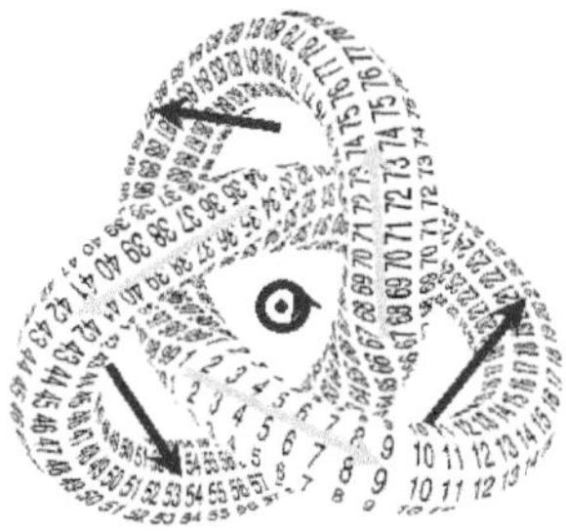

Figure 10.12. Once again at the same parameters of the FHN model 10.6 and 10.7 as in figures 10.8–10.11 and in the same generously sized box, the well-evolved compact left trefoil knot (2:3 torus knot) has perimeter $8.9\lambda_0$. This is about $3\lambda_0$ per penetration, as in figure 10.8. Tubes drawn at standard girth $= \lambda_0$ are almost tangent at three places but elsewhere have lots more 'elbowroom'. As in the prior figures, the tube here is embossed with equispaced station numbers 1...99 at which we evaluated differential geometry and motion; they look oppositely oriented in [12] only because the left knot in that different excitable medium is viewed from the other side. Its Wr, $w(s)$, $k(s)$, and motions (normalized to λ_0) are quite similar (contrary to the unaccountably mistaken assertion in [43] near the bottom of page 486. [The next page, incidentally, notes the beginnings what might be a long-term instability in this knot]). There is, of course, also a right knot, e.g., see figure 10.21 from earlier computations using a piecewise-linear kinetics rather than FHN.

vectors at many places along the filament is far from uniform during any one cycle, and not the same for V and p! This again indicates the intervention of a non-local influence.

Figures 10.12 and 10.13 present single rings that are knotted. They can be shown [47] to necessarily have zero full turns of $Wr + Tw$ (Tw being the ring integral of local twist, w) and sure enough, each does. They also have substantial, in fact equal and opposite, Wr; figure 10.12 for example resembles three planar semicircles connected by right-angle torsions, and resembles three turns of a helix, with $Wr = 3.3 \times 2\pi$ rad and so necessarily with the opposite (negative) amount of (left) twist, which is confirmed numerically. This twist is not uniformly distributed, as figure 10.21 will show, and this turns out to be essential for the dynamical stability of this knot. Figure 10.13 shows that more than one kind of knot can shrink to a stable configuration. The methods of [36] and [25] readily adapt to initializing every kind of torus knot, but I would love to know how to make other kinds. So here is our next question:

Question 10.6. *How can we write complex-valued IC functions for arbitrary knots, and which ones (all?) will prove viable as persistent particle-like solutions to the reaction–diffusion PDE of excitable media? The only two tried so far*

Figure 10.13. A stable left 2:5 torus knot, with perimeter $14.1\lambda_0$ (again approaching three wavelengths per self-linkage) and equal and opposite Tw, Wr near 5.7 full turns. It is slowly gliding toward you while precessing anticlockwise like all the foregoing. There is also a mirror-image version. I express organizing centre sizes in terms of λ_0 because of the belief that they would be the same in those terms in any other medium characterized by the same Q: equations (10.5) may not be pertinent to stable or persistent organizing centres, but they do suggest a uniform scaling with $Q\lambda_0$.

both persist and one persists in four rather different excitable media. But I only know how to do $m : n$ torus knots, and iterated torus knots (cable knots) [25]. Can 'random' initial conditions make links and knots, e.g. if the medium, while propagating shocks without phase singularities, were stirred or were diced and the cubes rotated at random or shuffled?

In the course of the studies that culminated in such pictures (1985–1994) my students and I inadvertently used plenty of topologically correct but quantitatively awkward initial conditions in a variety of distinct excitable media. In many such cases evolution under the reaction–diffusion PDE did *not* preserve the topological 'invariants'. Rather, bits of filament came too close together and cross-connected. Here question 10.2, 'What stops them?', arises again to spawn our next question:

Question 10.7. *What processes transpire if they* do not *stop? It has been known for two decades that topology and geometry provide no constraints against filaments joining to reconnect hyperbolically if pushed into contact [48] and that they also can just pull through one another. In numerical excitable media both have been observed to occur spontaneously. This leads back to Question 4: Are these solutions really asymptotically stable?*

10.2 Persistent organizing centres

Might the ostensible stability of the solutions shown in figures 10.8–10.13 be deceptive, maybe a residue of symmetric initial conditions? Or might it be as delicately dependent on peculiar parameters as the stability of the cylindrically symmetric scroll ring is known to be in the same medium [6]? To answer this question I gradually changed the parameters of the FHN excitable medium (μ_1 from 0.3 to 0.2, μ_2 from 0.7 to 0.9) until the 2D vortex ceased to rigidly rotate, but instead began to meander along a looped quasi-periodic path. The direction of this extravagant looping reflects the changing phase difference between the two frequencies. This difference and direction vary along the filament, partly because both frequencies are somewhat affected by local curvature and twist [42]. This would appear to test the stability of the formerly symmetric compact structures of figures 10.8–10.13, much as thermal agitation might. However, there is one important difference: a symmetric discrete computation is going to continue with exactly the same symmetry even if unstable, unless seeded with at least a small symmetry-breaking perturbation. This is generally provided by the discreteness of the numerical grid, but not if the initial conditions have a compatible symmetry, as they did in figures 10.9 and 10.11. Unless we invasively intrude by mapping all concentrations into a new array through some rotation or distortion, or introduce intermittent random noise[2], the only symmetry-breaking factor comes from the fact that the Laplacian is computed by a discrete algorithm that proceeds in one fixed direction through the cube, irrespective of symmetries in the initial conditions. Then the non-associativity of floating-point arithmetic in its terminal bits (see [43, p 491] and my web site [51]) can, in principle, start mirror-image computations diverging if their initial symmetry is unstable.

This peculiarity of floating-point arithmetic did not suffice to trigger symmetry-breaking in the case of figure 10.14. This shows two twice-linked rings started (at the parameters given in the caption, beyond the bifurcation to vortex meander) from the same initial conditions as figure 10.9 (there at parameters supporting only rigid rotation in 2D). These IC exactly replicate 3D blocks of numbers at eight locations in the array, but translated and rotated through multiples of 90 degrees. They would be handled identically but for the fact the Laplacian algorithm encounters them from different sides. This slight discrepancy did not trigger an instability in figure 10.9 and it evidently still does not at the more provocative parameters of figure 10.14: each ring remains 180-degree-symmetric through the vertical axis and the two rings remain the same but for a 90° rotation. An object with this same topology and symmetry arose on another occasion when I ineptly scaled the IC for a pair of rings mutually linked three times (as in a Star of David). This had 180-degree rotational symmetry about the z-axis. At two diametrically opposite points where these rings crossed at short range, they fused. Then there were two wiggly rings, one twice the length of

[2] Or massively perturb a short segment of the filament. I did this in [42] to test the notion that symmetry was very ready to break, if only it could get started. It did not, so I concluded there, incorrectly, that the even or odd number of rings is a decisive determinant of symmetry. My mistake lay in applying the perturbation too locally and not waiting long enough for the perturbed computation to diverge from the control, and in overlooking the preservation of symmetry in the three Borromean rings before I rotated them from body diagonal to upright. I thank Peter Ashwin for inducing me to reconsider this right after my talk.

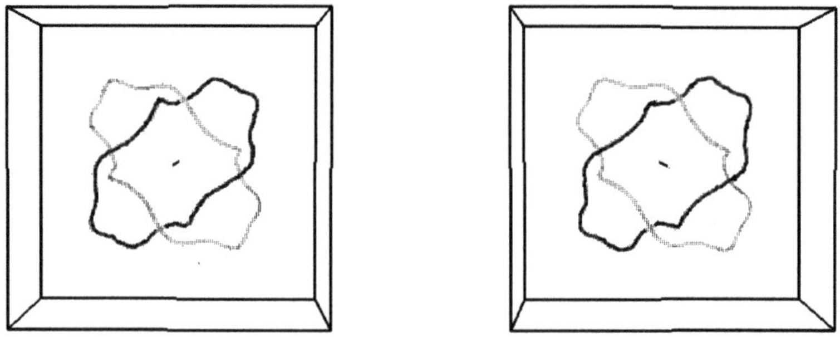

Figure 10.14. Like figure 10.9 but in a bigger box ($3.6\lambda_0$ wide) and in stereo for eyes crossed. (If you have difficulty fusing the two images, hold a pencil-point halfway between nose-bridge and page, and focus on it: this should direct your left eye to the right image, and right eye to left image. Adjust slightly so they snap into 3D fusion.) The computation was done at FHN parameters precluding rigid rotation around a phase singularity. For figures 10.14–10.17 parameter μ_1 of (10.6), (10.7) and figures 10.8–10.13 changed from 0.3 to 0.2 and parameter μ_2 changed from 0.7 to 0.9. Another difference: only the naked vortex line is depicted here, not sheathed in tubes as in figures 10.8–10.13. It is the intersection of a particular level surface of V with a particular level surface of p, though here it is picked out computationally by a sharp peak in the magnitude of $\nabla V \times \nabla p$ (which vector also points along the ring's tangent) [34]. One ring is distinguished by darkening. Forty vortex periods after changing parameters, you see strange ring shapes still preserving symmetries. A major perturbation of one ring did not quickly trigger symmetry-breaking instabilities (but maybe a different perturbation would have; or was a longer wait needed? Characteristic time L^2/D is hundreds of rotation periods.) See the PowerPoint lecture from [51] for a video. Reprinted from A T Winfree Persistent tangles of vortex rings in excitable media *Physica* D **84** 126–47 ©1995, with permission from Elsevier Science.

the other, mutually twice linked. Each preserved its perfect symmetry during scores of vortex periods. I presume that by applying an asymmetrical distortion or perturbation all along each ring, I could have induced this organizing centre to spontaneously diverge from symmetry (but I didn't think of it, 8 years ago). Figure 10.8 was not subjected to the parameter change, but it would presumably have had the same trouble relinquishing its initial symmetry.

In contrast, in figures 10.15, 10.16 and 10.17 the three rings have spontaneously evolved to become asymmetric and in distinctive ways. All are in cubes $3.6\lambda_0$ on edge; λ_0 is somewhat longer at these parameters, and the box was additionally relatively enlarged a bit because the wiggly rings take up more normalized volume.

Question 10.8. *Movies corresponding to figures 10.15–10.17 have been cited in support of the impression that, while the linkage and knottedness 'quantum numbers' seem to be constants of motion (apart from rare discrete transitions),*

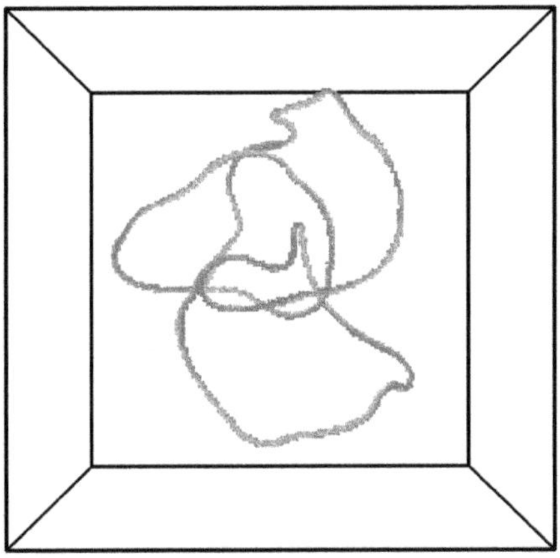

Figure 10.15. From initial conditions like figure 10.10, but in a bigger box ($3.6\lambda_0$ on edge) and 40 rotation periods after crossing the bifurcation to meander, naked filaments of finite thickness are colored red-green (see colour plate 1) across their diameter in the direction of the V gradient in this snapshot. Along the perimeter of any ring two alternations of color visible from any perspective reveal the two full twists required by co-oriented linkage with two other rings. The three have evolved distinctive wiggliness. See the PowerPoint from [51] for a video. Reprinted by permission from A T Winfree 1994 Persistent tangled vortex rings in generic excitable media *Nature* **371** 233–6 ©1994 Macmillan Magazines Ltd.

no exact shape is seen again in the same or another ring. This seems plausible, but is it strictly true, and if so how might it be interpreted? At each station s along the filament the two frequencies (of 2D vortex rotation and of its meander) are slightly modified in 3D by local k, w^2, and $\mathrm{d}w/\mathrm{d}s$ [42] but an un-checked possibility is that the filament approaches an asymptotic shape in which the local frequencies accordingly cease to change, and so local wiggliness is simply quasi-periodic.

10.3 Igniting resistance to curvature-induced contraction

The remainder of this chapter summarizes a conjecture about a mechanism that might be dominantly responsible for the persistence of these particle-like solutions to the nonlinear reaction–diffusion equation. It could be checked immediately by computation, but I have not done any since abandoning PDEs completely in 1995 to try my hand instead at instrument design and construction (and use) for observation of vortex filaments in chemical gels [44] and in heart muscle [49, 50]. And someone more skillful than I could probably check this notion by quantitative analytical modelling.

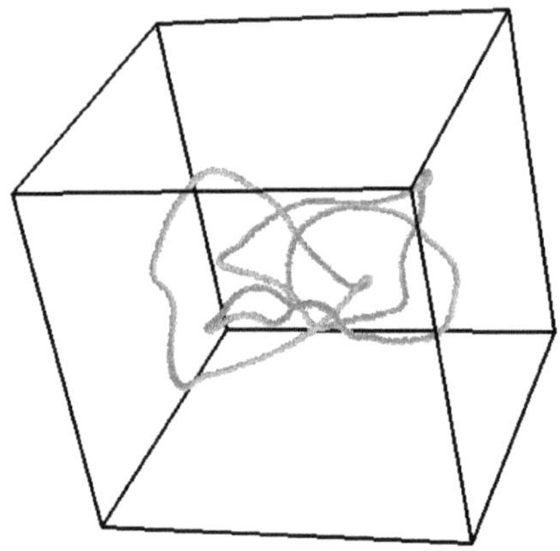

Figure 10.16. As in figure 10.15, but from figure 10.11 (Borromean rings). As in figures 10.9 and 10.14, the initial conditions had the same symmetry as the cubic numerical grid (e.g. see the three Robinson sculptures outside the Newton Institute in Cambridge [53] and they preserved that symmetry (while meandering) until I cut out a spherical region and rotated it to reorient the former body diagonal along the new grid's z-axis, interpolating values to fit the new cubic grid. Then the solution's three-fold rotation axis was no longer aligned to the cube's diagonal and three-fold axis, but to its edge and four-fold axis: the solution's identical parts thus became differently discretized, and the three rings divergently evolved much as they did in figure 10.15 without any such provocation; see also colour plate 1. Reprinted by permission from A T Winfree 1994 Persistent tangled vortex rings in generic excitable media *Nature* **371** 233–6 ©1994 Macmillan Magazines Ltd.

I think the key factor is twist. There is nothing to be twisted[3] in fluid vortex rings, nor have the suggested initial conditions [36, pp 243–4] been implemented in fluids, nor I think even in computations, to find out how they evolve. (This could be question 10.9, were I not trying to stop at a prime number.) So this story about linked and knotted vortex rings may apply *only* to the motionless media addressed here. It is observed in numerical experiments that rings generally do have non-uniform $w(s)$, even if no twist is topologically required (e.g. if the integer linking number, Lk is zero or if $Lk = Wr$ or both, as in figure 10.11, for example). Especially in linked or knotted rings, the ring integral of $w(s)$,

[3] There is in acoustic, electromagnetic, and quantum-mechanical vortex lines. Just as we do here, Berry and Dennis [4] use the crossed-gradient criterion (caption 10.14) and the topological principle connecting linkage and twist [35, 46] to infer an important role for twist in the acoustical and optical wave fields. But one difference should be noted: such singularities are passive consequences of phase circulation on a large scale, just as a tornado is a consequence of weather patterns, not their cause. In contrast, the phase singularities we deal with here are shock wave sources and causal organizing centres.

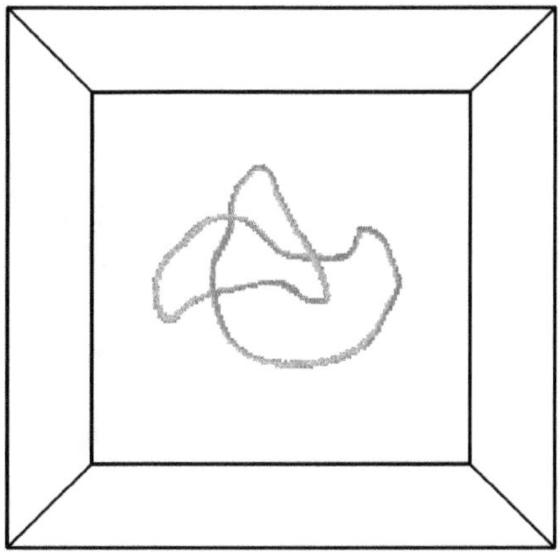

Figure 10.17. Seventy vortex periods after gradually changing parameters as before, the isolated trefoil has long ago quit changing its average shape, but its parts continue wiggling asymmetrically as it glides and precesses through the medium on a time scale two orders of magnitude more leisurely than shocks propagate and the vortex rotates in normal planes along the filament. See the PowerPoint lecture from [51] for a video; see also colour plate 1. Reprinted by permission from A T Winfree 1994 Persistent tangled vortex rings in generic excitable media *Nature* **371** 233–6 ©1994 Macmillan Magazines Ltd.

Tw, seems seldom if ever zero. Note that $Tw = Lk - Wr$, both of which global measures are size-independent if shape is conserved while a ring shrinks. Because each filament arc moves toward its local centre of curvature, the ring does shrink and so $w^2(s)$ generally increases along some arcs of the ring. Then three things happen, which I call Dynamics A, B and C:

10.3.1 Dynamic A: sproing

When a threshold of local twist, $w(s)$, is exceeded on a long enough arc of straight, uniformly twisted filament, that arc 'sproings' into an expanding helix with the same spatial periodicity (figure 10.18), just as observed in twisted elastic rods and in magnetohydrodynamic filaments. This threshold seems between 0.15 and 0.175 radians per space unit, which is 0.5 to 0.6 turns per wavelength at the parameters used in figures 10.8–10.13. This seems to be a maximum beyond which $w(s)$ as observed in all organizing centres depicted earlier never transgresses. Similar behaviour has been observed in a variety of other such media; the threshold is typically about 0.5 to 0.7 full turns per wavelength distance along the filament (figure 10.18), possibly becoming arbitrarily large in marginally excitable media with very large Q, and possibly going down to zero in media with the smallest possible quality factor, $Q = 6\pi$. (These thresholds

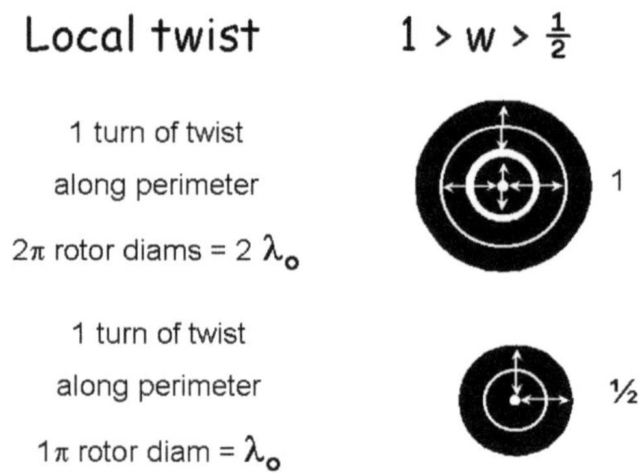

Figure 10.18. A perfectly circular vortex ring threaded by a single vortex line would have a total twist of one full turn. (Note the units: radians are used in the text.) If threaded perpendicularly, and distributing its local twist, w, uniformly, then $w = 1$ turn per wavelength of perimeter (lower, smaller diagram). If the threading tube has the same diameter $= \lambda_0/\pi$, then $w = 0.52\pi/(2\pi\lambda_0)$ (upper, larger diagram). This is the sproing threshold in the FHN medium used for figures 10.8–10.13. These diagrams suggest that the sproing threshold of 0.5 turn per wavelength discovered on straight filaments is not too high to be encountered on rings before they have shrunk to the minimum possible size. In reality this much twist also develops on bigger rings because twist, never uniform, causes gradients of spin rate and so causes increasing twist.

were measured on straight, uniformly twisted filaments; initially curved filaments probably have a somewhat offset threshold.) In each of the seven[4] topologically distinct, twisted, compact organizing centres known in that medium $|w(s)|$ comes almost up to that value along their most twisted arcs. Beyond that, a helical segment begins to expand and so opposes further curvature-induced contraction while it converts some Tw to Wr.

Sproing at a comparable threshold was confirmed in chemical gels [22] and in complex Ginzburg–Landau numerical fields [23, 28], where the transition proves to be made through a supercritical Hopf bifurcation. A corresponding derivation has not yet been attempted for excitable media, so we have our next question:

Question 10.9. *There is no theory of 'sproing' where first discovered, in excitable but non-oscillatory media. Is it a supercritical Hopf bifurcation there too? If the vortex meanders, so the twisted filament cannot be straight but must be initially helical already, is the threshold lower? Can it be zero or less?*

[4] The seven distinct organizing centres are those of figure 10.2 and a Star-of-David pair mutually thrice linked; there is also the cylindrically symmetric single circle but it has no twist, and the helix, but it is not compact.

For discussion of some beginnings of such theory, attempted in rapport with numerical experiments on the Oregonator model of chemical excitability by adding *ad hoc* nonlinear terms to the linear local theory of figure (10.4), see [16] and [41, section 5].

In figure 10.11 the knot could be drawn tighter as in [13, 14] were only a positive 'tension' of the filament acting against a hard tube, but that metaphor is evidently inappropriate to this case. It seems noteworthy that the area inside each ring threaded in figures 10.8–10.11 is at least *several times* the nominal area of that many tubes of unit wavelength in girth, and the multiple varies from case to case. It seems that the perimeter of the engirdling ring is *not* minimized by a tension in connection with the number of tubes it must accommodate, but has more to do with twist and sproing. Ring perimeter must accommodate a topologically required total Tw determined by the manner of knotting and/or the number and orientations of linking rings, while nowhere exceeding the sproing threshold of 0.5 times 2π radians per wavelength. In figures 10.8–10.11 things so arrange themselves that the average twist $\overline{w}(s)$ is 0.3 to 0.4 times 2π radians per wavelength. This makes all linked or knotted rings much bigger than needed to girdle the corresponding number of linking tubes. But what if a ring be linked by an even number of filaments with opposite orientations, so the induced twists cancel out to leave $Tw = 0$, as in the Borromean rings? It seems that where each such filament passes close to the engirdling ring its rotary phase field does impose a substantial twist on that nearby filament. Even though twist does sum to zero along the full perimeter, it also does persist locally, guaranteeing a certain minimum perimeter by the principle that sproing threshold is nowhere exceeded.

10.3.2 Dynamic B: non-local repulsion

The spin rate of the vortex depends on local twist, e.g. $\Delta\omega = (QD/8\pi)\, w^2$ (plus a much smaller $b_1 \frac{\mathrm{d}w}{\mathrm{d}s}$ with $b_1 = D$ in case of equal diffusion, plus another term, $c_1 k$, with $c_1 < 0$ for this medium and $c_1 = 0$ in media with equal diffusion). This means more twist, whether positive or negative, makes for higher-frequency shock trains radiating from that arc of filament. No counterexamples are known in any excitable medium, though it must be admitted that for the two FHN media of figures 10.8–10.17 there has been no attempt to demonstrate it numerically or analytically. Charging brazenly ahead anyway, the next essential observation is that twist generally does not become uniform along the filament and tops out somewhat under the nominal sproing threshold of π radians per wavelength. So the radiation frequency (initially: see later) should differ between twist-free arcs and maximally twisted arcs by no more than about $\Delta\omega = (QD/8\pi)\,(\pi/\lambda_0)^2$, with QD defined as $\lambda_0^2\omega_0/2\pi$, so $\Delta\omega = \omega_0/16$. (Contributions from $\frac{\mathrm{d}w}{\mathrm{d}s}$ are typically an order of magnitude smaller, and those from k slightly *de*crease spin rate.)

For every point along a closed-ring filament there must be another point sufficiently far away, and a curve between them inside the ring, where their two

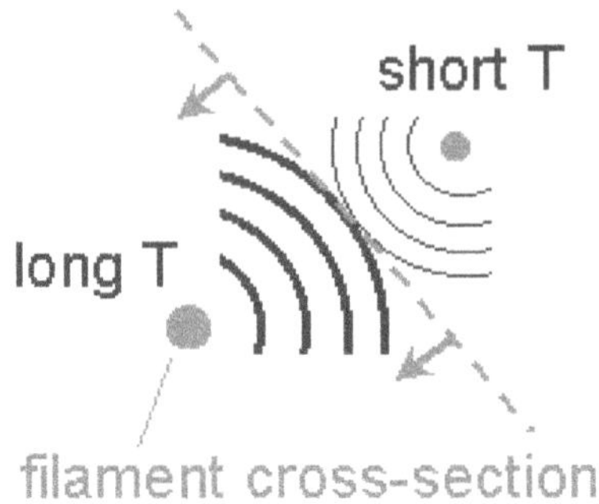

Figure 10.19. In 2D, periodic sources of relatively low (lower left) and high (upper right) frequency emit activation fronts that collide along the broken diagonal line. If speed were independent of frequency then the next collision would clearly occur midway between the next fronts in line to collide, thus on the side of the longer wavelength. For small frequency difference and speed dependence, it moves in the same direction at speed = fractional frequency difference × half the shock propagation speed. Something similar happens in 3D, between skew filament segments. The result is that shocks from the higher-frequency source impact the slower rotor at intervals shorter than its rotation, phase-shifting it and displacing it away from the source. With sources distributed along curves in 3D rather than localized at points as in 2D, things are a bit trickier in that not all of the collision interface's velocity projects directly along the line connecting filament arcs of extreme frequency.

radiations collide. Unlike the waves of linear physics, the shocks of nonlinear excitable media propagate without attenuation (replenishing their energy by local excitation like a foraging army or a grass fire), providing a distance-independent coupling across the field. They also collide and extinguish one another like grassfires. That collision interface inevitably moves away from the higher-frequency source (figure 10.19) at speed proportional to the frequency difference, $\delta\omega$: speed = $\mathit{shockspeed}/2 \times \delta\omega/\omega_0$, which is the same thing as $\delta\omega \times \lambda_0/4\pi$. It moves at least until it is directly slapping the lower-frequency source. When a 2D rotor is directly impacted by an excitation front, it is transiently disorganized, then reforms with a slight phase shift, displaced slightly at some fixed angle to the direction of propagation [19]. This leads to our next question:

Question 10.10. *How far and in what direction is this recurrent displacement, depending on the phase and spin rate differences, the 3D skew angle of attack, and the nature of the medium? See [43, pp 328, 490–2] for starter references.*

Each place s on a lower-frequency arc of filament is slapped outward by the impact of an unattenuated shock front from some other place s' on a higher-frequency arc. In 2D situations the displacement could be about a rotor diameter once per rotor period times the fractional frequency difference, $\delta\omega/\omega_0$. This happens to be $2/\pi$ of the collision interface speed, so I will estimate rotor displacement speed as the collision interface speed in section 10.6. In 3D this would suffice to oppose further curvature-induced contraction and so to keep skew filaments apart unless they are extremely curved toward one another. I attempt to quantify this under Dynamic C.

This Dynamic B is a bit peculiar. Previously [40,41] I called it 'non-local', in the sense that happenings along some arc of filament directly, and in a distance-independent way (but with a time delay proportional to distance, less than one half rotor period in the case of known persistent organizing centres), affect only some other disjoint arc of filament. This seems to be virgin territory for analytical modelling.

The foregoing interpretations stand in some want of evidence in as much as [11], from which figures 10.8–10.13 are adapted, contains no mention of 'slapping' nor of twist-dependent spin rates in that peculiar medium (but it is not clear that as little as 1 part in 16, as previously estimated, would easily be resolvable among radiations Doppler-shifted by filament motions observed to be comparably fast relative to shock propagation speed). Both *were* clearly involved in the rather different excitable medium used to make a stable knot [12] and the one used to make persistent meandering organizing centres of diverse sorts in [42], but these observations do not settle the question whether their involvement is indispensable.

Observation of the Borromean rings may be pertinent. They need have no twist, and because they are not linked, the arc-length integral of twist must be zero (more accurately, equal and opposite to the writhing, but that happens to be near zero). Yet the stable configuration develops as much twist (positive and negative equally) as the other organizing centres. I imagine that if there were less, the three rings would continue shrinking until they reconnect at symmetrically disposed cross-overs, develop substantial writhing, and convert that writhing to twist, ending perhaps as a stable trefoil knot.

Vinson [31, 32] confirmed both twist and slapping in a numerical model of the chemically excitable BZ reaction, but only in twist-free symmetric vortex rings, whose local spin rates differ not by twist but only because of an imposed parameter gradient.

It may be relevant that, in some media, a symmetry-breaking instability allows one of a pair of closely adjacent rotors to slap the other away at a steady pace, despite there being *no* difference between their native frequencies before recurrent slapping induces one by phase-shifting the slapped rotor. This continues in exactly the same way even at an arbitrarily large distance (e.g., see [37] about the expansion of helices by this mechanism, and [1, 29, 30]).

So, in short, I see the creation of persistent organizing centres as something like the formation of stars by igniting thermonuclear reactions: curvature impels ever-faster contraction[5] but only until the topologically locked-in total twist becomes compressed enough to nucleate a countervailing expansion by Dynamics A and/or B. Whether this is routinely followed on a longer time scale by 'supernova' has not yet been asked.

So we now have the first of three bonus questions, beyond the invited ten:

Question 10.11. *Why should this balance of influences promote stability (figures 10.8–10.13) or at least persistence (figures 10.14–10.17) of linked and/or knotted rings?*

First of all, it does not always. For example in the case of equal diffusion that so simplifies (10.4) (to $V_b = 0$, $V_n = D/R$), an equal-diffusion trefoil knotted ring blatantly violates that law by expanding without apparent limit after an initial shrinkage to critical twist intensity (supernova) [42]. So why should (10.4) plus Dynamics A and B *sometimes* promote persistence? I think mainly because the slap–jump mechanism of filament repulsion suffices to prevent skew intersections unless the segments involved have nearly the same spin frequency and sufficient curvature or other local-geometry factors noted in (10.4) to overcome the slight speed of slap–jump separation. This notion needs quantitative testing, which can be undertaken in earnest only after question 10.10 is answered in the form of a dynamical theory of slap–jump, and numerical means are created for observing the motion of collision interfaces.

And this brings us finally to equally speculative Dynamic C, without which also I think stabilization could not occur.

10.4 Dynamic C: synchronization

Rotors must eventually spin at the same pace all along every ring in any stable organizing centre. As previously mentioned and according to (10.4) they might do so if some fixed linear combination of curvature, twist and its arc-length derivative were realized at every point. But it is not: when the numerically evolved rings are plotted in $(k, w^2, \frac{\mathrm{d}w}{\mathrm{d}s})$ coordinates they do not lie on one plane, nor even near similar curved surfaces. What may happen instead is that highly twisted filament segments in the interior of the organizing centre come to be as close together as possible (about $0.5\lambda_0$), then something like the volume-averaged magnitude of concentration gradients is minimized by 'meshing gears' to synchrony. They no longer spin independently. The less twisted outer lobes of nearby rings are more widely spaced from nearest neighbours and might have become progressively more twisted due to arc-length gradients of spin rate, but instead they are forced to the same periodicity by recurrent shocks from the higher-frequency synchronous interior. Does this conjecture find support in the limited available numerical results?

[5] Again, except in certain marginally excitable media.

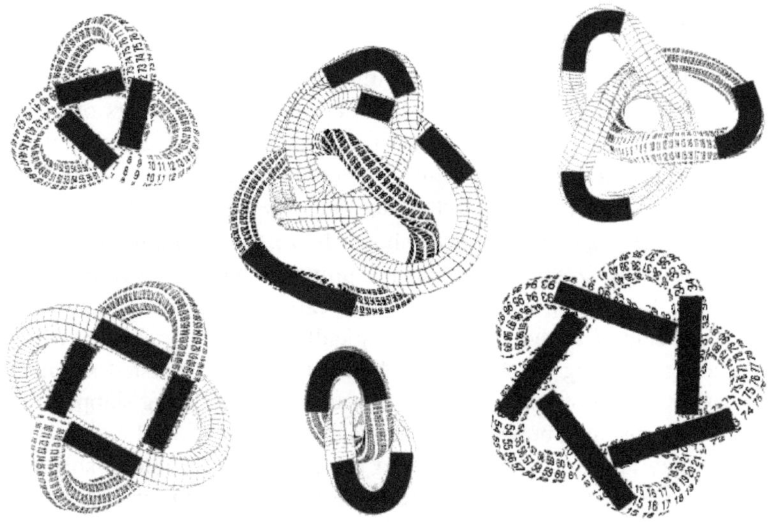

Figure 10.20. This is figure 10.2 with band-aids over the segments whose twist (whether positive or negative) approaches the observed extremum, near the sproing threshold measured on uniformly twisted straight filaments in this same medium. In three cases the band-aids are where 'expected', on relatively straight inner segments imagined to synchronize one another and drive more remote segments at this maximum frequency. However, in three other cases the inner segments have less twist and/or more curvature (which diminishes spin rate) than the outer arcs on which I therefore reluctantly place the band-aids. I hope I have made some mistake here.

Figure 10.20 repeats figure 10.2 with dark overlays where I conjecture, based on the $w(s)$ and $k(s)$ measured in [11], that the unmeasured spin rate would be greatest. In the three with interior straight segments (from figures 10.9, 10.12 and 10.13) these *are* the most twisted segments, but in three others (from figures 10.8, 10.10 and 10.11) curved segments on the *outer* arcs are more twisted, and so 'ought' to have the highest frequency, at least if these numerical outcomes with large w and k and no diffusion of p are anything like the equal-diffusion case of small w and k as analytically approximated in (10.4). Spin rates were not measured in these calculations of a decade ago, so we do not know whether figure 10.20 or the guesswork that motivated its construction is more correct. But if the sources of maximum-frequency shocks *are* really as far apart as they look in three cases here, it is hard to see how perfect synchrony could be maintained, and so hard to see how the structure could be stable. This conundrum might be called:

Question 10.12. *Is Dynamic C really an essential principle in stable organizing centres?*

10.5 Ball-park estimation of pertinent quantities

Can we exclude the kind of interpretation proffered here on grounds that the magnitudes of putatively balancing factors are necessarily orders of magnitude different, or that any such balance must be unstable? Let us try some rough estimates based on a simpler medium: with equal diffusion $c_1 = 0$ and $b_1 = D$ in (10.4). Then let us cheat by pretending that $D\frac{\mathrm{d}w}{\mathrm{d}s}$ is relatively slight (as it is, in fact, in all the organizing centres shown here), so w^2 alone determines the spin rate: $\Delta\omega = (QD/8\pi)w^2$ radians per unit time with w given in radians per unit distance. Suppose a non-circular ring bears non-uniform twist, $w(s)$, that sums to a constant Tw radians of topologically required total twist (implicitly approximating writhing $= 0$). If we think of the most twisted segment as a source of waves, and of some less twisted curved arc as being slapped by those waves, then the balance between outward slapping and inward curvature-driven contraction may appear to be unstable: if the curved arc is able to shrink then it becomes more curved and shrinks faster; if not, it is slapped further outward and becomes even less curved, and so is slapped away still faster.

But there is another way to think of it. Suppose these arcs are parts of one ring. Imagine the whole ring expanding or contracting to scale, with scale factor f. Total Tw remains unaffected but local twist $w(s)$ becomes $w(fs)/f$ and local radius of curvature $R(s) = 1/k(s)$ becomes $fR(fs)$. Suppose expansion or contraction reflects a balance between an arc around s_2 contracting inward at *inspeed* $= D/fR(s_2)$ while being slapped away by waves from fs_1 at higher frequency $w_{\max}$. The collision interface moves from s_1 toward s_2 (maybe not directly), eventually pushing arc s_2, at *outspeed* $= (\Delta\omega(s_1) - \Delta\omega(s_2))\lambda_0/4\pi$. This makes sense only until the interface reaches s_2 and starts slapping it away, but let us suppose that s_2 then backs off at that same speed. (If faster it would soon desist; but [7] argues that it might retreat more slowly in marginally excitable media.) And for simplicity take $\Delta\omega(s_2) = 0$ as though s_2 has no twist and take $\Delta\omega(s_1)$ at the $w_{\max}$ along this particular ring, i.e. s_2 is wherever $(\text{twist})^2$ is maximum. *Inspeed* and *outspeed* then balance if

$$\frac{D}{fR} = \frac{QD}{8\pi}\left(\frac{w_{\max}}{f}\right)^2 \frac{\lambda_0}{4\pi}.$$

If scale factor f fluctuates to enlarge the ring and dilute its twist then contraction inspeed $D/(fR)$ falls but the collision interface outspeed falls more, so a contraction ensues: the balance at $f = 1$

$$\frac{1}{R} = \frac{Q}{32\pi^2} w_{\max}^2 \lambda_0$$

can be stable at least to fluctuations of overall scale. Note that like (10.5) this again suggests that organizing centres should scale with λ_0 for diverse excitable media that all have the same quality factor, Q.[6] Is this plausible, given what

[6] In this connection I withdraw the jumbled sentence in [43, p 475, lines 9–12 from bottom].

little we know? We know that dimensionless Q is 41.6 for the medium used in figures 10.8–10.13 (and 51.4 in the next four figures) and that w_{max} cannot exceed sproing threshold π/λ_0 radians per space unit in that medium. We do not know that w_{max} peaks up all the way to the sproing threshold, but it comes close in the one medium for which extensive computations are available, so I adopt that value here. Brashly plugging in these numbers we see that $R = (32/Q)\lambda_0$ or $0.8\lambda_0$. In fact, the radius of curvature typical in those figures is in the range of $2\lambda_0$ down to $0.4\lambda_0$ (nowhere near as tight as the theoretical minimum, $\lambda_0/2\pi$), with some arc in all of them as sharply curved as $0.6\lambda_0$. So this estimate seems not implausibly outside the ball-park. Whether it has any other merit I do not yet know.

Simple as it is, this rough estimation has some testable implications. For example, if the internal traffic of shock waves should be disturbed, e.g. by interposing a slim barrier between two arcs of filament within a link, in the big empty spaces between tubes, then the organizing centre must be expected to rearrange itself dramatically. This would not occur were every arc of filament moving according to local geometry rules. And any medium in which slapped rotors jump *toward* the slapping source rather than away (as for example the marginally excitable medium computed in [7]) should be incapable of supporting stable organizing centres. Similarly, media with much larger dimensionless quality factor, Q, e.g. marginally excitable media such as used in [7], would either require shorter filament curvature radii than possible, and so not support stable organizing centres or (which I doubt) have a lower threshold for sproing in terms of turns per wavelength distance along the filament. Conversely, media with $Q = 6\pi$, the theoretical minimum for single-diffusion excitable media, would either have much larger organizing centres (in wavelength units) or exhibit higher sproing thresholds (the opposite of my impression from a few examples.)

Next, some comparable first estimate should be cobbled together for the two components of motion transverse to the movement of the collision interface, hopefully suggesting an answer to the mystery that all this complicated geometry ends up as a rigid-body precession and glide at fixed rates two orders of magnitude below vortex spin rate and shock propagation speed, respectively.

Now the skeptic speaks: Such loose back-of-envelope reasoning neglects many other degrees of freedom than mere scaling, f, of $R(s)$ and $w(s)$. And if it turns out that twist accumulates *all the way up* to a glass ceiling at sproing threshold, so that all organizing centres operate at the correspondingly shortened spin rate, then slight changes of scale would *not* affect collision interface speed and its stable balance against contraction speed would accordingly become an *in*stability. Were this so we would expect to see substantial arcs of filament pegged at that twist rate, which is not the case: $w(s)$ is smooth and typically peaks maybe 10% below threshold.

These are clearly only the roughest estimates, maybe even comically so, but being also the first, they leave us every reason to hope for substantial refinements.

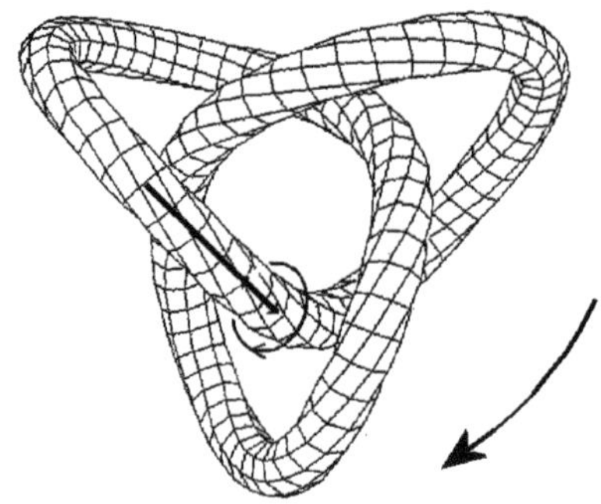

Figure 10.21. In a different excitable medium a narrow octagonal tube around a trefoil knot is oriented by the local gradient of one reactant (not by local filament curvature vector as in figures 10.8–10.13): it shows that the twist, $w(s)$, is not uniform in the steady state of this solution. The less curved inner segments of filament are more twisted, spinning slightly faster on that account, and so slapping away the outer lobes of the knot which would otherwise continue their curvature-driven contraction. Filament orientation and corresponding vortex rotation are indicated, and the two-orders-of-magnitude slower rigid precession of the whole knot. (Adapted from figure 26 of a summer 1988 plenary lecture [38].)

10.6 Passing in silence from 1994 to 2002

For a decade starting in 1985 I invested NSF support in (among other things) a quixotic effort to discover how local geometry guides the motions of vortex rings in excitable media. Even before (10.4) formalized a limiting case it was already clear from such numerical spectacles as figure 10.21 that something more is needed for understanding linked and knotted rings small enough to be computed. While this has been understood for a long time, no attempt has yet been made to record the evidence we now wish for in connection with an alternative conjectural mechanism for preservation of topological invariants of particle-like organizing centres in such media. Since 1994 there have been review articles but little further investigation along these lines. The dozen questions given here still seem speculations about imaginary objects, like medieval natural histories of the unicorn. But computer speed has improved many fold in the interim, so the story could at least be finished. And it now seems imaginable that laboratory research reports using tomographic optics [44] could appear, illustrating such particle-like organizing centres gliding through natural excitable media. If so, there would probably also be observations upon the consequences of their collision with a wall or even with other organizing centres. In general, fragmentation, transmutation and scattering would seem likely aftermaths. And in these events the two distinctively different arcs (twisted source arc, slapped expanding arc) of any ring would play distinctive roles. For example, the distant parts of an organizing centre striking a wall or another filament with its slapped arc might be little affected in parts remote from initial contact, but one striking with its high-frequency arc would immediately feel the repercussions in all parts. But before analysing any of those possibly complicated sequels, I should think it would be important to first establish whether (question 10.4) any organizing centres *actually have* a preferred shape to which they tend to rebound after modest perturbations. Even this basic principle has yet to be established by numerical experiments sufficiently prolonged after suitable perturbations or laboratory observation. None of these patterns have yet been pursued even as far as 100 rotation periods from their initial conditions. In equal-diffusion excitable media a cylindrically symmetric scroll ring of initial perimeter P wavelengths is guaranteed to shrink to nothing after $(Q/8\pi^2)P^2$ rotation periods, which in the examples shown is no more than about 20 periods, and this is observed exactly in computations. So 100 periods might be generously sufficient for observing the asymptotic outcome of other ring configurations of comparable diameter. But the slowest time constant for relaxation of the modes of Burgers' phase diffusion equation along a filament of perimeter P wavelengths is $(Q/\pi^2)P^2$ periods, which, with $P = n\pi d$ (n being a few links or loops of a knot) is $8n^2$ greater. So it remains possible that, where phase twist is importantly involved, asymptotic behaviour has not yet been fairly appraised. And so finally:

Question 10.13. *Is there any relation to the more recent apparition of linked and knotted vortex rings in Lagrangian field theory?*

Comparable numerical exploration of a soliton model of baryons starting from similar initial condition polynomials, but in an energy-conserving field rather than a dissipative reaction–diffusion field, reveals the possibility of stably twisted, linked and knotted vortex rings [2, 8, 10]. Such Skyrmions in Bose–Einstein condensates have also been computed [17]. Skyrmions are something like Kelvin–Helmholtz 'vortex atoms' [43, p 300] implemented by the sine-Gordon model. Rañada *et al* [26, 27] proposed knotted energy-minimizing magnetic vortices as an interpretation of ball lightning. Berry [3] shows linked and knotted singular filaments in the zeros of hydrogen atom wavefunctions. It might be that insights derived from such different field equations may illuminate the topologically similar solutions found in those of biological and chemical excitable media.

These conjectural Dynamics A, B and C in excitable media admittedly sound extravagantly fictional, too much like a Rube–Goldberg concatenation of gimmicks. It is as though the gracious simplicity of Maxwell's field equations had evolved backwards to the discrete field lines and ethereal gearworks of his and Michael Faraday's time. This unraveling of the simple idea that vortex motion is determined by local geometry was forced on us as early as 1987 [45] by numerical experiments intended to reveal those laws of motion. This chapter's explicit form of the alternative, however, arose only years after the last computations recorded everything *but* the measurements pertinent for testing the alternative. But this symposium was convened to address unanswered questions that offer the prospect of engendering new principles of nonlinear dynamics. I think we have some here.

Acknowledgments

Data underlying these interpretations and questions came from my unpublished notebooks of 1994, which are partly informed by an unpublished and incompletely documented dissertation [11]. Conversation with Peter Ashwin at Bristol altered my views on the maintenance of symmetry in some numerical experiments. Jose Luis Trueba, also at the meeting, helpfully critiqued the manuscript. I thank the US National Science Foundation for lifelong backing and current grant 9974334, and the Colston Research Society of Bristol for hospitality on a memorable early-summer weekend. I especially thank Alan Champneys and John Hogan for patient editing.

Addendum while in press

Shortly after the meeting reported in this book, Paul Sutcliffe undertook to protract the reported computations from $O(100)$ rotation periods to $O(1000)$. The reported plastic deformation after rebound from a collision and the spontaneous 'ballooning' of one lobe of some organizing centres [43, p 487] do not at all seem to be ephemeral anomalies. Rather, they were the harbingers of longer-term evolution now discovered on this generously prolonged time scale. Just as with motions of the planets and the moon, organizing centre motions resemble clockwork in the short term, but their long-term stability is far from evident. One imaginable outcome of what we have seen so far is that, in general, one lobe expands to become a large ring, slapped outward by volleys of waves from a more compact centre of knotting and linking. But why just one? And will the extreme twist in that compact centre itself diffuse outward along this long, little-curved extension? If so then the centre may no longer radiate at a higher frequency than the remote curved arc, and so no longer press a collision interface into intimate contact with it. If the ballooned arc can then force the collision interface away, it should resume contraction. The evolution of slight twist and slight curvature on this ballooned arc may finally provide the setting for which analytical theory (10.4) was designed so long ago.

References

[1] Aranson I S, Levine H and Tsimring L 1996 Spiral competition in three-component excitable media *Phys. Rev. Lett.* **76** 1170–3

[2] Battye R A and Sutcliffe P M 1999 Solitons, links, and knots *Phil. Trans. R. Soc.* A **455** 4305–31

[3] Berry M V 2002 Knotted zeros in the quantum states of hydrogen *Found. Phys.* preprint

[4] Berry M V and Dennis M R 2001 Knotted and linked phase singularities in monochromatic waves *Proc. R. Soc.* A preprint

[5] Biktashev V N, Holden A V and Zhang H 1994 Tension of organizing filaments of scroll waves *Phil. Trans. R. Soc.* A **347** 611–30

[6] Courtemanche M, Skaggs W and Winfree A T 1990 Stable three-dimensional action potential circulation in the FitzHugh–Nagumo model *Physica* D **41** 173–82

[7] Ermakova E A, Krinsky V I, Panfilov A V, and Pertsov A M 1986 Interaction of spiral and plane periodic autowaves in an active medium *Biofizika* **31** 318–23

[8] Faddeev L and Niemi A J 1997 Stable knot-like structures in classical field theory *Nature* **387** 58–61

[9] FitzHugh R 1961 Impulses and physiological states in theoretical models of nerve membrane *Biophys. J.* **1** 445–66

[10] Gladikowski J and Hellmund M 1997 Static solitons with nonzero Hopf number *Phys. Rev.* D **56** 5194–9

[11] Henze C E 1993 Stable organizing centers *PhD Dissertation* University of Arizona, University Microfilms #9333307

[12] Henze C E and Winfree A T 1991 A stable knotted singularity in an excitable medium *Int. J. Bif. Chaos* **1** 891–922

[13] Katrich V, Bednar J, Michoud D, Scharein R G, Dubochet J and Stasiak A 1996 Geometry and physics of knots *Nature* **384** 142–5

[14] Katrich V, Olson W K, Pieranski P, Dubochet J and Stasiak A 1997 Properties of ideal composite knots *Nature* **388** 148–51

[15] Keener J P 1988 The dynamics of three dimensional scroll waves in excitable media *Physica* D **31** 269–76

[16] Keener J P and Tyson J J 1992 The dynamics of scroll waves in excitable media *SIAM Rev.* **34** 1–39

[17] Khawaja U A and Stoof H 2001 Skyrmions in a ferromagnetic Bose–Einstein condensate *Nature* **411** 981–20

[18] Klapper I and Tabor M 1994 A new twist in the kinematics and elastic dynamics of thin filaments and ribbons *J. Phys. A: Math. Gen.* **27** 4919–24

[19] Krinsky V I and Aghadze K I 1983 Interaction of rotating waves in an active chemical medium *Physica* D **8** 50–6

[20] Margerit D and Barkley D 2001 Selection of twisted scroll waves in three-dimensional excitable media *Phys. Rev. Lett.* **86** 175–8

[21] Margerit D and Barkley D 2002 Cookbook asymptotics for spiral and scroll waves in excitable media *Preprint*

[22] Mironov S, Vinson M, Mulvey S and Pertsov A 1996 Destabilization of three-dimensional rotating chemical waves in an inhomogeneous BZ reaction *J. Phys. Chem.* **100** 1975–83

[23] Nam K, Ott E, Guzdar P N and Gabbay M 1998 Stability of spiral wave vortex filaments with phase twists *Phys. Rev.* E **58** 2580–5

[24] Panfilov A V and Pertsov A M 1984 Vortex ring in three-dimensional active medium described by reaction diffusion equation *Dokl. Akad. Nauk USSR* **274** 1500–3

[25] Poston T and Winfree A T 1988 Complex singularities and excitable media, unpublished manuscript

[26] Rañada A F, Soler M and Trueba J L 2000 Ball lightning as a force-free magnetic knot *Phys. Rev.* E **62** 7181–90

[27] Rañada A F and Trueba J L 2002 Topological electromagnetism with hidden nonlinearity *Adv. Chem. Phys.* **114** to appear

[28] Rousseau G, Chate H and Kapral R 1998 Coiling and supercoiling of vortex filaments in oscillatory media *Phys. Rev. Lett.* **80** 5671–4

[29] Ruiz-Villarreal M, Gomez-Gesteira M, Souto C, Munuzuri A P and Perez-Villar V 1996 Long-term vortex interaction in active media *Phys. Rev.* E **54** 2999–3002

[30] Schebesch I and Engel H 1999 Interaction of spiral waves in the Oregonator model of the light-sensitive Belousov–Zhabotinsky reaction *Phys. Rev.* E **60** 6429–34

[31] Vinson M 1998 Interactions of spiral waves in inhomogeneous excitable media *Physica* D **116** 313–24

[32] Vinson M and Pertsov A M 1999 Dynamics of scroll rings in a parameter gradient *Phys. Rev.* E **59** 2764–71

[33] Winfree A T 1973 Scroll-shaped waves of chemical activity in three dimensions *Science* **181** 937–9

[34] Winfree A T 1974 Rotating solutions to reaction/diffusion equations *SIAM-AMS Proceedings* vol 8, ed D Cohen (Providence, RI: American Mathematical Society) pp 13–31

[35] Winfree A T 1980 *The Geometry of Biological Time* 1st edn (Berlin: Springer)

[36] Winfree A T 1987 *When Time Breaks Down: The Three-dimensional Dynamics of Chemical Waves and Cardiac Arrhythmias* (Princeton, NJ: Princeton University Press)

[37] Winfree A T 1990 Vortices in motionless media *Appl. Mech. Rev.* **43** 297–309

[38] Winfree A T 1990 Stable particle-like solutions to the nonlinear wave equations of three-dimensional excitable media *SIAM Rev.* **32** 1–53

[39] Winfree A T 1991 Varieties of spiral wave behaviour in excitable media *Chaos* **1** 303–34

[40] Winfree A T 1993 The geometry of excitability *1992 Lectures in Complex Systems* vol 5, ed L Nadel and D Stein (Reading, MA: Addison-Wesley) pp 205–98

[41] Winfree A T 1995 Lingering mysteries about organizing centers in the Belousov–Zhabotinsky medium and its Oregonator model *Chemical Waves and Patterns* vol 1, ed R Kapral and K Showalter (Dordrecht: Kluwer) pp 3–56

[42] Winfree A T 1995 Persistent tangles of vortex rings in excitable media *Physica* D **84** 126–47

[43] Winfree A T 2001 *The Geometry of Biological Time* 2nd edn (Berlin: Springer)

[44] Winfree A T, Caudle S, Chen G, McGuire P and Szilagyi Z 1996 Quantitative optical tomography of chemical waves and their organizing centers *Chaos* **6** 617–26

[45] Winfree A T and Guilford W 1988 The dynamics of organizing centers: Experiments in differential geometry *Biomathematics and Related Computational Problems* ed L Ricciardi (Dordrecht: Kluwer Academic) pp 676–716

[46] Winfree A T and Strogatz S H 1983 Singular filaments organize chemical waves in three dimensions: 2. Twisted waves *Physica* D **9** 65–80

[47] Winfree A T and Strogatz S H 1983 Singular filaments organize chemical waves in three dimensions: 3. Knotted waves *Physica* D **9** 333–45

[48] Winfree A T and Strogatz S H 1984 Singular filaments organize chemical waves in three dimensions: 4. Wave taxonomy *Physica* D **13** 221–33

[49] Winfree A T and Witkowski F X 1998 Three-dimensional ventricular fibrillation? *Ann. Biomed. Eng.* **26** (supplement 1) S-16

[50] Witkowski F X, Leon L J, Penkoske P A, Giles W R, Spano M L, Ditto W L and Winfree A T 1998 Spatiotemporal evolution of ventricular fibrillation *Nature* **392** 78–82

[51] Powerpoint presentations of Art Winfree
http://eeb8.biosci.arizona.edu/art/2000_lectures/

[52] Symbolic sculpture and the Borromean rings
http://www.cpm.informatics.bangor.ac.uk/sculmath/borings.htm
web site: Centre for the Popularization of Mathematics, University of Bangor, UK

[53] Symbolic sculptures by John Robinson
http://www.newton.cam.ac.uk/art/sculptures.html
web site: Isaac Newton Institute for Mathematical Sciences, UK

Chapter 11

Spontaneous pattern formation in primary visual cortex

Paul C Bressloff
University of Utah

Jack D Cowan
University of Chicago

The primary visual cortex (V1) is the first cortical area to receive visual information transmitted by ganglion cells of the retina via the lateral geniculate nucleus (LGN) of the thalmus to the back of the brain; see figure 11.1. A fundamental property of the functional architecture of V1 is an orderly retinotopic mapping of the visual field onto the surface of cortex, with the left and right halves of visual field mapped onto the right and left cortices respectively. Except close to the fovea (centre of the visual field), this map can be approximated by the complex logarithm; see figure 11.2. Let $\mathbf{r}_\mathrm{R} = \{r_\mathrm{R}, \theta_\mathrm{R}\}$ be a point in the visual field represented in polar coordinates and let $\mathbf{r} = \{x, y\}$ be the corresponding point in the cortex given in Cartesian coordinates. Under the retino-cortical map, $\mathbf{r} = \{\log \mathbf{r}_\mathrm{R}, \theta_\mathrm{R}\}$. Evidently, if we introduce the complex representation of $\mathbf{r}_\mathrm{R}$, $z_\mathrm{R} = \mathbf{r}_\mathrm{R} \mathrm{e}^{\mathrm{i}\theta_\mathrm{R}}$ then $z = \log z_\mathrm{R} = \log r_\mathrm{R} + \mathrm{i}\theta_\mathrm{R} = x + \mathrm{i}y$ generates the complex cortical representation. One of the interesting properties of the retino-cortical map is that the action of rotations and dilatations in the visual field correspond to translations in the x and y directions, respectively, in the cortex.

Superimposed upon the retinotopic map are a number of additional feature maps reflecting the fact that neurons respond preferentially to stimuli with particular features [37, 42, 54]. For example, most cortical cells signal the local *orientation* of a contrast edge or bar—they are tuned to a particular local orientation [32]. Cells also show a left/right eye preference known as ocular dominance and some are also direction selective. The latter is illustrated in figure 11.1 where the response of a cell to a moving bar is shown. In recent

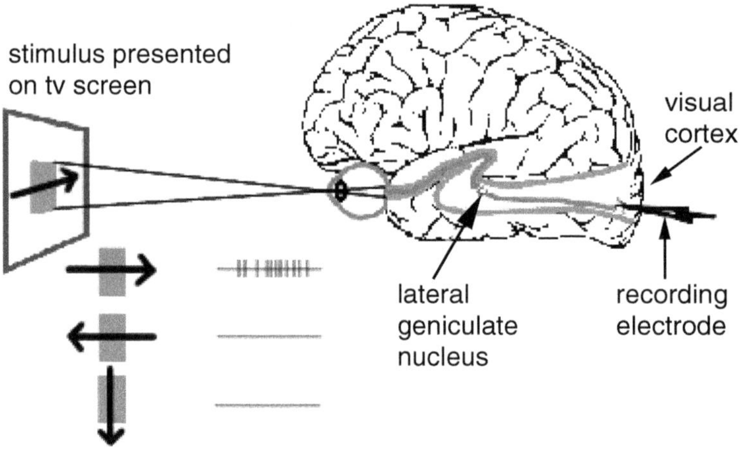

Figure 11.1. The visual pathway.

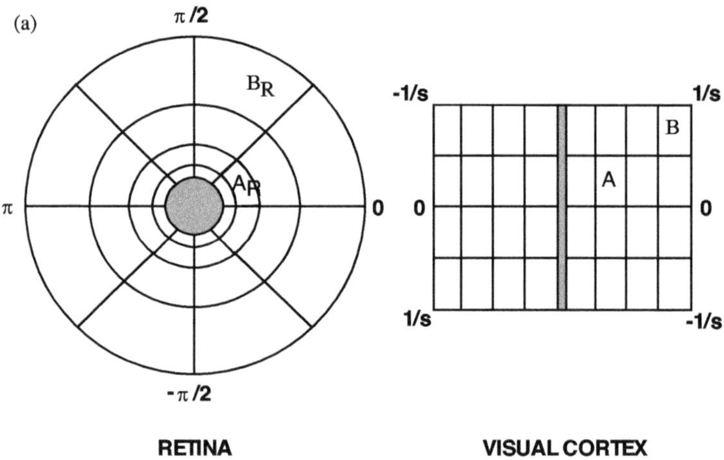

Figure 11.2. The retino-cortical map generated by the complex logarithm.

years much information has accumulated about the distribution of orientation selective cells in V1 [26]. In figure 11.3 a typical arrangement of such cells, obtained via microelectrodes implanted in cat V1, is given. The first panel shows how orientation preferences rotate smoothly over the surface of V1, so that approximately every 300 μm the same preference reappears, i.e. the distribution is π-periodic in the orientation preference angle. The second panel shows the receptive fields of the cells, and how they change with V1 location. The third panel shows more clearly the rotation of such fields with translation across V1.

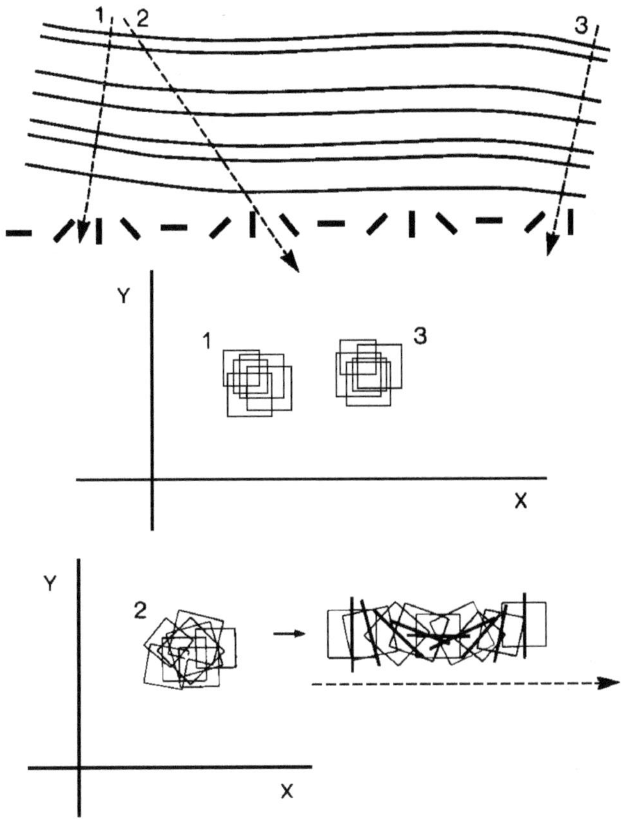

Figure 11.3. Orientation tuned cells in layers of V1 which is shown in cross section. Note the constancy of orientation preference at each cortical location (electrode tracks 1 and 3), and the rotation of orientation preference as cortical location changes (electrode track 2). Reprinted from C D Gilbert Horizontal integration and cortical dynamics *Neuron* **9** 1–13 ©1992 with permission from Elsevier Science.

A more complete picture of the two–dimensional distribution[1] of both orientation preference and ocular dominance has been obtained using optical imaging techniques [5–7]. The basic experimental procedure involves shining light directly on to the surface of the cortex. The degree of light absorption within each patch of cortex depends on the local level of activity. Thus, when an oriented

[1] The cortex is, of course, three-dimensional since it has non-zero thickness with a distinctive layered structure. However, one find that cells with similar feature preferences tend to arrange themselves in vertical columns so that, to a first approximation, the layered structure of cortex can be ignored. For example, electrode track 1 in figure 11.3 is a vertical penetration of the cortex that passes through a single column of cells with the same orientation preference and ocular dominance.

Figure 11.4. Iso-orientation contours in a hypercolumn. There are two ocular dominance columns corresponding to left (L) and right (R) eye preference. Each ocular dominance column contains two orientation singularities or pinwheels. A broken ring is drawn around one orientation singularity.

image is presented across a large part of the visual field, the regions of cortex that are particularly sensitive to that stimulus will be differentiated. (An example of optical imaging data is shown in figure 11.5). The basic topography revealed by these methods has a number of characteristic features [42]:

(i) Orientation preference changes continuously as a function of cortical location except at singularities (or pinwheels).
(ii) There exist linear regions, approximately 800 μm $\times$ 800 μm in area (in macaque monkeys), bounded by singularities, within which iso-orientation regions form parallel slabs.
(iii) Iso-orientation slabs tend to cross the borders of ocular dominance stripes at right angles.

Singularities tend to align with the centres of ocular dominance stripes. These experimental findings suggest that there is an underlying periodicity in the microstructure of V1 with a period of approximately 1 mm (in cats and primates). The fundamental domain of this periodic tiling of the cortical plane is the hypercolumn [33, 37], which contains two sets of orientation preferences $\phi \in [0, \pi)$, one for each eye, organized around a set of four singularities; see figure 11.4.

Given the existence of a regularly repeating set of orientation and ocular dominance maps, how does such a periodic structure manifest itself anatomically?

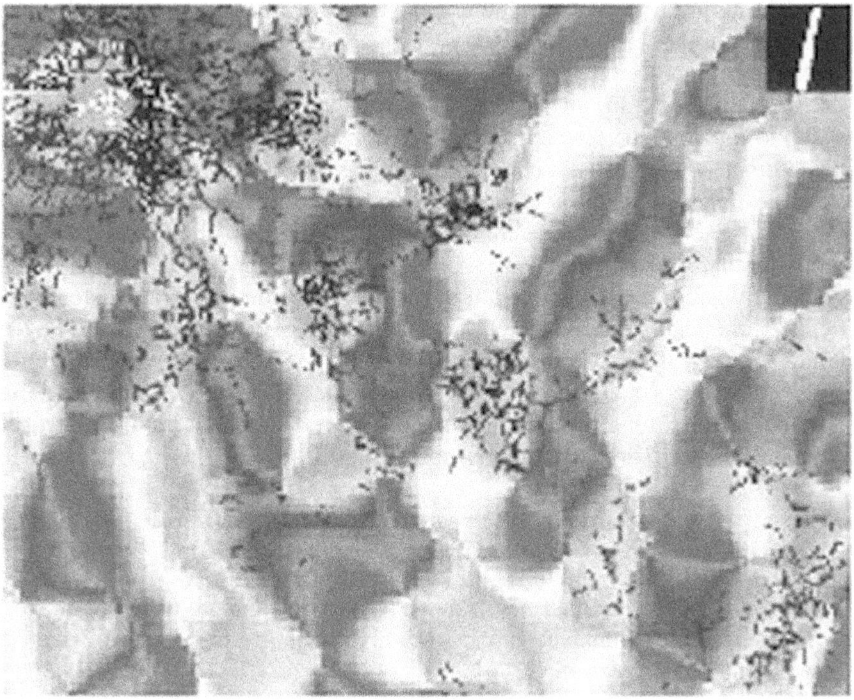

Figure 11.5. Lateral connections made by a cells in Tree Shrew V1. A radioactive tracer is used to show the locations of all terminating axons from cells in a central injection site, superimposed on an orientation map obtained by optical imaging. (Patches with the same coarse-grained orientation preference are shown in the same colour—this is purely for visualization purposes). The patchy distribution of the lateral connections is clearly seen, linking regions of like orientation preference along a particular visuotopic axis. The local axonal field, however, is isotropic and connects all neurons within a neighbourhood ($\approx$0.7 mm) of the injection site; see also colour plate 4. Reproduced from W H Bosking, Y Zhang, B Schofield and D Fitzpatrick 1997 Orientation selectivity and the arrangement of horizontal connections in tree shrew striate cortex *J. Neurosci.* **17** 2112–27 ©1997 by the Society for Neuroscience.

Two cortical circuits have been fairly well characterized. There is a local circuit operating at sub-hypercolumn dimensions in which cells make connections with most of their neighbours in a roughly isotropic fashion [20]. The other circuit operates between hypercolumns, connecting cells with similar functional properties separated by several millimetres of cortical tissue. Optical imaging combined with labelling techniques has generated considerable information concerning the pattern of connections both within and between hypercolumns [5, 6,9,39,65]. A particularly striking result concerns the intrinsic lateral connections of V1. The axons of these connections make terminal arbors only every 0.7 mm or so along their tracks [25,46], and they seem to connect mainly to cells with similar orientation preferences [9, 39, 65]. In addition, as shown in figure 11.5, there is a pronounced anisotropy of the pattern of such connections: its long axis runs parallel to a patch's preferred orientation [9, 25]. Thus differing iso-orientation patches connect to patches in neighbouring hypercolumns in differing directions. Ongoing studies of feedback connections from points in extrastriate areas back to area V1 [1], show that these connectional fields are also distributed in highly regular geometric patterns, having a topographic spread of up to 13 mm that is significantly larger than the spread of intrinsic lateral connections. Stimulation of a hypercolumn via lateral or feedback connections modulates rather than initiates spiking activity [30, 56]. Thus this long-range connectivity is ideally structured to provide local cortical processes with contextual information about the global nature of stimuli. As a consequence the lateral connections have been invoked to explain a wide variety of context-dependent visual processing phenomena [16, 23, 27].

From the perspective of nonlinear dynamics, there are two very distinct questions one can ask about the large-scale structure of cortex:

(i) How did the feature maps and connectivity patterns first develop?
(ii) What types of spontaneous and stimulus-driven spatio-temporal dynamics arise in the mature cortex?

It appears that in both cases the Turing mechanism for spontaneous pattern formation plays a crucial role.

11.1 The Turing mechanism and its role in cooperative cortical dynamics

In 1952 Turing [57] introduced an important set of new ideas concerning spontaneous pattern formation. The details are well known, but we will re-state them here by way of providing a context for the rest of this chapter. Turing considered the development of animal coat markings as a problem of pattern formation. He started by introducing the idea that chemical markers in the skin comprise a system of diffusion-coupled chemical reactions among substances called *morphogens*. Turing introduced the following two-component reaction–diffusion system:

$$\frac{\partial \mathbf{c}}{\partial t} = \mathbf{f}(\mathbf{c}) + D\nabla^2\mathbf{c} \tag{11.1}$$

where $\mathbf{c}$ is a vector of morphogen concentrations, $\mathbf{f}$ is (in general) a nonlinear vector function representing the reaction kinetics and D is the (diagonal) matrix of positive diffusion coefficients. What Turing showed was that two different reactions can exist such that in the absence of diffusion ($D = 0$), $\mathbf{c}$ tends to a linearly stable homogeneous state, and when $D \neq 0$, $D_1 \neq D_2$, the homogeneous state becomes unstable and $\mathbf{c}$ tends to a spatially inhomogeneous state. This was the now famous *diffusion-driven instability*.

Wilson and Cowan introduced exactly the same mechanism in a neural context [63, 64]. Here we briefly summarize their formulation. Let $a_{\rm E}(\mathbf{r}, t)$ be the activity of excitatory neurons in a given volume element of a slab of neural tissue located at $\mathbf{r} \in \mathbb{R}^2$, and $a_{\rm I}(\mathbf{r}, t)$ the correspond activity of inhibitory neurons. $a_{\rm E}$ and $a_{\rm I}$ can be interpreted as local spatio-temporal averages of the membrane potentials or voltages of the relevant neural populations. In case neuron activation rates are low they can be shown to satisfy nonlinear evolution equations of the form:

$$\begin{aligned}
\tau \frac{\partial a_{\rm E}(\mathbf{r}, t)}{\partial t} &= -a_{\rm E}(\mathbf{r}, t) + \tau \int_{\mathbb{R}^2} w_{EE}(\mathbf{r}|\mathbf{r}')\sigma_{\rm E}[a_{\rm E}(\mathbf{r}', t)]\, d\mathbf{r}' \\
&\quad - \tau \int_{\mathbb{R}^2} w_{EI}(\mathbf{r}|\mathbf{r}')\sigma_{\rm I}[a_{\rm I}(\mathbf{r}', t)]\, d\mathbf{r}' + h_{\rm E}(\mathbf{r}, t) \\
\tau \frac{\partial a_{\rm I}(\mathbf{r}, t)}{\partial t} &= -a_{\rm I}(\mathbf{r}, t) + \tau \int_{\mathbb{R}^2} w_{IE}(\mathbf{r}|\mathbf{r}')\sigma_{\rm E}[a_{\rm E}(\mathbf{r}', t)]\, d\mathbf{r}' \\
&\quad - \tau \int_{\mathbb{R}^2} w_{II}(\mathbf{r}|\mathbf{r}')\sigma_{\rm I}[a_{\rm I}(\mathbf{r}', t)]\, d\mathbf{r}' + h_{\rm I}(\mathbf{r}, t) \qquad (11.2)
\end{aligned}$$

where $w_{lm}(\mathbf{r}|\mathbf{r}') = w_{lm}(|\mathbf{r} - \mathbf{r}'|)$ gives the weight per unit volume of all synapses to the lth population from neurons of the mth population a distance $|\mathbf{r} - \mathbf{r}'|$ away, $\sigma_{\rm E}$ and $\sigma_{\rm I}$ are taken to be smooth output functions

$$\sigma(x) = \frac{1}{\tau} \frac{1}{1 + e^{-\eta(x-\kappa)}} \qquad (11.3)$$

where η determines the slope or sensitivity of the input–output characteristics of the population and κ is a threshold, $h_{\rm E}$ and $h_{\rm I}$ are external stimuli, and τ is the membrane time constant.

Equations (11.2) can be rewritten in the more compact form:

$$\tau \frac{\partial a_l(\mathbf{r}, t)}{\partial t} = -a_l(\mathbf{r}, t) + \tau \sum_{m=E,I} \int_{\mathbb{R}^2} w_{lm}(|\mathbf{r} - \mathbf{r}'|)\sigma[a_m(\mathbf{r}', t)]\, d\mathbf{r}' + h_l(\mathbf{r}, t) \qquad (11.4)$$

Note that $w_{l\rm E} \geq 0$ and $w_{l\rm I} \leq 0$.

In the case of a constant external input, $h_l(\mathbf{r}) = \bar{h}_l$, there exists at least one fixed point solution $a_l(\mathbf{r}) = \bar{a}_l$ of equation (11.4), where

$$\bar{a}_l = \tau \sum_{m=E,I} W_{lm}\sigma(\bar{a}_m) + \bar{h}_l \qquad (11.5)$$

and $W_{lm} = \int_{\mathbb{R}^2} w_{lm}(\mathbf{r})\, d\mathbf{r}$. If $\bar{h}_l$ is sufficiently small relative to the threshold κ then this fixed point is unique and stable. Under the change of coordinates

$a_l \to a_l - \bar{h}_l$, it can be seen that the effect of $\bar{h}_l$ is to shift the threshold by the amount $-\bar{h}_l$. Thus there are two ways to increase the excitability of the network and thus destabilize the fixed point: either by increasing the external input $\bar{h}_l$ or reducing the threshold κ. The latter can occur through the action of drugs on certain brain stem nuclei which, as we shall see, provides a mechanism for generating geometric visual hallucinations [11, 12, 14, 21].

The local stability of $(\bar{a}_{\mathrm{E}}, \bar{a}_{\mathrm{I}})$ is found by linearization:

$$\frac{\partial b_l(\mathbf{r}, t)}{\partial t} = -b_l(\mathbf{r}, t) + \mu \sum_{m=E,I} \int_{\mathbb{R}^2} w_{lm}(|\mathbf{r} - \mathbf{r}'|) b_m(\mathbf{r}', t)\, \mathrm{d}\mathbf{r}' \tag{11.6}$$

where $b_l(\mathbf{r}, t) = a_l(\mathbf{r}, t) - \bar{a}_l$ and we have performed a rescaling of the local weights $\tau\sigma'(\bar{a}_l) w_{lm} \to \mu w_{lm}$ with μ a measure of the degree of network excitability. We have also rescaled t in units of the membrane time constant τ. Assuming solutions of the form $b_l(\mathbf{r}, t) = b_l(\mathbf{r})\mathrm{e}^{-\lambda t}$ we are left with the eigenvalue problem:

$$\lambda b_l(\mathbf{k}) = -b_l(\mathbf{k}) + \mu \sum_m W_{lm}(|\mathbf{k}|^2) b_m(\mathbf{k}) \tag{11.7}$$

where $b_l(\mathbf{k})$ and $W_{lm}(|\mathbf{k}|^2)$ are, respectively, the Fourier coefficients of $b_l(\mathbf{r})$ and $w_{lm}(\mathbf{r})$. This leads to the matrix *dispersion relation* for λ as a function of $q = |\mathbf{k}|$ given by solutions of the characteristic equation

$$\det([\lambda + 1]I - \mu \mathbf{W}(q)) = 0 \tag{11.8}$$

where $\mathbf{W}$ is the matrix of Fourier coefficients of the w_{lm}. One can actually simplify the formulation by reducing equations (11.4) to an equivalent one-population model:

$$\tau \frac{\partial a(\mathbf{r}, t)}{\partial t} = -a_l(\mathbf{r}, t) + \tau \int_{\mathbb{R}^2} w(|\mathbf{r} - \mathbf{r}'|) \sigma[a(\mathbf{r}', t)]\, \mathrm{d}\mathbf{r}' + h(\mathbf{r}, t) \tag{11.9}$$

from which we obtain the dispersion relation $\lambda = -1 + \mu W(q) \equiv \lambda(q)$, with $W(q)$ the Fourier transform of $w(\mathbf{r})$.

In either case it is relatively straightforward to set up the conditions under which the homogeneous state first loses its stability at $\mu = \mu_{\mathrm{c}}$ and at a wavevector with $q = q_{\mathrm{c}} \neq 0$. In the case of equation (11.9) the condition is that $W(q)$ be *bandpass*. This can be achieved with the 'Mexican hat' function (see figure 11.6):

$$w(|\mathbf{r}|) = \left(\frac{a_+}{\sigma_+}\right) \mathrm{e}^{-r^2/\sigma_+^2} - \left(\frac{a_-}{\sigma_-}\right) \mathrm{e}^{-r^2/\sigma_-^2} \tag{11.10}$$

the Fourier transform of which is:

$$W(q) = \tfrac{1}{2}(a_+ \mathrm{e}^{-\frac{1}{4}\sigma_+^2 q^2} - a_- \mathrm{e}^{-\frac{1}{4}\sigma_-^2 q^2}). \tag{11.11}$$

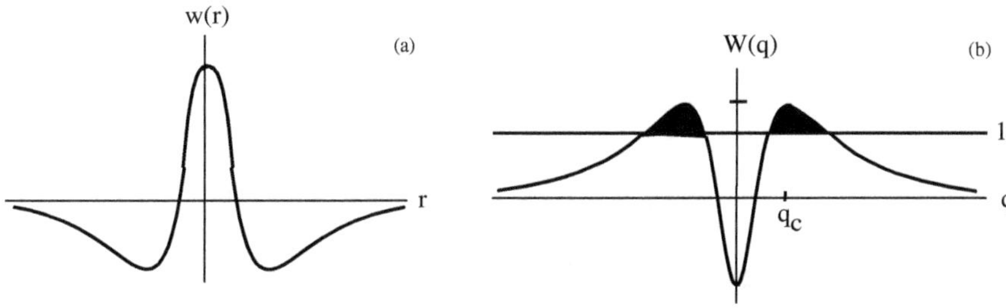

Figure 11.6. Neural basis of the Turing mechanism. Mexican hat interaction function showing short-range excitation and long-range inhibition (*a*). Fourier transform $W(q)$ of Mexican hat function (*b*). There exists a critical parameter $\mu_c = W(q_c)^{-1}$ where $W(q_c) = [\max_q\{W(q)\}]$ such that for $\mu_c < \mu < \infty$ the homogeneous fixed point is unstable.

Evidently $W(0) = \frac{1}{2}(a_+ - a_-)$ and $W(\infty) = 0$. It is simple to establish that λ passes through zero at the critical value μ_c signalling the growth of spatially periodic patterns with wavenumber q_c, where $W(q_c) = \max_q\{W(q)\}$. Close to the bifurcation point these patterns can be represented as linear combinations of plane waves

$$b(\mathbf{r}) = \sum_n (c_n e^{i\mathbf{k}_n\cdot\mathbf{r}} + c_n^* e^{-i\mathbf{k}_n\cdot\mathbf{r}})$$

where the sum is over all wavevectors with $|\mathbf{k}_n| = q_c$ and n can be bounded by restricting the solutions to doubly-periodic patterns in $\mathbb{R}^2$. Depending on the boundary conditions various patterns of stripes or spots can be obtained as solutions. Figure 11.7 shows, for example, a late stage in the development of stripes [61]. Amplitude equations for the coefficients c_n can then be obtained in the usual fashion to determine the linear stability of the various solutions. This analysis of the Wilson–Cowan equations was first carried out by Ermentrout and Cowan as part of their theory of visual hallucinations [21], and is an exact parallel of Turing's original analysis, although he did not develop amplitude equations for the various solutions.

Essentially the same analysis can be applied to a variety of problems concerning the neural development of the various feature maps and connectivity patterns highlighted at the beginning of the chapter. Consider, for example, the development of topographic maps from eye to brain [60, 62]. Such maps develop by a process which involves both genetic and epigenetic factors. Thus the actual growth and decay of connections is epigenetic, involving synaptic plasticity. However, the final solution is constrained by genetic factors, which act, so to speak, as boundary conditions. The key insight was provided by von der Malsburg [59] who showed that pattern formation can occur in a developing neural network whose synaptic connectivity or weight matrix is activity dependent and modifiable, provided some form of *competition* is present. Thus Häussler and

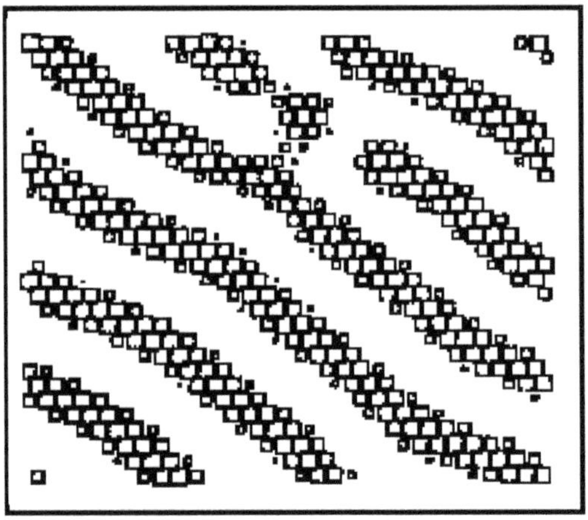

Figure 11.7. A late stage in the spontaneous formation of stripes of neural activity. See text for details.

von der Malsburg [29] formulated the topographic mapping problem (in the case of a one-dimensional cortex) as follows. Let w_{rs} be the weight of connections from the retinal point r to the cortical point s, and $\mathbf{w}$ the associated weight matrix. An evolution equation for $\mathbf{w}$ embodying synaptic plasticity and competition can then be written as

$$\frac{\mathrm{d}\mathbf{w}}{\mathrm{d}t} = \alpha\mathbf{J} + \beta\mathbf{w}\cdot C(\mathbf{w}) - \mathbf{w}\cdot B(\alpha\mathbf{J} + \beta\mathbf{w}\cdot C(\mathbf{w})) \tag{11.12}$$

where $\mathbf{J}$ is a matrix with all elements equal to unity, $C_{rs}(\mathbf{x}) = \sum_{r's'} c(r-r', s-s')x_{r's'}$, and

$$B_{rs}(\mathbf{x}) = \frac{1}{2}\left(\frac{1}{N}\sum_{r'} x_{r's} + \frac{1}{N}\sum_{s'} x_{rs'}\right).$$

One can easily show that $\mathbf{w} = \mathbf{J}$ is an unstable fixed point of equation (11.12). Linearizing about this fixed point leads to the linear equation:

$$\frac{\mathrm{d}\mathbf{v}}{\mathrm{d}t} = \alpha\mathbf{v} + C(\mathbf{v}) - B(\mathbf{v}) - B[C(\mathbf{v})] \tag{11.13}$$

where $\mathbf{v} = \mathbf{w} - \mathbf{J}$. Since B and C are linear operators, we can rewrite equation (11.13) in the form:

$$\tau\frac{\mathrm{d}\mathbf{v}}{\mathrm{d}t} = -\mathbf{v} + \tau(I - B)[(I + C)(\mathbf{v})] \tag{11.14}$$

where the time constant $\tau = (1-\alpha)^{-1}$. It is not too difficult to see that the term $(I - B)[(I + C)(\mathbf{v})]$ is equivalent to the action of an effective convolution kernel of the form:

$$w(\mathbf{r}) = w_{+}(\mathbf{r}) - w_{-}(\mathbf{r})$$

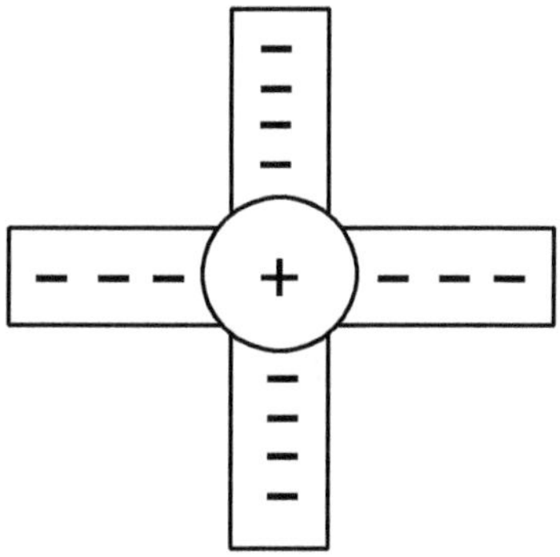

Figure 11.8. Structure of the weight kernel $w(r, s)$.

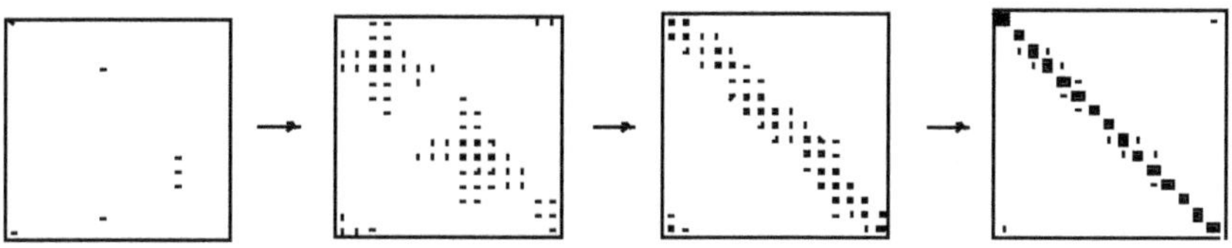

Figure 11.9. Stages in the development of an ordered retinotopic map. A single stripe develops in the (r, s)-plane.

so that equation (11.14) can be rewritten in the familiar form:

$$\tau \frac{\partial \mathbf{v}(\mathbf{r}, t)}{\partial t} = -\mathbf{v}(\mathbf{r}, t) + \tau \int_{\mathbb{R}^2} w(\mathbf{r} - \mathbf{r}')\mathbf{v}(\mathbf{r}', t)\, \mathrm{d}\mathbf{r}' \tag{11.15}$$

where in this case $\mathbf{r} = \{r, s\}$ and $\mathbf{v}$ is a matrix. Once again there is a dispersion relation of the form $\lambda = -1 + \mu W(\mathbf{k}) \equiv \lambda(\mathbf{k})$, where $\mathbf{k} = \{k, l\}$ and, as in the previous examples, assuming appropriate boundary conditions—in this case periodic—it is the Fourier transform $W(\mathbf{k})$ that determines which of the eigenmodes

$$\sum_{kl} c_{kl} \exp\left(\mathrm{i}\frac{2\pi}{N}(kr + ls)\right)$$

emerges at the critical wavenumber $\mathbf{k}_\mathrm{c} = \{k_\mathrm{c}, l_\mathrm{c}\}$. Figure 11.8 shows the form of $w(r, s)$ in the (r, s)-plane. It will be seen that it is similar to the Mexican Hat except that the inhibitory surround is in the form of a cross. This forces the eigenmodes that emerge from the Turing instability to be diagonal in the (r, s)-plane. If the wavenumber is selected so that only one wave is present, this corresponds to an ordered retino-cortical map. Figure 11.9 shows details of the emergence of the required mode.

A second example is to be found in models for the development of ocular dominance maps [52]. Let $n_\mathrm{R}(\mathbf{r}, t)$ and $n_\mathrm{L}(\mathbf{r}, t)$ be, respectively, the (normalized) right and left eye densities of synaptic connections to the visual cortex modelled as a two–dimensional sheet. Such densities are presumed to evolve according to an evolution equation of the form:

$$\frac{\partial u_l(\mathbf{r},t)}{\partial t} = \sum_{m=\mathrm{R,L}} \int_{\mathbb{R}^2} w_{lm}(|\mathbf{r}-\mathbf{r}'|)\sigma[u_m(\mathbf{r}',t)]\,\mathrm{d}\mathbf{r}' \tag{11.16}$$

where $u_l = \log(n_l/(1-n_l))$ such that $\sigma(u_l) = n_l$ and the coupling matrix $\mathbf{w}$ is given by

$$\mathbf{w}(r) = w(r)\begin{pmatrix} +1 & -1 \\ -1 & +1 \end{pmatrix}.$$

With the additional constraint $n_\mathrm{R} + n_\mathrm{L} = 1$, equation (11.16) reduces to the one-dimensional form:

$$\frac{\partial u_\mathrm{R}(\mathbf{r},t)}{\partial t} = 2\int_{\mathbb{R}^2} w(|\mathbf{r}-\mathbf{r}'|)\sigma[u_\mathrm{R}(\mathbf{r}',t)]\,\mathrm{d}\mathbf{r}' - \int_{\mathbb{R}^2} w(|\mathbf{r}'|)\,\mathrm{d}\mathbf{r}'. \tag{11.17}$$

which can be rewritten in terms of the variable $n_\mathrm{R}(\mathbf{r},t)$ as:

$$\begin{aligned}\frac{\partial n_\mathrm{R}(\mathbf{r},t)}{\partial t} &= n_\mathrm{R}(\mathbf{r},t)(1-n_\mathrm{R}(\mathbf{r},t)) \\ &\quad \times \left[2\int_{\mathbb{R}^2} w(|\mathbf{r}-\mathbf{r}'|)n_\mathrm{R}(\mathbf{r}',t)\,\mathrm{d}\mathbf{r}' - \int_{\mathbb{R}^2} w(|\mathbf{r}'|)\,\mathrm{d}\mathbf{r}'\right].\end{aligned} \tag{11.18}$$

The fixed points of this equation are easily seen to be $n_\mathrm{R}(\mathbf{r}) = 0, 1$ and $n_\mathrm{R}(\mathbf{r}) = \frac{1}{2}$. The first two fixed points are stable. However, the third fixed point is unstable to small perturbations. Linearizing about this fixed point we find the dispersion relation $\lambda = \frac{1}{2}W(|\mathbf{k}|)$. Once again the Fourier transform of the interaction kernel $w(|\mathbf{r}|)$ controls the emergence of the usual eigenmodes, in this case plane waves of the form $\mathrm{e}^{\mathrm{i}\mathbf{k}\cdot\mathbf{r}}$ in the cortical plane. Note that the fixed point $n_\mathrm{R} = n_\mathrm{L} = \frac{1}{2}$ corresponds to the fixed point $u_\mathrm{R} = u_\mathrm{L} = 0$ which is a point of reflection symmetry for the function $\sigma[u]$. It is this additional symmetry which results in the emergence of stripes rather than spots or blobs when the fixed point destabilizes.

There are many other examples of the role of the Turing instability in visual neuroscience such as the Marr–Poggio model of stereopsis [40] and the Swindale model for the development of iso-orientation patches [53]. However, all of the neural models involve the same basic mechanism of competition between excitation and inhibition (the Mexican hat form of interaction; see figure 11.6), and most have some underlying symmetry that plays a crucial role in the selection and stability of the resulting patterns. In what follows, we shall develop these ideas further by considering in detail our own recent work on spontaneous pattern formation in primary visual cortex [11, 12, 14]. In this work we have investigated how correlations between the pattern of patchy lateral connections and the underlying orientation map within V1 (as highlighted in the introduction to the chapter) affect the large-scale dynamics of V1 idealized as a continuous two-dimensional sheet of interacting hypercolumns [11, 12, 14]. We have shown that the patterns of lateral connection are invariant under the so-called *shift-twist* action of the planar Euclidean group E(2) acting on the product space $\mathbb{R}^2 \times \mathbf{S}^1$. By virtue of the anisotropy of the lateral connections (see figure 11.5), this shift-twist symmetry supports distinct scalar and pseudoscalar group representations of E(2) [8], which characterize the type of cortical activity patterns that arise through spontaneous symmetry breaking [11]. Following on from the original work of Ermentrout and Cowan [21], we have used our continuum model to develop a theory for the generation of geometric visual hallucinations, based on the idea that some disturbance such as a drug or flickering light can destabilize V1 inducing a spontaneous pattern of cortical activity that reflects the underlying architecture of V1. These activity patterns are seen as hallucinatory images in the visual field, whose spatial scale is determined by the range of lateral connections and the cortical-retinotopic map. Four examples of common hallucinatory images that are reproduced by our model [11] are shown in figure 11.10. Note the contoured nature of the third and fourth images, which could not have been generated in the original Ermentrout–Cowan model [21]. Our results suggest that the circuits in V1 that are normally involved in the detection of oriented edges and in the formation of contours, are also responsible for the generation of simple hallucinations.

11.2 A continuum model of V1 and its intrinsic circuitry

Consider a local population of excitatory (E) and inhibitory (I) cells at cortical position $\mathbf{r} \in \mathbb{R}^2$ with orientation preference ϕ. We characterize the state of the population at time t by the real-valued activity variable $a_l(\mathbf{r}, \phi, t)$ with $l = E, I$. As in section 11.1, V1 is treated as an (unbounded) continuous two-dimensional sheet of nervous tissue with the additional simplifying assumption that ϕ and $\mathbf{r}$ are independent variables—all possible orientations are represented at every position. Hence, one interpretation of our model would be that it is a continuum version of a lattice of hypercolumns. An argument for the validity of this continuum model is to note that the separation of two points in the visual field—visual *acuity*—(at a given retinal eccentricity of $\mathbf{r}^\circ$), corresponds to hypercolumn spacing [33], and so to each location in the visual field there corresponds a representation in V1 of that location with finite resolution and all possible orientations. Our large-scale model of V1 takes the form

$$\begin{aligned}\frac{\partial a_l(\mathbf{r}, \phi, t)}{\partial t} = & -a_l(\mathbf{r}, \phi, t) + h_l(\mathbf{r}, \phi, t) \\ & + \sum_{m \equiv E, I} \int_{\mathbb{R}^2} \int_0^\pi w_{lm}(\mathbf{r}, \phi | \mathbf{r}', \phi') \sigma[a_m(\mathbf{r}', \phi', t)] \frac{\mathrm{d}\phi'}{\pi} \, \mathrm{d}\mathbf{r}'\end{aligned} \tag{11.19}$$

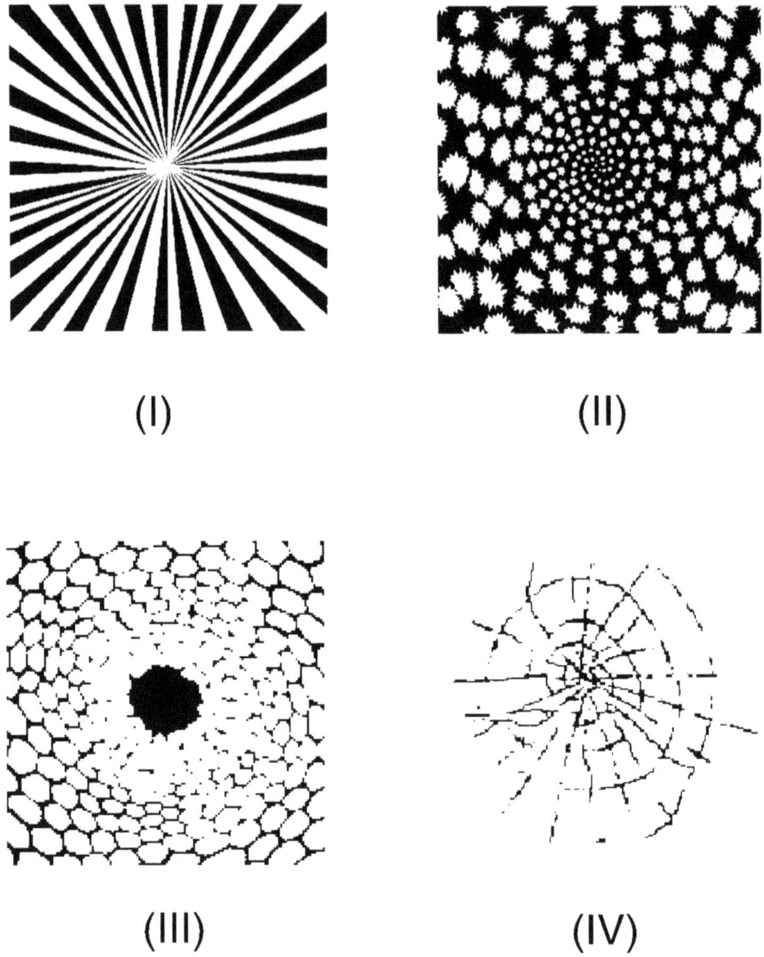

Figure 11.10. Hallucinatory form constants: funnel (I) and spiral images (II) seen following ingestion of LSD; honeycomb generated by marihuana (III); and cobweb petroglyph (IV). Reproduced from P C Bressloff, J D Cowan, M Golubitsky, P J Thomas and M C Weiner Geometric visual hallucinations, Euclidean symmetry and the functional architecture of striate cortex *Phil. Trans. R. Soc.* B **356** 299–330 © 2001 The Royal Society.

which is a generalized version of the Wilson–Cowan equations of nerve tissue introduced in section 11.1, with t measured in units of τ. The distribution $w_{lm}(\mathbf{r}, \phi|\mathbf{r}', \phi')$ represents the strength or *weight* of connections from the iso-orientation patch ϕ' at cortical position $\mathbf{r}'$ to the orientation patch ϕ at position $\mathbf{r}$.

Motivated by experimental observations concerning the intrinsic circuitry of V1 (see the introduction to the chapter), we decompose w in terms of *local* connections from elements within the same hypercolumn, and patchy excitatory *lateral* connections from elements in other hypercolumns:

$$w_{lm}(\mathbf{r}, \phi|\mathbf{r}', \phi') = w_{\text{loc}}(\phi|\phi')\delta(\mathbf{r} - \mathbf{r}') + \varepsilon w_{\text{lat}}(\mathbf{r}, \phi|\mathbf{r}', \phi')\delta_{m,\text{E}}\beta_l \qquad (11.20)$$

where ε is a parameter that measures the weight of lateral relative to local connections. Observations in [30] suggest that ε is small and, therefore, that the lateral connections modulate rather than drive V1 activity. Note that although the lateral connections are excitatory [25,46], 20% of the connections in layers II and III of V1 end on inhibitory interneurons, so the overall action of the lateral connections can become inhibitory, especially at high levels of activity [30]. The relative strengths of the lateral inputs into local excitatory and inhibitory populations are represented by the factors β_l.

The local weight distribution is taken to be homogeneous, that is,

$$w_{\text{loc}}(\phi|\phi') = W(\phi - \phi') \qquad (11.21)$$

for some π-periodic, even function W. It follows that an isolated hypercolumn (zero lateral interactions) has internal O(2) symmetry corresponding to rotations and reflections within the ring. In order to incorporate the anisotropic nature of the lateral connections, we further decompose w_{lat} as [12]

$$w_{\text{lat}}(\mathbf{r}, \phi|\mathbf{r}', \phi') = J(T_{-\phi}(\mathbf{r} - \mathbf{r}'))\delta(\phi - \phi') \qquad (11.22)$$

where

$$J(\mathbf{r}) = \int_{-\pi/2}^{\pi/2} p(\eta) \int_{-\infty}^{\infty} g(s)\delta(\mathbf{r} - s\mathbf{e}_\eta)\, \mathrm{d}s\, \mathrm{d}\eta \qquad (11.23)$$

with $\mathbf{e}_\eta = (\cos(\eta), \sin(\eta))$, $p(-\eta) = p(\eta)$ and T_ϕ the rotation matrix

$$T_\phi \begin{pmatrix} x \\ y \end{pmatrix} = \begin{pmatrix} \cos\phi & -\sin\phi \\ \sin\phi & \cos\phi \end{pmatrix} \begin{pmatrix} x \\ y \end{pmatrix}.$$

Such a distribution links neurons with the same orientation and spatial frequency label, with the function $p(\eta)$ determining the degree of spatial spread (anisotropy) in the pattern of connections relative to the direction of their common orientation preference. The weighting function $g(s)$ specifies how the strength of interaction varies with the distance of separation. A simplified schematic representation of the pattern of lateral connections is illustrated for our coupled hypercolumn model in figure 11.11.

Substituting equations (11.20) and (11.22) back into equation (11.19) leads to the evolution equation

$$\begin{aligned} \frac{\partial a_l(\mathbf{r}, \phi, t)}{\partial t} = & -a_l(\mathbf{r}, \phi, t) + \sum_m \int W_{lm}(\phi - \phi')\sigma[a_m(\mathbf{r}', \phi', t)] \frac{\mathrm{d}\phi'}{\pi} \\ & + \varepsilon\beta_l \int_{-\pi/2}^{\pi/2} p(\eta) \int_{-\infty}^{\infty} g(s)\sigma[a_\text{E}(\mathbf{r} + s\mathbf{e}_{\eta+\phi}, \phi, t)]\, \mathrm{d}\eta\, \mathrm{d}s \\ & + h_l(\mathbf{r}, \phi, t). \end{aligned} \qquad (11.24)$$

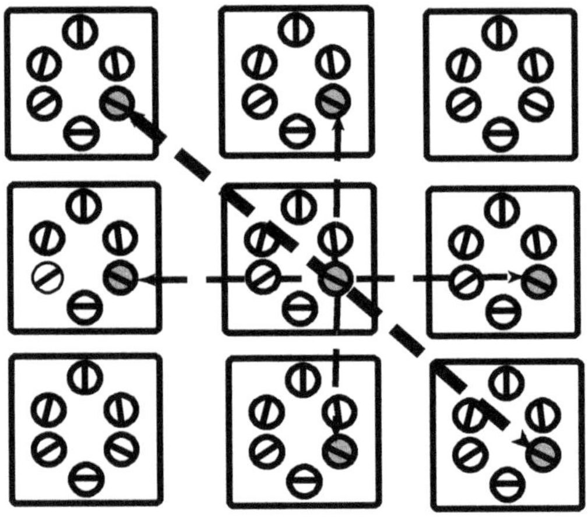

Figure 11.11. Schematic diagram of a coupled hypercolumn model of V1. It is assumed that there are isotropic local interactions within a hypercolumn, and anisotropic lateral interactions between hypercolumns. The latter connect iso-orientation patches located within some angular distance from the visuotopic axis parallel to their (common) orientation preference (as illustrated for the shaded patches).

If $p(\eta) = 1/\pi$ for all η then the weight distribution is isotropic and the system (11.24) is equivariant with respect to $E(2) \times O(2)$, where E(2) denotes the Euclidean group of translations, rotations and reflections in the cortex, and O(2) is the internal symmetry group of an isolated hypercolumn. It is important to emphasize that cortical rotations are distinct from rotations in the visual field (which correspond to vertical translations in cortex), as well as from internal rotations with respect to orientation. When $p(\eta)$ is non-uniform, the resulting anisotropy breaks both cortical and internal O(2) symmetries. However, full Euclidean symmetry, $E(2) = \mathbb{R}^2 \dot{+} O(2)$, is recovered by considering the combined Euclidean action on $\{\mathbf{r}, \phi\}$, which introduces a form of *shift-twist* symmetry in the plane [11, 12, 17, 66]. More specifically, the anisotropic weight distribution (11.22) is invariant with respect to the following action of the Euclidean group:

$$
\begin{aligned}
\mathbf{s} \cdot (\mathbf{r}, \phi) &= (\mathbf{r} + \mathbf{s}, \phi) \qquad && \mathbf{s} \in \mathbb{R}^2 \\
\xi \cdot (\mathbf{r}, \phi) &= (T_\xi \mathbf{r}, \phi + \xi) \qquad && \xi \in S^1 \\
\kappa \cdot (\mathbf{r}, \phi) &= (\kappa \mathbf{r}, -\phi)
\end{aligned}
\tag{11.25}
$$

where κ is the reflection $(x_1, x_2) \mapsto (x_1, -x_2)$. The corresponding group action on a function $a : \mathbb{R}^2 \times S \to \mathbb{R}$ is given by

$$
\gamma \cdot a(P) = a(\gamma^{-1} \cdot P) \qquad \text{for all } \gamma \in \mathbb{R}^2 \dot{+} O(2) \tag{11.26}
$$

and the invariance of $w_{\rm lat}(P|P')$ is expressed as

$$\gamma \cdot w_{\rm lat}(P|P') = w_{\rm lat}(\gamma^{-1} \cdot P|\gamma^{-1} \cdot P') = w_{\rm lat}(P|P').$$

It can be seen that the rotation operation comprises a translation or shift of the orientation preference label ϕ to $\phi + \xi$, together with a rotation or twist of the position vector $\mathbf{r}$ by the angle ξ. Such an operation provides a novel way to generate the Euclidean group E(2) of rigid motions in the plane. The fact that the weighting functions are invariant with respect to this form of E(2) has important consequences for the global dynamics of V1 in the presence of anisotropic lateral connections [11, 12].

11.3 Orientation tuning and O(2) symmetry

In the absence of lateral connections ($\varepsilon = 0$) each hypercolumn is independently described by the so-called ring model of orientation tuning [3, 4, 10, 41, 51], in which the internal structure of a hypercolumn is idelaized as a ring of orientation selective cells. That is, equation (11.19) reduces to

$$\frac{\partial a_l}{\partial t} = -a_l + \sum_{m=E,I} W_{lm} * \sigma(a_m) + h_l \tag{11.27}$$

where $*$ indicates a convolution operation

$$W * f(\phi) = \int_{-\pi/2}^{\pi/2} W(\phi - \phi') f(\phi') \frac{\mathrm{d}\phi'}{\pi}. \tag{11.28}$$

Just as in section 11.1 the local stability of $(\bar{a}_{\rm E}, \bar{a}_{\rm I})$ is found by linearization about the fixed points $\bar{a}_l$:

$$\frac{\partial b_l}{\partial t} = -b_l + \mu \sum_m W_{lm} * b_m \tag{11.29}$$

where $b_l(\mathbf{r}, \phi, t) = a_l(\mathbf{r}, \phi, t) - \bar{a}_l$. Equation (11.29) has solutions of the form

$$b_l(\mathbf{r}, \phi, t) = B_l \mathrm{e}^{\lambda t}[z(\mathbf{r})\mathrm{e}^{2in\phi} + \bar{z}(\mathbf{r})\mathrm{e}^{-2in\phi}] \tag{11.30}$$

where $z(\mathbf{r})$ is an arbitrary (complex) amplitude with complex conjugate $\bar{z}(\mathbf{r})$, and λ satisfies the eigenvalue equation

$$(1 + \lambda)\mathbf{B} = \mu \widetilde{\mathbf{W}}(n)\mathbf{B}. \tag{11.31}$$

Here $\widetilde{W}_{lm}(n)$ is the nth Fourier coefficient in the expansion of the π-periodic weights kernels $W_{lm}(\phi)$:

$$W_{lm}(\phi) = \widetilde{W}_{lm}(0) + 2\sum_{n=1}^{\infty} \widetilde{W}_{lm}(n) \cos(2n\phi) \qquad l, m \equiv E, I \tag{11.32}$$

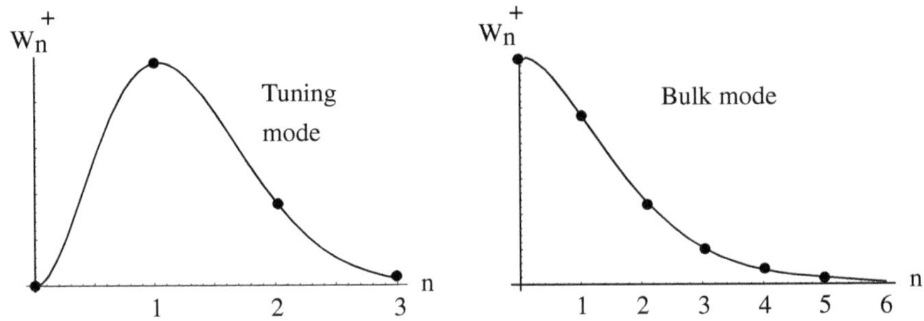

Figure 11.12. Spectrum W_n^+ of local weight distribution with a maximum at $n = 1$ (tuning mode) (*a*) and a maximum at $n = 0$ (bulk mode) (*b*).

It follows that

$$\lambda_n^\pm = -1 + \mu W_n^\pm \tag{11.33}$$

for integer n, where

$$W_n^\pm = \tfrac{1}{2}[\widetilde{W}_{EE}(n) + \widetilde{W}_{II}(n) \pm \Sigma(n)] \tag{11.34}$$

are the eigenvalues of the weight matrix with

$$\Sigma(n) = \sqrt{[\widetilde{W}_{EE}(n) - \widetilde{W}_{II}(n)]^2 + 4\widetilde{W}_{EI}(n)\widetilde{W}_{IE}(n)}. \tag{11.35}$$

The corresponding eigenvectors (up to an arbitrary normalization) are

$$\mathbf{B}_n^\pm = \begin{pmatrix} -\widetilde{W}_{EI}(n) \\ \tfrac{1}{2}[\widetilde{W}_{EE}(n) - \widetilde{W}_{II}(n) \mp \Sigma(n)] \end{pmatrix}. \tag{11.36}$$

Equation (11.33) implies that, for sufficiently small μ (low network excitability), $\lambda_n^\pm < 0$ for all n and the homogeneous resting state is stable. However, as μ increases an instability can occur leading to the spontaneous formation of an orientation tuning curve.

For the sake of illustration, suppose that the Fourier coefficients are given by the Gaussians

$$\widetilde{W}_{lm}(n) = \alpha_{lm}\mathrm{e}^{-n^2\xi_{lm}^2/2} \tag{11.37}$$

with ξ_{lm} determining the range of the axonal fields of the excitatory and inhibitory populations. We consider two particular cases.

Case A

If $\widetilde{W}_{EE}(n) = \widetilde{W}_{IE}(n)$ and $\widetilde{W}_{II}(n) = \widetilde{W}_{EI}(n)$ for all n, then $W_n^- = 0$ and

$$W_n^+ = \alpha_{EE}\mathrm{e}^{-n^2\xi_{EE}^2/2} - \alpha_{II}\mathrm{e}^{-n^2\xi_{II}^2/2} \tag{11.38}$$

Suppose that $\xi_{II} > \xi_{EE}$ and $0 < \alpha_{II} < \alpha_{EE}$. As we described in section 11.1 the resulting combination of short-range excitation and longer-range inhibition generates a Turing instability. Of particular relevance to orientation tuning is the case where W_n^+ has a unique (positive) maximum at $n = 1$; see figure 11.12(*a*). The homogeneous state then destabilizes at the critical point $\mu = \mu_c \equiv 1/W_1^+$ due to excitation of the eigenmodes $\mathbf{b}(\mathbf{r}, \phi, t) = \mathbf{B}a(\mathbf{r}, \phi, t)$ with $\mathbf{B} = (1, 1)^{\mathrm{T}}$ and

$$a(\mathbf{r}, \phi, t) = z(\mathbf{r})\mathrm{e}^{2\mathrm{i}\phi} + \bar{z}(\mathbf{r})\mathrm{e}^{-2\mathrm{i}\phi} = |z(\mathbf{r})| \cos(2[\phi - \phi^*(\mathbf{r})]) \tag{11.39}$$

with $z(\mathbf{r}) = |z(\mathbf{r})|\mathrm{e}^{-2\mathrm{i}\phi^*(\mathbf{r})}$. Since these modes have a single maximum in the interval $(-90°, 90°)$, each hypercolumn supports an activity profile consisting of a solitary peak centred about $\phi^*(\mathbf{r}) = \arg z(\mathbf{r})$. It can be shown that the saturating nonlinearities of the system stabilize the tuning curves beyond the critical point μ_c [10, 22]; see also section 11.4.2. The location of the peak $\phi^*(\mathbf{r})$ of the tuning curve at $\mathbf{r}$ is arbitrary in the presence of constant inputs, reflecting the hidden O(2) symmetry of a hypercolumn. However, the inclusion of an additional small amplitude input $\Delta h_l(\mathbf{r}, \phi) \sim \cos[2(\phi - \Phi(\mathbf{r}))]$ breaks this symmetry, and locks the location of the tuning curve at each point $\mathbf{r}$ to the orientation corresponding to the peak of the local stimulus, that is, $\phi^*(\mathbf{r}) = \Phi(\mathbf{r})$. As one moves further away from the point of instability, the amplitude of the tuning curve increases and sharpening occurs due to the nonlinear effects of the firing rate function (11.3). This is illustrated in figure 11.13, where the input and output (normalized) firing rate of the excitatory population of a single hypercolumn are shown. Thus the local intracortical connections within a hypercolumn serve both to amplify and sharpen a weakly oriented input signal from the LGN [4,51]. However, if the local level of inhibition is reduced such that $\alpha_{II} \ll \alpha_{EE}$, then W_n^+ is a monotonically decreasing function of $|n|$ (see figure 11.12(*b*)), and the homogeneous fixed point undergoes a bulk instability resulting in broadening of the tuning curve. This is consistent with experimental data demonstrating a loss of stable orientation tuning in cats with blocking of intracortical inhibition [44][2].

Case B

If $\Sigma(n)$ is pure imaginary, $\Sigma(n) = \mathrm{i}\Omega(n)$, then

$$W_n^{\pm} = \alpha_{EE}\mathrm{e}^{-n^2\xi_{EE}^2/2} - \alpha_{II}\mathrm{e}^{-n^2\xi_{II}^2/2} \pm \mathrm{i}\Omega(n). \tag{11.40}$$

Assume, as in case A, that the difference of Gaussians has a maximum at $n = 1$. Then an instability will occur at the critical point $\mu_c = 1/\mathrm{Re}(W_1^+)$ due to

[2] The idea that local cortical interactions play a role in orientation tuning is still controversial. The classical model of Hubel and Wiesel [31] proposes a very different mechanism, in which both the orientation preference and tuning of a cell arise primarily from the geometrical alignment of the receptive fields of the LGN neurons projecting to it. This has also received recent experimental support [24]. However, intracellular measurements indicate that direct inputs from the LGN to neurons in layer 4 of V1 provide only a fraction of the total excitatory inputs relevant to orientation tuning [58].

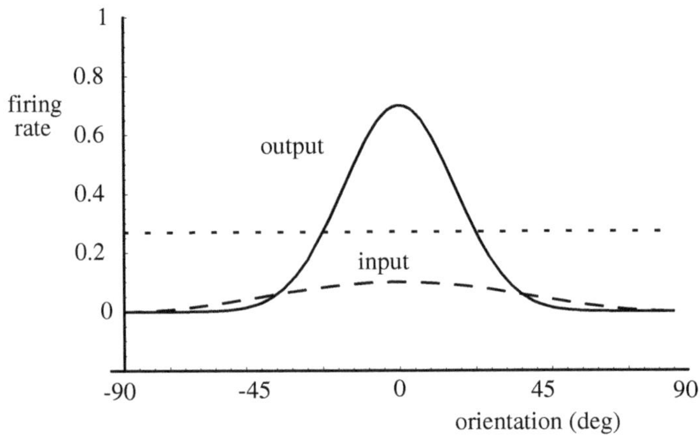

Figure 11.13. Sharp orientation tuning curve in a single hypercolumn. Local recurrent excitation and inhibition amplifies a weakly modulated input from the LGN. The dotted line is the base-line output without orientation tuning.

excitation of the oscillatory eigenmodes

$$\mathbf{b}(\mathbf{r}, \phi, t) = [z_{\mathrm{L}}(\mathbf{r})\mathrm{e}^{\mathrm{i}(\Omega_0 t - 2\phi)} + z_{\mathrm{R}}(\mathbf{r})\mathrm{e}^{\mathrm{i}(\Omega_0 t + 2\phi)}]\mathbf{B} + \text{c.c.} \tag{11.41}$$

where $\Omega_0 = \mu_c \Omega(1)$ and $\mathbf{B} = \mathbf{B}_1^+$. It is then possible for rotating tuning curves to be generated spontaneously within a hypercolumn [4].

11.4 Amplitude equation for interacting hypercolumns

An isolated hypercolumn exhibits spontaneous O(2) symmetry breaking leading to the formation of an orientation tuning curve. How is this process modulated by anisotropic lateral interactions between hypercolumns? In this section we use perturbation theory to derive a dynamical equation for the complex amplitude $z(\mathbf{r})$ for orientation tuning in the presence of lateral interactions. We will then use this amplitude equation to show how the lateral interactions induce correlations between $z(\mathbf{r})$ at different points in the cortex, leading to spatially periodic patterns of activity across V1; see section 11.5. These patterns reproduce the commonly found types of geometric visual hallucinations when mapped back into visual field coordinates under the retino-cortical map of figure 11.2; see section 11.5.4. Our basic assumptions in the derivation of the amplitude equation are as follows:

(i) each hypercolumn is close to a bifurcation point signalling the onset of sharp orientation tuning; and
(ii) the interactions between hypercolumns are *weak*.

11.4.1 Cubic amplitude equation: stationary case

Let us perform a Taylor expansion of equation (11.24) with $b_l(\mathbf{r}, \phi, t) = a_l(\mathbf{r}, \phi, t) - \bar{a}_l$:

$$\begin{aligned}\frac{\partial b_l}{\partial t} &= -b_l + \sum_{m=E,I} W_{lm} * [\mu b_m + \gamma_m b_m^2 + \gamma_m' b_m^3 + \cdots] + \Delta h_l \\ &\quad + \varepsilon \beta_l w_{\text{lat}} \circ ([\bar{\sigma}_{\text{E}} + \mu b_{\text{E}} + \cdots]) \end{aligned} \tag{11.42}$$

where $\Delta h_l = h_l - \bar{h}_l$ and $\mu = \sigma'(\bar{a}_{\text{E}})$, $\gamma_l = \sigma''(\bar{a}_l)/2$, $\gamma_l' = \sigma'''(\bar{a}_l)/6$. The convolution operation $*$ is defined by equation (11.28)) and

$$[w_{\text{lat}} \circ f](\mathbf{r}, \phi) = \int_{-\pi/2}^{\pi/2} \int_{-\infty}^{\infty} p(\eta) g(s) f(\mathbf{r} + s\mathbf{e}_{\eta+\phi}, \phi)\, \mathrm{d}s\, \mathrm{d}\eta \tag{11.43}$$

for an arbitrary function $f(\mathbf{r}, \phi)$ and w_{lat} given by equation (11.22). Suppose that the system is ε-close to the point of marginal stability of the homogeneous fixed point associated with excitation of the modes $\mathrm{e}^{\pm 2\mathrm{i}\phi}$. That is, take $\mu = \mu_{\text{c}} + \varepsilon \Delta\mu$ where $\mu_{\text{c}} = 1/W_1^+$, see equation (11.33). Substitute into equation (11.42) the perturbation expansion

$$b_m = \varepsilon^{1/2} b_m^{(1)} + \varepsilon b_m^{(2)} + \varepsilon^{3/2} b_m^{(3)} + \cdots. \tag{11.44}$$

Finally, introduce a slow time scale $\tau = \varepsilon t$ and collect terms with equal powers of ε. This leads to a hierarchy of equations of the form (up to $\mathcal{O}(\varepsilon^{3/2})$)

$$[\mathcal{L}\mathbf{b}^{(1)}]_l = 0 \tag{11.45}$$

$$\begin{aligned}[\mathcal{L}\mathbf{b}^{(2)}]_l &= v_l^{(2)} \\ &\equiv \sum_{m\equiv E,I} \gamma_m W_{lm} * [b_m^{(1)}]^2 + \beta_l \bar{\sigma}_{\text{E}} w_{\text{lat}} \circ 1 \end{aligned} \tag{11.46}$$

$$\begin{aligned}[\mathcal{L}\mathbf{b}^{(3)}]_l &= v_l^{(3)} \\ &\equiv -\frac{\partial b_l^{(1)}}{\partial \tau} + \sum_{m\equiv E,I} W_{lm} * [\Delta\mu b_m^{(1)} + \gamma_m' [b_m^{(1)}]^3 + 2\gamma_m b_m^{(1)} b_m^{(2)}] \\ &\quad + \mu_{\text{c}} \beta_l w_{\text{lat}} \circ b_{\text{E}}^{(1)} + \Delta h_l \end{aligned} \tag{11.47}$$

with the linear operator $\mathcal{L}$ defined according to

$$[\mathcal{L}\mathbf{b}]_l = b_l - \mu_{\text{c}} \sum_{m=E,I} W_{lm} * b_m. \tag{11.48}$$

We have also assumed that the modulatory external input is $\mathcal{O}(\varepsilon^{3/2})$ and rescaled $\Delta h_l \to \varepsilon^{3/2} \Delta h_l$

The first equation in the hierarchy, equation (11.45), has solutions of the form

$$\mathbf{b}^{(1)}(\mathbf{r},\phi,\tau) = (z(\mathbf{r},\tau)\mathrm{e}^{2\mathrm{i}\phi} + \bar{z}(\mathbf{r},\tau)\mathrm{e}^{-2\mathrm{i}\phi})\mathbf{B} \tag{11.49}$$

with $\mathbf{B} \equiv \mathbf{B}_1^+$ defined in equation (11.36). We obtain a dynamical equation for the complex amplitude $z(\mathbf{r},\tau)$ by deriving solvability conditions for the higher-order equations. We proceed by taking the inner product of equations (11.46) and (11.47) with the dual eigenmode $\widetilde{\mathbf{b}}(\phi) = \mathrm{e}^{2\mathrm{i}\phi}\widetilde{\mathbf{B}}$ where

$$\widetilde{\mathbf{B}} = \begin{pmatrix} W_{IE}(1) \\ -\frac{1}{2}[W_{EE}(1) - W_{II}(1) - \Sigma(1)] \end{pmatrix} \tag{11.50}$$

so that

$$[\mathcal{L}^{\mathrm{T}}\widetilde{\mathbf{b}}]_l \equiv \widetilde{b}_l - \mu_{\mathrm{c}} \sum_{m\equiv E,I} W_{ml} * \widetilde{b}_m = 0.$$

The inner product of any two vector-valued functions of ϕ is defined as

$$\langle \mathbf{u}|\mathbf{v}\rangle = \int_0^\pi [\bar{u}_{\mathrm{E}}(\phi)v_{\mathrm{E}}(\phi) + \bar{u}_{\mathrm{I}}(\phi)v_{\mathrm{I}}(\phi)]\frac{\mathrm{d}\phi}{\pi}. \tag{11.51}$$

With respect to this inner product, the linear operator $\mathcal{L}$ satisfies $\langle\widetilde{\mathbf{b}}|\mathcal{L}\mathbf{b}\rangle = \langle\mathcal{L}^{\mathrm{T}}\widetilde{\mathbf{b}}|\mathbf{b}\rangle = 0$ for any $\mathbf{b}$. Since $\mathcal{L}\mathbf{b}^{(p)} = \mathbf{v}^{(p)}$, we obtain a hierarchy of solvability conditions $\langle\widetilde{\mathbf{b}}|\mathbf{v}^{(p)}\rangle = 0$ for $p = 2, 3, \ldots$.

It can be shown from equations (11.43), (11.46) and (11.49) that the first solvability condition is identically satisfied (provided that the system is bifurcating from a uniform state). The solvability condition $\langle\widetilde{\mathbf{b}}|\mathbf{v}^{(3)}\rangle = 0$ generates a cubic amplitude equation for $z(\mathbf{r},\tau)$. As a further simplification we set $\gamma_m = 0$, since this does not alter the basic structure of the amplitude equation. Using equations (11.43), (11.47) and (11.49) we then find that (after rescaling τ)

$$\begin{aligned}\frac{\partial z(\mathbf{r},\tau)}{\partial\tau} &= z(\mathbf{r},\tau)(\Delta\mu - A|z(\mathbf{r},\tau)|^2) + f(\mathbf{r}) \\ &\quad + \beta\int_0^\pi w_{\mathrm{lat}} \circ [z(\mathbf{r},\tau) + \bar{z}(\mathbf{r},\tau)\mathrm{e}^{-4i\phi}]\frac{\mathrm{d}\phi}{\pi}\end{aligned} \tag{11.52}$$

where

$$f(\mathbf{r}) = \mu_{\mathrm{c}} \sum_{l\equiv E,I} \widetilde{B}_l \int_0^\pi \mathrm{e}^{-2\mathrm{i}\phi}\Delta h_l(\mathbf{r},\phi)\frac{\mathrm{d}\phi}{\pi} \tag{11.53}$$

and

$$\beta = \frac{\mu_{\mathrm{c}}^2 B_{\mathrm{E}}}{\widetilde{\mathbf{B}}^{\mathrm{T}}\mathbf{B}} \sum_{l\equiv E,I} \beta_l\widetilde{B}_l \qquad A = -\frac{3}{\widetilde{\mathbf{B}}^{\mathrm{T}}\mathbf{B}} \sum_{l\equiv E,I} \widetilde{B}_l\gamma_l' B_l^3. \tag{11.54}$$

Equation (11.52) is our reduced model of weakly interacting hypercolumns. It describes the effects of anisotropic lateral connections and modulatory inputs from the LGN on the dynamics of the (complex) amplitude $z(\mathbf{r},\tau)$. The latter determines the response properties of the orientation tuning curve associated with the hypercolumn at cortical position $\mathbf{r}$. The coupling parameter β is a linear combination of the relative strengths of the lateral connections innervating excitatory neurons and those innervating inhibitory neurons with D_E, D_I determined by the local weight distribution. Since $D_E > 0$ and $D_I < 0$, we see that the effective interactions between hypercolumns have both an excitatory and an inhibitory component.

11.4.2 Orientation tuning revisited

In the absence of lateral interactions, equation (11.52) reduces to

$$\frac{\partial z(\mathbf{r},\tau)}{\partial \tau} = z(\mathbf{r},\tau)(\Delta\mu - A|z(\mathbf{r},\tau)|^2) + f(\mathbf{r}). \tag{11.55}$$

For the nonlinear output function (11.3), we find that $A > 0$. Hence, if $f(\mathbf{r}) = 0$ then there exist (marginally) stable time-independent solutions of the form $z(\mathbf{r}) = \sqrt{\Delta\mu/A}\mathrm{e}^{-\mathrm{i}\phi(\mathbf{r})}$ where $\phi(\mathbf{r})$ is an arbitrary phase that determines the location of the peak of the tuning curve at position $\mathbf{r}$. Now consider the effects of a weakly biased input from the LGN $h_l(\mathbf{r},\phi,\tau) = C(\mathbf{r})\cos(2[\phi - \omega\tau])$. This represents a slowly rotating stimulus with angular velocity ω and contrast $C(\mathbf{r}) = \mathcal{O}(\varepsilon^{3/2})$. Equation (11.53) implies that $f(\mathbf{r}) = C(\mathbf{r})\mathrm{e}^{-2\mathrm{i}\omega\tau}$. Writing $z = v\mathrm{e}^{-2\mathrm{i}(\phi+\omega\tau)}$ we obtain from (11.55) the pair of equations

$$\begin{aligned}\dot{v} &= v(\mu - \mu_c + Av^2) + C\cos(2\phi)\\ \dot{\phi} &= -\omega - \frac{C}{2v}\sin(2\phi).\end{aligned} \tag{11.56}$$

Thus, provided that ω is sufficiently small, equation (11.56) will have a stable fixed point solution $\{v^*(\mathbf{r}), \phi^*(\mathbf{r})\}$ in which the peak of the tuning curve is entrained to the signal. That is, writing $\mathbf{b}^{(1)}(\mathbf{r},\phi) = \mathbf{B}a(\mathbf{r},\phi)$,

$$a(\mathbf{r},\phi) = v^*(\mathbf{r})\cos(2[\phi - \omega\tau - \phi^*(\mathbf{r})]) \tag{11.57}$$

with $\phi^*(\mathbf{r}) = 0$ when $\omega = 0$.

It is also possible to repeat our bifurcation analysis in the case where each hypercolumn undergoes a bulk instability. This occurs, for example, when the spectrum of local connections is as in figure 11.12(*b*). The amplitude equation (11.52) now takes the form

$$\frac{\partial a(\mathbf{r},\tau)}{\partial \tau} = a(\mathbf{r},\tau)(\Delta\mu - Aa(\mathbf{r},\tau)^2) + f_0(\mathbf{r}) + \beta\int_0^{\pi} w_{\mathrm{lat}} \circ a(\mathbf{r},\tau)\frac{\mathrm{d}\phi}{\pi} \tag{11.58}$$

with a real and f_0 the ϕ-averaged LGN input. It follows that, in the absence of lateral interactions, each hypercolumn bifurcates to a ϕ-independent state whose amplitude $a(\mathbf{r})$ is a root of the cubic

$$a(\mathbf{r})(\Delta\mu - Aa(\mathbf{r})^2) + f_0(\mathbf{r}) = 0. \tag{11.59}$$

11.4.3 Cubic amplitude equation: oscillatory case

In our derivation of the amplitude equation (11.52) we assumed that the local cortical circuit generates a stationary orientation tuning curve. However, as shown in section 11.3, it is possible for a time-periodic tuning curve to occur when $\mathrm{Im}(W_1^+) \neq 0$. Taylor expanding (11.24) as before leads to the hierarchy of equations (11.45)–(11.47) except that the linear operator $\mathcal{L} \to \mathcal{L}_t = \mathcal{L} + \partial/\partial t$. The lowest-order solution (11.49) now takes the form

$$\mathbf{b}^{(1)}(\mathbf{r}, \phi, t, \tau) = [z_{\mathrm{L}}(\mathbf{r}, \tau)\mathrm{e}^{\mathrm{i}(\Omega_0 t - 2\phi)} + z_{\mathrm{R}}(\mathbf{r}, \tau)\mathrm{e}^{\mathrm{i}(\Omega_0 t + 2\phi)}]\mathbf{B} + \mathrm{c.c.} \quad (11.60)$$

where z_{L} and z_{R} represent the complex amplitudes for anticlockwise (L) and clockwise (R) rotating waves (around the ring of a single hypercolumn), and $\Omega_0 = \mu_{\mathrm{c}}\,\mathrm{Im}(\Sigma(1))$. Introduce the generalized inner product

$$\langle \mathbf{u}|\mathbf{v}\rangle = \lim_{T\to\infty} \frac{1}{T}\int_{-T/2}^{T/2}\int_0^{\pi} [\overline{u}_{\mathrm{E}}(\phi, t)v_{\mathrm{E}}(\phi, t) + \overline{u}_{\mathrm{I}}(\phi, t)v_{\mathrm{I}}(\phi, t)]\frac{\mathrm{d}\phi}{\pi}\,\mathrm{d}t \quad (11.61)$$

and the dual vectors $\widetilde{\mathbf{b}}_{\mathrm{L}} = \widetilde{\overline{\mathbf{B}}}\mathrm{e}^{\mathrm{i}(\Omega_0 t - 2\phi)}$, $\widetilde{\mathbf{b}}_{\mathrm{R}} = \widetilde{\overline{\mathbf{B}}}\mathrm{e}^{\mathrm{i}(\Omega_0 t + 2\phi)}$. Using the fact that $\langle \widetilde{\mathbf{b}}_{\mathrm{L}}|\mathcal{L}_t b\rangle = \langle \widetilde{\mathbf{b}}_{\mathrm{R}}|\mathcal{L}_t \mathbf{b}\rangle = 0$ for arbitrary $\mathbf{b}$ we obtain the pair of solvability conditions $\langle \widetilde{\mathbf{b}}_{\mathrm{L}}|\mathbf{v}^{(p)}\rangle = \langle \widetilde{\mathbf{b}}_{\mathrm{R}}|\mathbf{v}^{(p)}\rangle = 0$ for each $p \geq 2$.

As in the stationary case, the $p = 2$ solvability conditions are identically satisfied. The $p = 3$ solvability conditions then generate cubic amplitude equations for z_{L}, z_{R} of the form

$$\begin{aligned}\frac{\partial z_{\mathrm{L}}(\mathbf{r}, \tau)}{\partial \tau} &= (1 + \mathrm{i}\Omega_0)z_{\mathrm{L}}(\mathbf{r}, \tau)(\Delta\mu - A|z_{\mathrm{L}}(\mathbf{r}, \tau)|^2 - 2A|z_{\mathrm{R}}(\mathbf{r}, \tau)|^2)\\ &\quad + \beta\int_0^{\pi} w_{\mathrm{lat}} \circ [z_{\mathrm{L}}(\mathbf{r}, \tau) + z_{\mathrm{R}}(\mathbf{r}, \tau)\mathrm{e}^{4\mathrm{i}\phi}]\frac{\mathrm{d}\phi}{\pi}\end{aligned} \quad (11.62)$$

and

$$\begin{aligned}\frac{\partial z_{\mathrm{R}}(\mathbf{r}, \tau)}{\partial \tau} &= (1 + \mathrm{i}\Omega_0)z_{\mathrm{R}}(\mathbf{r}, \tau)(\Delta\mu - A|z_{\mathrm{R}}(\mathbf{r}, \tau)|^2 - 2A|z_{\mathrm{L}}(\mathbf{r}, \tau)|^2)\\ &\quad + \beta\int_0^{\pi} w_{\mathrm{lat}} \circ [z_{\mathrm{R}}(\mathbf{r}, \tau) + z_{\mathrm{L}}(\mathbf{r}, \tau)\mathrm{e}^{-4\mathrm{i}\phi}]\frac{\mathrm{d}\phi}{\pi}\end{aligned} \quad (11.63)$$

where

$$f_{\pm}(\mathbf{r}) = \lim_{T\to\infty}\frac{\mu_{\mathrm{c}}}{T}\int_{-T/2}^{T/2}\int_0^{\pi} \mathrm{e}^{-\mathrm{i}(\Omega_0 t \pm 2\phi)}\sum_{l \equiv E, I}\widetilde{B}_l \Delta h_l(\mathbf{r}, \phi, t)\frac{\mathrm{d}\phi}{\pi}\,\mathrm{d}t. \quad (11.64)$$

Note that the amplitudes only couple to time-dependent inputs from the LGN.

11.5 Cortical pattern formation and E(2) symmetry

We now use the amplitude equations derived in section 11.4 to investigate how O(2) symmetry breaking within a hypercolumn is modified by the presence of anisotropic lateral interactions, and show how it leads to the formation of spatially periodic activity patterns across the cortex that break the underlying E(2) symmetry. We begin by considering the case of stationary patterns. Oscillatory patterns will be considered in section 11.5.5.

11.5.1 Linear stability analysis

Since we are focusing on spontaneous pattern formation, we shall assume that there are no inputs from the LGN, $f(\mathbf{r}) = 0$. Equation (11.52) then has the trivial solution $z = 0$. Linearizing about this solution gives

$$\frac{\partial z(\mathbf{r}, \tau)}{\partial \tau} = \Delta\mu\, z(\mathbf{r}, \tau) + \beta \int_0^{\pi} w_{\text{lat}} \circ [z(\mathbf{r}, \tau) + \bar{z}(\mathbf{r}, \tau)\mathrm{e}^{-4\mathrm{i}\phi}] \frac{\mathrm{d}\phi}{\pi}. \tag{11.65}$$

If we ignore boundary effects by treating V1 as an unbounded two dimensional sheet, then equation (11.65) has two classes of solution, $z_{\pm}$, of the form

$$z_+(\mathbf{r}, \tau) = \mathrm{e}^{\lambda_+ \tau} \mathrm{e}^{-2\mathrm{i}\varphi} [c\mathrm{e}^{\mathrm{i}\mathbf{k}\cdot\mathbf{r}} + \bar{c}\mathrm{e}^{-\mathrm{i}\mathbf{k}\cdot\mathbf{r}}] \tag{11.66}$$
$$z_-(\mathbf{r}, \tau) = \mathrm{i}\mathrm{e}^{\lambda_- \tau} \mathrm{e}^{-2\mathrm{i}\varphi} [c\mathrm{e}^{\mathrm{i}\mathbf{k}\cdot\mathbf{r}} + \bar{c}\mathrm{e}^{-\mathrm{i}\mathbf{k}\cdot\mathbf{r}}] \tag{11.67}$$

where $\mathbf{k} = q(\cos(\varphi), \sin(\varphi))$ and c is an arbitrary complex amplitude. Substitution into equation (11.65) and using equation (11.43) leads to the eigenvalue equation

$$\lambda_{\pm} = \Delta\mu + \beta \int_0^{\pi} \left[\int_{-\infty}^{\infty} g(s) \mathrm{e}^{iqs\cos(\phi)}\, \mathrm{d}s \right] (1 \pm \chi \mathrm{e}^{-4\mathrm{i}\phi}) \frac{\mathrm{d}\phi}{\pi} \tag{11.68}$$

where

$$\chi = \int_{-\pi/2}^{\pi/2} p(\eta) \mathrm{e}^{-4\mathrm{i}\eta}\, \mathrm{d}\eta. \tag{11.69}$$

Using an expansion in terms of Bessel functions

$$\mathrm{e}^{\mathrm{i}x\cos(\phi)} = \sum_{n=-\infty}^{\infty} (-\mathrm{i})^n J_n(x) \mathrm{e}^{in\phi} \tag{11.70}$$

the eigenvalue equation reduces to the more compact form

$$\lambda_{\pm} = \Delta\mu + \beta G_{\pm}(q) \tag{11.71}$$

with

$$G_{\pm}(q) = G_0(q) \pm \chi G_2(q) \tag{11.72}$$

and

$$G_n(q) = (-1)^n \int_{-\infty}^{\infty} g(s) J_{2n}(qs)\, \mathrm{d}s. \tag{11.73}$$

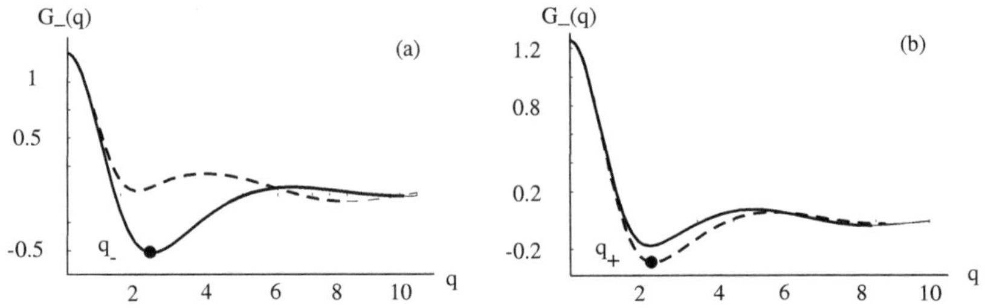

Figure 11.14. Plot of functions $G_-(q)$ (full curve) and $G_+(q)$ (broken curve) in the case $\chi = 1$ (strong anisotropy) and $g(s)$ defined by (11.74) for $\xi = 1$ and $s_0 = 1$ (*a*). The critical wavenumber for spontaneous pattern formation is q_-. The marginally stable eigenmodes are odd functions of ϕ. Similarly, with $\chi = \sin 4\eta_0/4\eta_0$ (*b*) with lateral spread of width $\eta_0 = \pi/3$. The marginally stable eigenmodes are now even functions of ϕ.

Before using equation (11.71) to determine how the lateral interactions modify the condition for marginal stability, we need to specify the form of the weight distribution $g(s)$. From experimental data based on tracer injections it appears that the patchy lateral connections extend several mm on either side of a hypercolumn and the field of axon terminals within a patch tends to diminish in size the further away it is from the injection site [25,39,46,65]. The total extent of the connections depends on the particular species under study. In our continuum model we assume that

$$g(s) = \mathrm{e}^{-(s-s_0)^2/2\xi^2}\Theta(s - s_0) \tag{11.74}$$

where ξ determines the range and s_0 the minimum distance of the (non-local) lateral connections. Recall that there is growing experimental evidence to suggest that lateral connections tend to have an inhibitory effect in the presence of high-contrast visual stimuli but an excitatory effect at low contrasts [27]. It is possible that during the experience of hallucinations there are sufficient levels of activity within V1 for the inhibitory effects of the lateral connections to predominate. Many subjects who have taken LSD and similar hallucinogens report seeing bright white light at the centre of the visual field which then explodes into a hallucinatory image in about 3 s, corresponding to a propagation velocity in V1 of about 2.5 cm s^{-1} suggestive of slowly moving epileptiform activity [49]. In light of this, we assume that $\beta < 0$ during the experience of a visual hallucination.

In figure 11.14(*a*) we plot $G_\pm(q)$ as a function of q for the given weight distribution (11.74) and the spread function $p(\eta) = \delta(\eta)$ for which $\chi = 1$. It can be seen that $G_\pm(q)$ has a unique minimum at $q = q_\pm \neq 0$ and $G_-(q_-) < G_+(q_+)$. Since $\beta < 0$ it follows that the homogeneous state $z(\mathbf{r}, \tau) = 0$ becomes marginally stable at the modified critical point $\mu_c' = \mu_c - \varepsilon\beta G_-(q_-)$. The corresponding marginally stable modes are given by combining equations (11.49) and (11.67) for $\lambda_- = 0$. Writing $\mathbf{b}^{(1)}(\mathbf{r}, \phi) = a(\mathbf{r}, \phi)\mathbf{B}$ we have

$$a(\mathbf{r}, \phi) = \sum_{n=1}^{N} c_n \mathrm{e}^{\mathrm{i}\mathbf{k}_n \cdot \mathbf{r}} \sin(\phi - \varphi_n) + \text{c.c.} \tag{11.75}$$

where $\mathbf{k}_n = q_-(\cos\varphi_n, \sin\varphi_n)$ and c_n is a complex amplitude. These modes will be recognized as linear combinations of plane waves modulated by *odd* (phase-shifted) π-periodic functions $\sin[2(\phi - \varphi_n)]$. The infinite degeneracy arising from rotation invariance means that all modes lying on the circle $|\mathbf{k}| = q_-$ become marginally stable at the critical point. However, this can be reduced to a finite set of modes by restricting solutions to be doubly periodic functions as explained in section 11.5.2.

The solutions (11.75) are precisely the lowest-order odd eigenfunctions derived using the perturbation methods of [11][3]. It is also possible for even (+) eigenmodes to destabilize first when there is a sufficient spread in the distribution of lateral connections about the visuotopic axis as shown in figure 11.11. More specifically, if we take $p(\eta) = \Theta(|\eta| - \eta_0)/2\eta_0$, then

$$G_\pm(q) = G_0(q) \pm \frac{\sin(4\eta_0)}{4\eta_0} G_2(q) \tag{11.76}$$

such that $G_+(q_+) < G_-(q_-)$ when $\eta_0 > \pi/4$, which is illustrated in figure 11.14(*b*). It follows that the homogeneous state now becomes marginally stable at the critical point $\mu_c' = \mu_c - \varepsilon\beta G_+(q_+)$ due to excitation of the even modes given by equations (11.49) and (11.66) for $\lambda_+ = 0$:

$$a(\mathbf{r}, \phi) = \sum_{n=1}^{N} c_n \mathrm{e}^{\mathrm{i}\mathbf{k}_n \cdot \mathbf{r}} \cos(\phi - \varphi_n) + \text{c.c.} \tag{11.77}$$

where $\mathbf{k}_n = q_+(\cos(\varphi_n), \sin(\varphi_n))$.

A third class of solution can occur when each hypercolumn undergoes a bulk instability, as described by the amplitude equation (11.58). Repeating the previous linear analysis, we find that there are now only even eigenmodes, which are ϕ-independent (to leading order), and take the form

$$a(\mathbf{r}) = \sum_{n=1}^{N} [c_n \mathrm{e}^{\mathrm{i}\mathbf{k}_n \cdot \mathbf{r}} + \bar{c}_n \mathrm{e}^{-\mathrm{i}\mathbf{k}_n \cdot \mathbf{r}}]. \tag{11.78}$$

The corresponding eigenvalue equation is

$$\lambda = \Delta\mu + G_0(q) \tag{11.79}$$

[3] Note that in [11] we used a different perturbation scheme in which the strength of lateral connections ε and the distance from the bifurcation point $\mu - \mu_c$ were taken to be two independent parameters. The linearized equations were first solved using a perturbation expansion in the coupling. Amplitude equations for the linear modes were then derived by carrying out a Poincaré–Linstedt expansion with respect to $\mu - \mu_c$. This approach is particularly suitable for studying the role of symmetries in the spontaneous formation of cortical activity patterns underlying visual hallucinations.

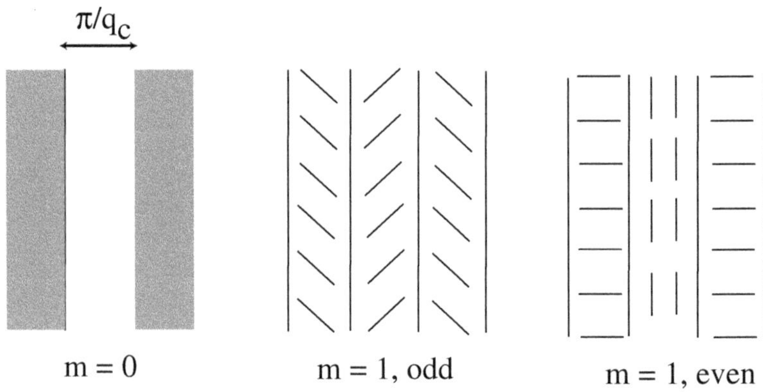

Figure 11.15. Three classes of rolls found in cortical pattern formation.

with $G_0(q)$ defined in equation (11.73). Thus $|\mathbf{k}_n| = q_0$ where q_0 is the minimum of $G_0(q)$.

It follows from our analysis that there are three classes of eigenmode that can bifurcate from the resting state. These are represented, respectively, by linear combinations of one of the three classes of roll pattern shown in figure 11.15. The $m = 0$ roll corresponds to modes of the form (11.78), and consists of alternating regions of high and low cortical activity in which individual hypercolumns do not amplify any particular orientation: the resulting patterns are said to be *non-contoured*. The $m = 1$ rolls correspond to the odd- and even-oriented modes of equations (11.75) and (11.77). These are constructed using a winner-takes-all rule in which only the orientation with maximal response is shown at each point in the cortex (after some coarse-graining). The resulting patterns are said to be *contoured*. The particular class that is selected depends on the detailed structure of the local and lateral weights. The $m = 0$ type will be selected when the local inhibition within a hypercolumn is sufficiently weak, whereas the $m = 1$ type will occur when there is strong local inhibition, with the degree of anisotropy in the lateral connections determining whether the patterns are even or odd.

11.5.2 Doubly-periodic planforms

Rotation symmetry implies that the space of marginally stable modes is infinite-dimensional. That is, all plane-waves with wavevectors $\mathbf{k}$ lying on the critical circle $|\mathbf{k}| = q_c$ are allowed, with $q_c = q_-$ for odd modes and $q_c = q_+$ for even modes. (For concreteness, we focus on the contoured eigenmodes.) However, translation symmetry means that we can restrict the space of solutions to that of doubly periodic functions corresponding to regular tilings of the plane. The associated space of marginally stable eigenmodes is then finite-dimensional. A finite set of specific functions can then be identified as candidate *planforms* in

Table 11.1. Generators for the planar lattices and their dual lattices.

Lattice	$\boldsymbol{\ell}_1$	$\boldsymbol{\ell}_2$	$\hat{\boldsymbol{\ell}}_1$	$\hat{\boldsymbol{\ell}}_2$
Square	$(1, 0)$	$(0, 1)$	$(1, 0)$	$(0, 1)$
Hexagonal	$(1, \frac{1}{\sqrt{3}})$	$(0, \frac{2}{\sqrt{3}})$	$(1, 0)$	$\frac{1}{2}(-1, \sqrt{3})$
Rhombic	$(1, -\cot\eta)$	$(0, \operatorname{cosec}\eta)$	$(1, 0)$	$(\cos\eta, \sin\eta)$

the sense that they approximate time-independent solutions of equation (11.24) sufficiently close to the critical point where the homogeneous state loses stability. These planforms consist of finite linear combinations of one of the three types of stripe pattern shown in figure 11.15.

Let $\mathcal{L}$ be a planar lattice; that is, choose two linearly independent vectors $\boldsymbol{\ell}_1$ and $\boldsymbol{\ell}_2$ and let

$$\mathcal{L} = \{2\pi m_1 \boldsymbol{\ell}_1 + 2\pi m_2 \boldsymbol{\ell}_2 : m_1, m_2 \in \mathbb{Z}\}.$$

Note that $\mathcal{L}$ is a subgroup of the group of planar translations. A function $f :\to \mathbb{R}$ is *doubly periodic with respect to* $\mathcal{L}$ if

$$f(x + \boldsymbol{\ell}, \phi) = f(x, \phi)$$

for every $\boldsymbol{\ell} \in \mathcal{L}$. Let θ be the angle between the two basis vectors $\boldsymbol{\ell}_1$ and $\boldsymbol{\ell}_2$. We can then distinguish three types of lattice according to the value of θ: square lattice ($\theta = \pi/2$), rhombic lattice ($0 < \theta < \pi/2$, $\theta \neq \pi/3$) and hexagonal ($\theta = \pi/3$). After rotation, the generators of the planar lattices are given in table 11.1. Also shown are the generators of the *dual lattice* satisfying $\hat{\boldsymbol{\ell}}_i.\boldsymbol{\ell}_j = \delta_{i,j}$ with $|\hat{\boldsymbol{\ell}}_i| = 1$.

Imposing double periodicity means that the original Euclidean symmetry group is restricted to the symmetry group $\Gamma_{\mathcal{L}}$ of the lattice $\mathcal{L}$. In particular, there are only a finite number of shift-twists and reflections to consider for each lattice (modulo an arbitrary rotation of the whole plane), which correspond to the so-called holohedries of the plane; see figure 11.16. Consequently the corresponding space of marginally stable modes is now finite-dimensional—we can only rotate eigenfunctions through a finite set of angles (for example, multiples of $\pi/2$ for the square lattice and multiples of $\pi/3$ for the hexagonal lattice). The marginally stable modes for each of the lattices are given in table 11.2.

11.5.3 Selection and stability of patterns

It remains to determine the amplitudes c_n of the doubly-periodic solutions that bifurcate from the homogeneous state; see table 11.2. We proceed by applying the perturbation method of section 11.4.1 to the amplitude equation (11.52). First, introduce a small parameter ξ determining the distance from the point

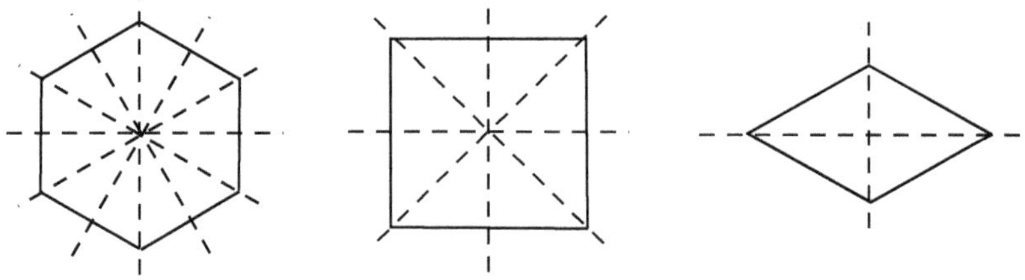

Figure 11.16. Holohedries of the plane.

Table 11.2. Eigenmodes corresponding to shortest dual wavevectors $\mathbf{k}_i = q_c\hat{\boldsymbol{\ell}}_i$. Here $u(\phi) = \cos(2\phi)$ for even modes and $u(\phi) = \sin(2\phi)$ for odd modes.

Lattice	$a(\mathbf{r}, \phi)$
Square	$c_1 u(\phi)\mathrm{e}^{\mathrm{i}\mathbf{k}_1\cdot\mathbf{r}} + c_2 u(\phi - \frac{1}{2}\pi)\mathrm{e}^{\mathrm{i}\mathbf{k}_2\cdot\mathbf{r}} + \mathrm{c.c.}$
Hexagonal	$c_1 u(\phi)\mathrm{e}^{\mathrm{i}\mathbf{k}_1\cdot\mathbf{r}} + c_2 u(\phi - \frac{2}{3}\pi)\mathrm{e}^{\mathrm{i}\mathbf{k}_2\cdot\mathbf{r}} + c_3 u(\phi + \frac{2}{3}\pi)\mathrm{e}^{-\mathrm{i}(\mathbf{k}_1+\mathbf{k}_2)\cdot\mathbf{r}} + \mathrm{c.c.}$
Rhombic	$c_1 u(\phi)\mathrm{e}^{\mathrm{i}\mathbf{k}_1\cdot\mathbf{r}} + c_2 u(\phi - \eta)\mathrm{e}^{\mathrm{i}\mathbf{k}_2\cdot\mathbf{r}} + \mathrm{c.c.}$

of marginal stability according to $\Delta\mu - \Delta\mu_c = \xi^2$ with $\Delta\mu_c = -\beta G_-(q_-)$ ($\Delta\mu_c = -\beta G_+(q_+)$) if odd (even) modes are marginally stable. Note that the parameter ξ is independent of the coupling parameter ε. Also introduce a second slow time scale $\hat{\tau} = \xi^2\tau$. Next substitute the series expansion

$$z(\mathbf{r}, \hat{\tau}) = \xi z^{(1)}(\mathbf{r}, \hat{\tau}) + \xi^2 z^{(2)}(\mathbf{r}, \hat{\tau}) + \xi^3 z^{(3)}(\mathbf{r}, \hat{\tau}) + \cdots \tag{11.80}$$

into equation (11.52) and collect terms with equal powers of ξ. This generates a hierarchy of equations of the form (up to $\mathcal{O}(\xi^3)$)

$$\mathcal{M}z^{(1)} = 0 \tag{11.81}$$

$$\mathcal{M}z^{(2)} = 0 \tag{11.82}$$

$$\mathcal{M}z^{(3)} = z^{(1)}[1 - A|z^{(1)}|^2] - \frac{\mathrm{d}z^{(1)}}{\mathrm{d}\hat{\tau}} \tag{11.83}$$

where, for any complex function z,

$$\mathcal{M}z = -\Delta\mu_c z - \beta\int_0^\pi w_{\mathrm{lat}} \circ [z + \bar{z}\mathrm{e}^{-4\mathrm{i}\phi}]\frac{\mathrm{d}\phi}{\pi}. \tag{11.84}$$

The first equation in the hierarchy has solutions of the form

$$z^{(1)}(\mathbf{r}, \hat{\tau}) = \Gamma\sum_{n=1}^{N} \mathrm{e}^{-2\mathrm{i}\varphi_n}[c_n(\hat{\tau})\mathrm{e}^{\mathrm{i}\mathbf{k}_n\cdot\mathbf{r}} + \bar{c}_n(\hat{\tau})\mathrm{e}^{-\mathrm{i}\mathbf{k}_n\cdot\mathbf{r}}] \tag{11.85}$$

where $\Gamma = 1$ for even (+) modes and $\Gamma = \mathrm{i}$ for odd (−) modes (see equations (11.66) and (11.67)). Here $N = 2$ for the square or rhombic lattice and $N = 3$ for the hexagonal lattice. Also $\mathbf{k}_n = q_c\hat{\boldsymbol{\ell}}_n$ for $n = 1, 2$ and $\mathbf{k}_3 = -\mathbf{k}_1 - \mathbf{k}_2$. A dynamical equation for the amplitudes $c_n(\tau)$ can then be obtained as a solvability condition for the third-order equation (11.83). Define the inner product of two arbitrary doubly-periodic functions $f(\mathbf{r})$ and $g(\mathbf{r})$ by

$$\langle f|g\rangle = \int_\Lambda \bar{f}(\mathbf{r})g(\mathbf{r})\,\mathrm{d}\mathbf{r} \tag{11.86}$$

where Λ is a fundamental domain of the periodically tiled plane (whose area is normalized to unity). Taking the inner product of the left-hand side of equation (11.83) with $\tilde{f}_n(\mathbf{r}) = \mathrm{e}^{\mathrm{i}\mathbf{k}_n\cdot\mathbf{r}}$ leads to the following solvability condition

$$\langle \tilde{f}_n|\mathrm{e}^{2\mathrm{i}\varphi_n}\mathcal{M}z^{(3)} + \Gamma^2\mathrm{e}^{-2\mathrm{i}\varphi_n}\overline{\mathcal{M}z^{(3)}}\rangle = 0. \tag{11.87}$$

The factor $\Gamma^2 = \pm 1$ ensures that the appropriate marginal stability condition $\lambda_\pm = 0$ is satisfied by equation (11.71). Finally, we substitute for $\mathcal{M}z^{(3)}$ using the right-hand side of equation (11.83) to obtain an amplitude equation for c_n, which turns out to be identical for both odd and even solutions:

$$\frac{\mathrm{d}c_n}{\mathrm{d}\hat{\tau}} = c_n\left[1 - \gamma(0)|c_1|^2 - 2\sum_{p\neq n}\gamma(\varphi_n - \varphi_p)|c_p|^2\right] \tag{11.88}$$

where

$$\gamma(\varphi) = [2 + \cos(4\varphi)]A. \tag{11.89}$$

We consider solutions of these amplitude equations for each of the basic lattices.

Square or rhombic lattice

First, consider planforms corresponding to a bimodal structure of the square or rhombic type ($N = 2$). That is, take $\mathbf{k}_1 = q_c(1, 0)$ and $\mathbf{k}_2 = q_c(\cos(\theta), \sin(\theta))$, with $\theta = \pi/2$ for the square lattice and $0 < \theta < \pi/2$, $\theta \neq \pi/3$ for a rhombic lattice. The amplitudes evolve according to a pair of equations of the form

$$\frac{\mathrm{d}c_1}{\mathrm{d}\hat{\tau}} = c_1[1 - \gamma(0)|c_1|^2 - 2\gamma(\theta)|c_2|^2] \tag{11.90}$$

$$\frac{\mathrm{d}c_2}{\mathrm{d}\hat{\tau}} = c_2[1 - \gamma(0)|c_2|^2 - 2\gamma(\theta)|c_1|^2]. \tag{11.91}$$

Since $\gamma(\theta) > 0$, three types of steady state are possible for arbitrary phases ψ_1, ψ_2:

(i) the homogeneous state, $c_1 = c_2 = 0$;
(ii) rolls, $c_1 = \sqrt{1/\gamma(0)}\mathrm{e}^{\mathrm{i}\psi_1}$, $c_2 = 0$ or $c_1 = 0$, $c_2 = \sqrt{1/\gamma(0)}\mathrm{e}^{\mathrm{i}\psi_2}$; and
(iii) squares or rhombics, $c_n = \sqrt{1/[\gamma(0) + 2\gamma(\theta)]}\mathrm{e}^{\mathrm{i}\psi_n}$, $n = 1, 2$.

A standard linear stability analysis shows that if $2\gamma(\theta) > \gamma(0)$ then rolls are stable whereas the square or rhombic patterns are unstable. The opposite holds if $2\gamma(\theta) < \gamma(0)$. Note that here stability is defined with respect to perturbations with the same lattice structure. Using equation (11.89) we deduce that, in the case of a rhombic lattice of angle $\theta \neq \pi/2$, rolls are stable if $\cos(4\theta) > -1/2$ whereas θ-rhombics are stable if $\cos(4\theta) < -\frac{1}{2}$, that is, if $\pi/6 < \theta < \pi/3$; rolls are stable and square patterns unstable on a square lattice.

Hexagonal lattice

Next consider planforms on a hexagonal lattice with $N = 3$, $\varphi_1 = 0$, $\varphi_2 = 2\pi/3$, $\varphi_3 = -2\pi/3$. The cubic amplitude equations take the form

$$\frac{dc_n}{d\hat{\tau}} = c_n[1 - \gamma(0)|c_n|^2 - 2\gamma(2\pi/3)(|c_{n+1}|^2 + |c_{n-1}|^2)] \qquad (11.92)$$

where $n = 1, 2, 3 \bmod 3$. Unfortunately, equation (11.92) is not sufficient to determine the selection and stability of the steady-state solutions bifurcating from the homogeneous state. One has to carry out an *unfolding* of the amplitude equation that includes higher-order terms (quartic and quintic) in $z, \bar{z}$. One could calculate this explicitly by carrying out a double expansion in the parameters ε and ξ, which is equivalent to the perturbation approach used by [11]. In addition to generating higher-order terms, one finds that there is an $\mathcal{O}(\varepsilon)$ contribution to the coefficients $\gamma(\varphi)$ such that $2\gamma(2\pi/3) - \gamma(0) = \mathcal{O}(\varepsilon)$ and, in the case of even planforms, an $\mathcal{O}(\varepsilon)$ contribution to the right-hand side of equation (11.92) of the form $\eta\bar{c}_{n-1}\bar{c}_{n+1}$.

Considerable information about the bifurcating solutions can be obtained using group-theoretic methods. First, one can use an important result from bifurcation theory in the presence of symmetries, namely, the *equivariant branching lemma* [28]: when a symmetric dynamical system goes unstable, new solutions emerge that (generically) have symmetries corresponding to the axial subgroups of the underlying symmetry group. A subgroup Σ is *axial* if the dimension of the space of solutions that are fixed by Σ is equal to one. Thus one can classify the bifurcating solutions by finding the axial subgroups of the symmetry group of the lattice (up to conjugacy). This has been carried out elsewhere for the particular shift-twist action of the Euclidean group described at the end of section 11.2 [11, 12]. The results are listed in table 11.3.

It can be seen that major differences emerge between the even and odd cases. Second, symmetry arguments can be used to determine the general form of higher-order contributions to the amplitude equation (11.92) and this leads to the bifurcation diagrams shown in figure 11.17 [11, 12]. It turns out that stability depends crucially on the sign of the $\mathcal{O}(\varepsilon)$ coefficient $2\gamma(2\pi/3) - \gamma(0)$, which is assumed to be positive in figure 11.17. The subcritical nature of the bifurcation to hexagonal patterns in the case of even patterns is a consequence of an additional quadratic term appearing on the right-hand side of (11.92).

Table 11.3. Even and odd planforms for hexagonal lattice.

Even planform	(c_1, c_2, c_3)	Odd planform	(c_1, c_2, c_3)
0-hexagon	$(1, 1, 1)$	Hexagon	$(1, 1, 1)$
π-hexagon	$(1, 1, -1)$	Triangle	(i, i, i)
Roll	$(1, 0, 0)$	Roll	$(1, 0, 0)$
		Patchwork quilt	$(0, 1, 1)$

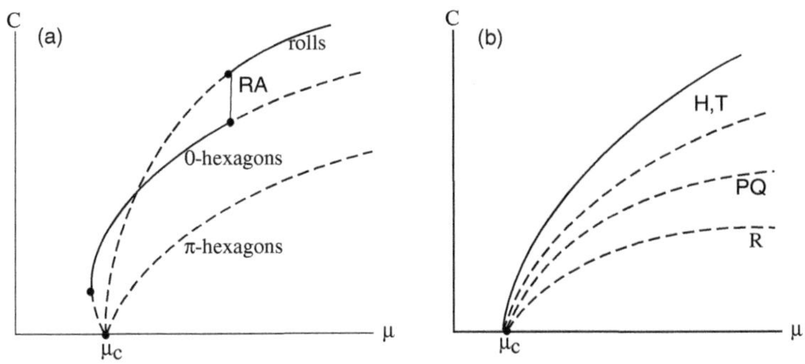

Figure 11.17. Bifurcation diagram showing the variation of the amplitude C with the parameter μ for patterns on a hexagonal lattice. Full and broken curves indicate stable and unstable solutions respectively. Even patterns (*a*): stable hexagonal patterns are the first to appear (subcritically) beyond the bifurcation point. Subsequently the stable hexagonal branch exchanges stability with an unstable branch of roll patterns due to a secondary bifurcation that generates rectangular patterns RA. Higher-order terms in the amplitude equation are needed to determine its stability. Odd patterns (*b*): Either hexagons (H) or triangles (T) are stable (depending on higher-order terms in the amplitude equation) whereas patchwork quilts (PQ) and rolls (R) are unstable. Secondary bifurcations (not shown) may arise from higher-order terms.

11.5.4 From cortical patterns to geometric visual hallucinations

We have now identified the stable planforms that are generated as primary bifurcations from a homogeneous, low-activity state of the continuum model (11.24). These planforms consist of certain linear combinations of the roll patterns shown in figure 11.15 and can thus be classified into non-contoured ($m = 0$) and contoured ($m = 1$ even or odd) patterns. Given a particular activity state in cortex, we can infer what the corresponding image in visual coordinates is like by applying the inverse of the retino-cortical map shown in figure 11.2. (In the case of contoured patterns, one actually has to specify the associated tangent map as detailed in [11].) Some examples of stable V1 planforms are presented in

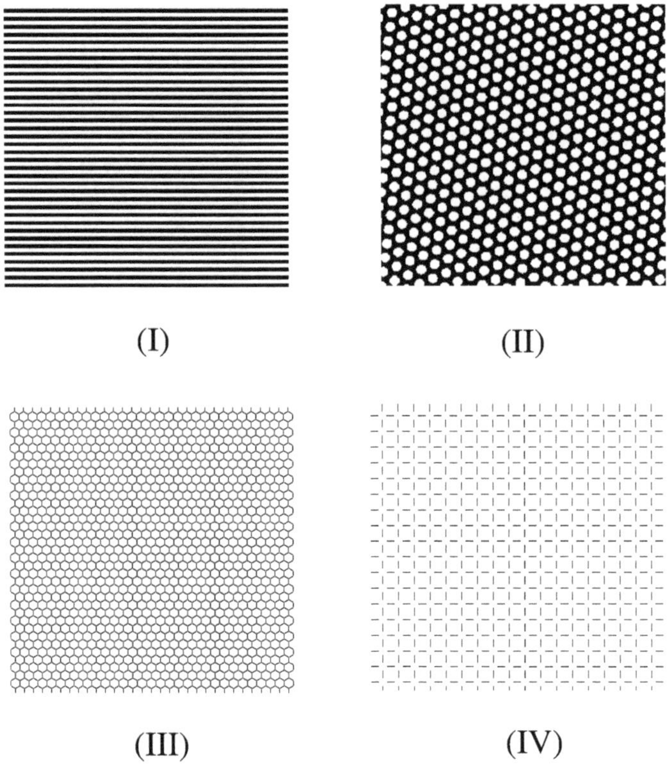

Figure 11.18. V1 planforms: non-contoured roll (I), non-contoured hexagon (II), even contoured hexagon (III) and even contoured square (IV).

figure 11.18 and the associated visual images are presented in figure 11.19. It will be seen that the two non-contoured planforms correspond to the type (I) and (II) Kluver form constants, as originally proposed by Ermentrout and Cowan [21], whereas the two contoured planforms reproduce the type (III) and (IV) form constants; see figure 11.10.

11.5.5 Oscillating patterns

It is also possible for the cortical model to spontaneously form oscillating patterns. This will occur if, in the absence of any lateral connections, each hypercolumn undergoes a Hopf bifurcation to a time-periodic tuning curve along the lines described in section 11.4.3. (It is not possible for the lateral connections to induce oscillations when the individual hypercolumns exhibit stationary tuning curves. This is a consequence of the fact that the lateral connections are modulatory and originate only from excitatory neurons.) Linearizing equations (11.62) and

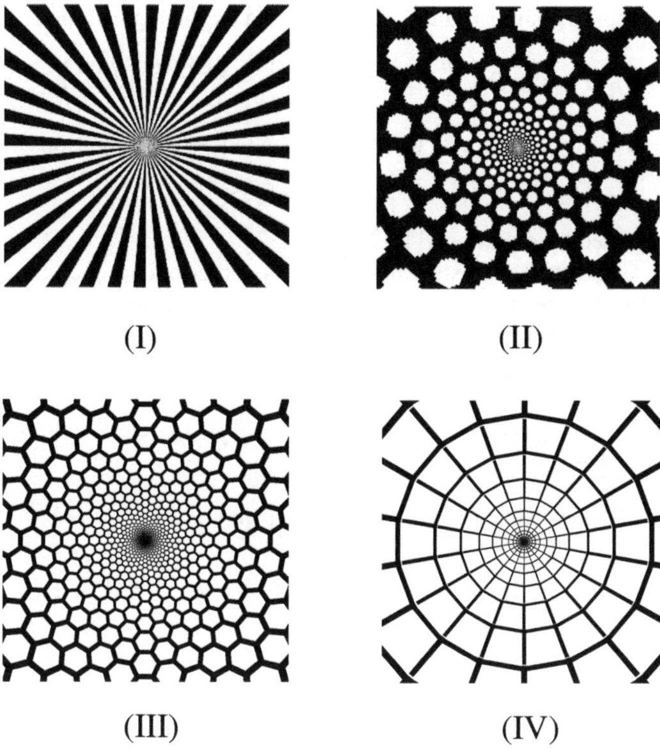

Figure 11.19. Visual field images of V1 planforms shown in figure 11.18.

(11.63) about the zero state $z_R = z_L = 0$ (assuming zero inputs from the LGN) gives

$$\frac{\partial z_L(\mathbf{r},\tau)}{\partial \tau} = \Delta\mu(1+\mathrm{i}\Omega_0)z_L(\mathbf{r},\tau) + \beta\int_0^\pi w_{\mathrm{lat}} \circ [z_L(\mathbf{r},\tau) + z_R(\mathbf{r},\tau)\mathrm{e}^{4\mathrm{i}\phi}]\frac{\mathrm{d}\phi}{\pi} \tag{11.93}$$

$$\frac{\partial z_R(\mathbf{r},\tau)}{\partial \tau} = \Delta\mu(1+\mathrm{i}\Omega_0)z_R(\mathbf{r},\tau) + \beta\int_0^\pi w_{\mathrm{lat}} \circ [z_R(\mathbf{r},\tau) + z_L(\mathbf{r},\tau)\mathrm{e}^{-4\mathrm{i}\phi}]\frac{\mathrm{d}\phi}{\pi}. \tag{11.94}$$

Equation (11.93) and (11.94) have solutions of the form

$$z_L(\mathbf{r},\tau) = u\mathrm{e}^{\lambda\tau}\mathrm{e}^{2\mathrm{i}\varphi}\mathrm{e}^{\mathrm{i}\mathbf{k}\cdot\mathbf{r}} \qquad z_R(\mathbf{r},\tau) = v\mathrm{e}^{\lambda\tau}\mathrm{e}^{-2\mathrm{i}\varphi}\mathrm{e}^{\mathrm{i}\mathbf{k}\cdot\mathbf{r}} \tag{11.95}$$

where $\mathbf{k} = q(\cos(\varphi), \sin(\varphi))$ and λ is determined by the eigenvalue equation

$$[\lambda - \Delta\mu(1 + \mathrm{i}\Omega_0)] \begin{pmatrix} cu \\ v \end{pmatrix} = \beta \begin{pmatrix} G_0(q) & \chi G_2(q) \\ \chi G_2(q) & G_0(q) \end{pmatrix} \begin{pmatrix} cu \\ v \end{pmatrix} \tag{11.96}$$

with $G_n(q)$ given by equation (11.73). Equation (11.96) has solutions of the form

$$\lambda_\pm = \Delta\mu(1 + \mathrm{i}\Omega_0) + \beta G_\pm(q) \qquad v = \pm u \tag{11.97}$$

for $G_\pm(q)$ defined by equation (11.72).

Hence, as in the case of stationary patterns, either odd or even time-periodic patterns will bifurcate from the homogeneous state depending on the degree of spread in the lateral connections. Close to the bifurcation point these patterns are approximated by the eigenfunctions of the linear theory according to

$$a(\mathbf{r}, \phi, t) = \sum_{n=1}^{N} \cos(2[\phi - \varphi_n])[c_n \mathrm{e}^{\mathrm{i}(\Omega t + \mathbf{k}_n \cdot \mathbf{r})} + d_n \mathrm{e}^{\mathrm{i}(\Omega t - \mathbf{k}_n \cdot \mathbf{r})} + \text{c.c.}] \tag{11.98}$$

with $\mathbf{k}_n = q_+(\cos\varphi_n, \sin\varphi_n)$ for even solutions and

$$a(\mathbf{r}, \phi, t) = \sum_{n=1}^{N} \sin(2[\phi - \varphi_n])[c_n \mathrm{e}^{\mathrm{i}(\Omega t + \mathbf{k}_n \cdot \mathbf{r})} + d_n \mathrm{e}^{\mathrm{i}(\Omega t - \mathbf{k}_n \cdot \mathbf{r})} + \text{c.c.}] \tag{11.99}$$

with $\mathbf{k}_n = q_-(\cos\varphi_n, \sin\varphi_n)$ for odd solutions, and where $\Omega = \Omega_0(1 + \varepsilon\Delta\mu)$. These should be compared with the corresponding solutions (11.75) and (11.77) of the stationary case. An immediate consequence of our analysis is that the oscillating patterns form standing waves within a single hypercolumn, that is, with respect to the orientation label ϕ. However, it is possible for travelling waves to propagate across V1 if there exist solutions for which $c_n \neq 0, d_n = 0$ or $c_n = 0, d_n \neq 0$. In order to investigate this possibility, we carry out a perturbation analysis of equations (11.62) and (11.63) along the lines of section 11.5.3 (after the restriction to doubly periodic solutions).

First, introduce a second slow time variable $\hat{\tau} = \xi^2\tau$ where $\xi^2 = \Delta\mu - \Delta\mu_c$ and take $z_{L,R} = z_{L,R}(\mathbf{r}, \tau, \hat{\tau})$. Next, substitute into equations (11.62) and (11.63) the series expansions

$$z_{L,R} = \xi z_{L,R}^{(1)} + \xi^2 z_{L,R}^{(2)} + \xi^3 z_{L,R}^{(3)} + \cdots \tag{11.100}$$

and collect terms with equal powers of ξ. This generates a hierarchy of equations of the form (up to $\mathcal{O}(\xi^3)$)

$$[\mathcal{M}_\tau \mathbf{z}^{(p)}]_L = 0 \qquad [\mathcal{M}_\tau \mathbf{z}^{(p)}]_R = 0 \tag{11.101}$$

for $p = 1, 2$ and

$$[\mathcal{M}_\tau \mathbf{z}^{(3)}]_L = (1 + \mathrm{i}\Omega_0) z_L^{(1)} [1 - A|z_L^{(1)}|^2 - 2A|z_R^{(1)}|^2] - \frac{\partial z_L^{(1)}}{\partial \hat{\tau}} \tag{11.102}$$

$$[\mathcal{M}_\tau \mathbf{z}^{(3)}]_R = (1 + \mathrm{i}\Omega_0) z_R^{(1)} [1 - A|z_R^{(1)}|^2 - 2A|z_L^{(1)}|^2] - \frac{\partial z_R^{(1)}}{\partial \hat{\tau}} \tag{11.103}$$

where, for any $\mathbf{z} = (z_L, z_R)$,

$$[\mathcal{M}_\tau \mathbf{z}]_L = \frac{\partial z_L}{\partial \tau} - (1 + i\Omega_0)\Delta\mu_c z_L - \beta \int_0^\pi w_{lat} \circ [z_L + z_R e^{+4i\phi}] \frac{d\phi}{\pi} \tag{11.104}$$

$$[\mathcal{M}_\tau \mathbf{z}]_R = \frac{\partial z_L}{\partial \tau} - (1 + i\Omega_0)\Delta\mu_c z_R - \beta \int_0^\pi w_{lat} \circ [z_R + z_L e^{-4i\phi}] \frac{d\phi}{\pi}. \tag{11.105}$$

The first equation in the hierarchy has solutions of the form

$$z_L^{(1)}(\mathbf{r}, \tau, \hat{\tau}) = e^{i\delta\Omega\tau} \sum_{n=1}^{N} e^{2i\varphi_n} [c_n(\hat{\tau}) e^{i\mathbf{k}_n \cdot \mathbf{r}} + d_n(\hat{\tau}) e^{-i\mathbf{k}_n \cdot \mathbf{r}}] \tag{11.106}$$

$$z_R^{(1)}(\mathbf{r}, \tau, \hat{\tau}) = \Gamma e^{i\delta\Omega\tau} \sum_{n=1}^{N} e^{-2i\varphi_n} [c_n(\hat{\tau}) e^{i\mathbf{k}_n \cdot \mathbf{r}} + d_n(\hat{\tau}) e^{-i\mathbf{k}_n \cdot \mathbf{r}}] \tag{11.107}$$

where $\Gamma = 1$ for even modes and $\Gamma = -1$ for odd modes (see equations (11.95) and (11.97)), and $\delta\Omega = \Omega_0 \Delta\mu_c$. As in the stationary case we restrict ourselves to doubly-periodic solutions with $N = 2$ for the square or rhombic lattice and $N = 3$ for the hexagonal lattice. A dynamical equation for the complex amplitudes $c_n(\hat{\tau})$ and $d_n(\hat{\tau})$ can be obtained as a solvability condition for the third-order equations (11.102) and (11.103). Define the inner product of two arbitrary doubly-periodic vectors $\mathbf{f}(\mathbf{r}, \tau) = (f_L(\mathbf{r}, \tau), f_R(\mathbf{r}, \tau))$ and $\mathbf{g}(\mathbf{r}, \tau) = (g_L(\mathbf{r}, \tau), g_R(\mathbf{r}, \tau))$ by

$$\langle \mathbf{f} | \mathbf{g} \rangle = \lim_{T \to \infty} \frac{1}{T} \int_{-T/2}^{T/2} \int_\Lambda [f_L(\mathbf{r}, \tau) g_L(\mathbf{r}, \tau) + f_R(\mathbf{r}, \tau) g_R(\mathbf{r}, \tau)] \, d\mathbf{r} \, d\tau \tag{11.108}$$

where Λ is a fundamental domain of the periodically tiled plane (whose area is normalized to unity). Taking the inner product of the left-hand side of equation (11.83) with the vectors $\tilde{\mathbf{f}}_n(\mathbf{r}, \tau) = e^{i\mathbf{k}_n \cdot \mathbf{r}} e^{i\delta\Omega\tau} (e^{2i\varphi_n}, \Gamma e^{-2i\varphi_n})$ and $\tilde{\mathbf{g}}_n(\mathbf{r}, \tau) = e^{-i\mathbf{k}_n \cdot \mathbf{r}} e^{i\delta\Omega\tau} (e^{2i\varphi_n}, \Gamma e^{-2i\varphi_n})$ leads to the following pair of solvability conditions

$$\langle \tilde{\mathbf{f}}_n | \mathcal{M}_\tau \mathbf{z}^{(3)} \rangle = 0 \qquad \langle \tilde{\mathbf{g}}_n | \mathcal{M}_\tau \mathbf{z}^{(3)} \rangle = 0. \tag{11.109}$$

Finally, we substitute for $\mathcal{M}_\tau \mathbf{z}^{(3)}$ using equations (11.102) and (11.103) to obtain amplitude equations for c_n and d_n (which at this level of approximation are the same for odd and even solutions):

$$\begin{aligned}\frac{\mathrm{d}c_n}{\mathrm{d}\hat{\tau}} = &-4(1+\mathrm{i}\Omega_0)\bar{d}_n \sum_{p\neq n} \gamma(\varphi_n - \varphi_p) c_p d_p + (1+\mathrm{i}\Omega_0) c_n \\ &\times \left[1 - 2\gamma(0)(|c_n|^2 + 2|d_n|^2) - 4\sum_{p\neq n} \gamma(\varphi_n - \varphi_p)(|c_p|^2 + |d_p|^2)\right]\end{aligned} \tag{11.110}$$

$$\begin{aligned}\frac{\mathrm{d}d_n}{\mathrm{d}\hat{\tau}} = &-4(1+\mathrm{i}\Omega_0)\bar{c}_n \sum_{p\neq n} \gamma(\varphi_n - \varphi_p) c_p d_p + (1+\mathrm{i}\Omega_0) d_n \\ &\times \left[1 - 2\gamma(0)(|d_n|^2 + 2|c_n|^2) - 4\sum_{p\neq n} \gamma(\varphi_n - \varphi_p)(|c_p|^2 + |d_p|^2)\right]\end{aligned} \tag{11.111}$$

with $\gamma(\varphi)$ given by equation (11.89).

The analysis of the amplitude equations (11.110) and (11.111) is considerably more involved than for the corresponding stationary problem. We discuss only the square lattice here ($N = 2$) with $\mathbf{k}_1 = q_c(1,0)$ and $\mathbf{k}_2 = q_c(0,1)$. The four complex amplitudes (c_1, c_2, d_1, d_2) evolve according to the set of equations of the form

$$\frac{\mathrm{d}c_1}{\mathrm{d}\hat{\tau}} = (1+\mathrm{i}\Omega_0)(c_1[1 - \kappa(|c_1|^2 + 2|d_1|^2) - 2\kappa(|c_2|^2 + |d_2|^2)] - 2\kappa \bar{d}_1 c_2 d_2) \tag{11.112}$$

$$\frac{\mathrm{d}c_2}{\mathrm{d}\hat{\tau}} = (1+\mathrm{i}\Omega_0)(c_2[1 - \kappa(|c_2|^2 + 2|d_2|^2) - 2\kappa(|c_1|^2 + |d_1|^2)] - 2\kappa \bar{d}_2 c_1 d_1) \tag{11.113}$$

$$\frac{\mathrm{d}d_1}{\mathrm{d}\hat{\tau}} = (1+\mathrm{i}\Omega_0)(d_1[1 - \kappa(|c_1|^2 + 2|d_1|^2) - 2\kappa(|c_2|^2 + |d_2|^2)] - 2\kappa \bar{c}_1 c_2 d_2) \tag{11.114}$$

$$\frac{\mathrm{d}d_2}{\mathrm{d}\hat{\tau}} = (1+\mathrm{i}\Omega_0)(d_2[1 - \kappa(|c_2|^2 + 2|d_2|^2) - 2\kappa(|c_1|^2 + |d_1|^2)] - 2\kappa \bar{d}_2 c_1 d_1) \tag{11.115}$$

where $\kappa = 6A$. These equations have the same structure as the cubic equations obtained for the standard Euclidean group action using group-theoretic methods [50]. An identical set of equations has been obtained for oscillatory activity patterns in the Ermentrout–Cowan model [55]. It can be shown that there exist five possible classes of solution that can bifurcate from the homogeneous state. We list examples from each class:

(i) travelling rolls (TR): $c_1 \neq 0$, $c_2 = d_1 = d_2 = 0$ with $|c_1|^2 = 1/\kappa$;
(ii) travelling squares (TS): $c_1 = c_2 \neq 0$, $d_1 = d_2 = 0$ with $|c_1|^2 = 1/3\kappa$;

(iii) standing rolls (SR): $c_1 = d_1, c_2 = d_2 = 0$ with $|c_1|^2 = 1/3\kappa$;
(iv) standing squares (SS): $c_1 = d_1 = c_2 = d_2$ with $|c_1|^2 = 1/9\kappa$; and
(v) Aalternating rolls (AR): $c_1 = -ic_2 = d_1 = -id_2$ with $|c_1|^2 = 1/5\kappa$.

Up to arbitrary phase-shifts the corresponding planforms are

$$\begin{aligned}
&\text{(TR)} \quad a(\mathbf{r}, \phi, t) = |c_1| u(\phi) \cos(\Omega t + q_c x); \\
&\text{(TS)} \quad a(\mathbf{r}, \phi, t) = |c_1| [u(\phi) \cos(\Omega t + q_c x) + u(\phi - \pi/2) \cos(\Omega t + q_c y)]; \\
&\text{(SR)} \quad a(\mathbf{r}, \phi, t) = |c_1| u(\phi) \cos(\Omega t) \cos(q_c x); \\
&\text{(SS)} \quad a(\mathbf{r}, \phi, t) = |c_1| \cos(\Omega t)[u(\phi) \cos(q_c x) + u(\phi - \pi/2) \cos(q_c y)]; \text{ and} \\
&\text{(AR)} \quad a(\mathbf{r}, \phi, t) = |c_1| [\cos(\Omega t) u(\phi) \cos(q_c x) + \sin(\Omega t) u(\phi - \pi/2) \cos(q_c y)].
\end{aligned}$$

(In contrast to the more general case considered in [50], our particular system does not support standing cross-roll solutions of the form $c_1 = d_1, c_2 = d_2$ with $|c_1| \neq |c_2|$.)

Linear stability analysis shows that (to cubic order) the TR solution is stable, the AR solution is marginally stable and the other solutions are unstable [50, 55]. In order to resolve the degeneracy of the AR solution one would need to carry out a double expansion in the parameters ξ, ε and include higher-order terms in the amplitude equation. The situation is even more complicated in the case of the hexagonal lattice where such a double expansion is expected to yield additional contributions to equations (11.110) and (11.111). As in the stationary case, group-theoretic methods can be used to determine generic aspects of the bifurcating solutions. First note that, as with other Hopf bifurcation problems, the amplitude equations have an extra phase-shift symmetry in time that was not in the original problem. This takes the form $\psi : (c_n, d_n) \to (e^{i\psi} c_n, e^{i\psi} d_n)$ for $n = 1, \ldots, N$ with $\psi \in \mathbf{S}^1$. Thus the full spatio-temporal symmetry is $\Gamma_{\mathcal{L}} \times \mathbf{S}^1$ for a given lattice $\mathcal{L}$. One can then appeal to the *equivariant Hopf theorem* [28] which guarantees the existence of primary bifurcating branches that have symmetries corresponding to the isotropy subgroups of $\Gamma_{\mathcal{L}} \times \mathbf{S}^1$ with two-dimensional fixed-point subspaces. (In the case of a square lattice this generates the five solutions listed earlier.) The isotropy subgroups for the standard Euclidean group action have been calculated elsewhere [19, 45, 50]. As in the stationary bifurcation problem, the isotropy subgroups of the shift-twist action may differ in a non-trivial way. This particular issue will be explored elsewhere.

11.6 Spatial frequency tuning and SO(3) symmetry

One of the simplifications of our large-scale cortical model has been to treat V1 as a continuum of interacting hypercolumns in which the internal structure of each hypercolumn is idealized as a ring of orientation selective cells. This reduction can be motivated in part by the fact that there exists a physical ring of orientation domains around each pinwheel, as illustrated by the circle in

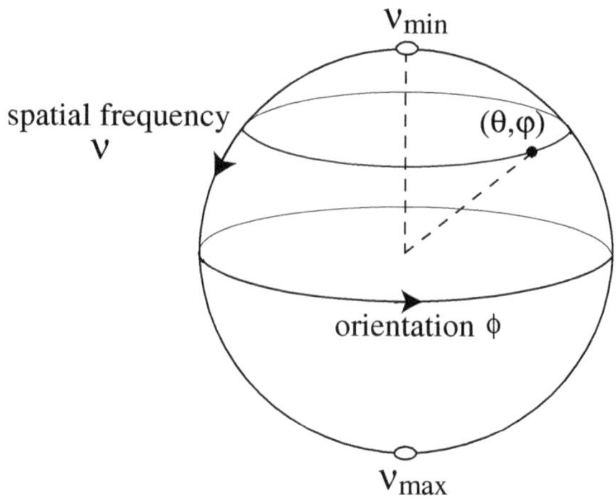

Figure 11.20. Spherical model of orientation and spatial frequency tuning.

figure 11.4. However, even if one restricts attention to the single eye case, there still exist two pinwheels per ocular dominance column. Moreover, the ring model does not take into account the fact that within each pinwheel region there is a broad distribution of orientation preferences so that the average orientation selectivity is weak. A fuller treatment of the two-dimensional structure of a hypercolumn can be carried out by incorporating another internal degree of freedom within the hypercolumn, which reflects the fact that cortical cells are also selective to the spatial frequency of a stimulus. (In the case of a grating stimulus, this would correspond to the inverse of the wavelength of the grating.) Indeed, recent optical imaging data suggest that the two pinwheels per hypercolumn are associated with high and low spatial frequencies respectively [7,35,36]. Recently, we have proposed a generalization of the ring model that takes into account this structure by incorporating a second internal degree of freedom corresponding to (log) spatial frequency preference [15]. Here we show how this new model can be used to extend our theory of cortical pattern formation to include both orientation and spatial frequency preferences.

11.6.1 The spherical model of a hypercolumn

Each hypercolumn (when restricted to a single ocular dominance column) is now represented by a sphere with the two orientation singularities identified as the north and south poles respectively; see figure 11.20. Following recent optical imaging results [7, 35, 36], the singularities are assumed to correspond to the two extremes of (log) spatial frequency within the hypercolumn. In terms of spherical polar coordinates (r, θ, φ) with $r = 1, \theta \in [0, \pi)$ and $\varphi \in [0, 2\pi)$, we thus define the orientation preference ϕ and (log) spatial frequency ν according to

$$\nu = \nu_{\min} + \frac{\theta}{\pi}[\nu_{\max} - \nu_{\min}] \qquad \phi = \varphi/2. \tag{11.116}$$

Note that we consider $\nu = \log p$ rather than spatial frequency p as a cortical label. This is motivated by the observation that dilatations in visual field coordinates correspond to horizontal translations in cortex; see the introduction of the chapter. Using certain scaling arguments it can then be shown that all hypercolumns have approximately the same bandwidth in ν even though there is broadening with respect to lower spatial frequencies as one moves towards the periphery of the visual field [15].

It is important to emphasize that the sphere describes the network topology of the local weight distribution expressed in terms of the internal labels for orientation and spatial frequency. It is not, therefore, expected to match the actual spatial arrangement of cells within a hypercolumn. Given this spherical topology, the natural internal symmetry of a hypercolumn is now SO(3) rather than O(2). In particular, the local weight distribution is assumed to be SO(3) invariant. Imposing such a symmetry is not just a mathematical convenience, but actually reflects the anatomical observation that the pattern of local interactions within a hypercolumn depends on cortical separation rather than feature separation [18]. When the weight distribution is expressed in terms of cortical coordinates, this naturally leads to Euclidean symmetry. However, we are describing the interactions in terms of cortical labels for orientation and spatial frequency. Hence, the natural measure of spatial separation is now in terms of geodesics or great circles on the sphere, and the group that preserves geodesic separation is SO(3).

Having introduced the spherical model of a single hypercolumn, it is straightforward to write down a generalization of our large-scale cortical model given by equation (11.19):

$$\begin{aligned}\frac{\partial a(\mathbf{r}, P, t)}{\partial t} &= -a(\mathbf{r}, P, t) + \int w(P|P')\sigma[a(\mathbf{r}, P', t)]\,\mathrm{d}P' \\ &\quad + \varepsilon \int w_{\mathrm{lat}}(\mathbf{r}, P|\mathbf{r}', P')\sigma[a(\mathbf{r}', P', t)]\,\mathrm{d}\mathbf{r}'\,\mathrm{d}P' + h(\mathbf{r}, P, t)\end{aligned} \tag{11.117}$$

where we have introduced the compact notation $P = \{\theta, \varphi\}$ and $\mathrm{d}P = \sin\theta\,\mathrm{d}\theta\,\mathrm{d}\varphi$. (For ease of notation, we consider here a one-population model by identifying the states $a_{\mathrm{E}} = a_{\mathrm{I}}$. Note that such a state arose previously in case A of section 11.3). In equation (11.117), $w(P|P')$ denotes the distribution of local connections within a hypercolumn, whereas $w_{\mathrm{lat}}(\mathbf{r}, P|\mathbf{r}', P')$ denotes the distribution of horizontal connections between the hypercolumns at $\mathbf{r}$ and $\mathbf{r}'$. In the following we focus on the spontaneous dynamical behaviour of the model by fixing $h(\mathbf{r}, P, t) = h_0$ such that $a(\mathbf{r}, P, t) = a_0$ is a homogeneous fixed-point solution of equation (11.117).

An SO(3) invariant local weight distribution can be constructed in terms of spherical harmonics [15]:

$$w(\theta,\varphi|\theta',\varphi') = \mu \sum_{n=0}^{\infty} W_n \sum_{m=-n}^{n} \overline{Y_n^m}(\theta',\varphi') Y_n^m(\theta,\varphi) \tag{11.118}$$

with W_n real. The functions $Y_n^m(\theta,\varphi)$ constitute the angular part of the solutions of Laplace's equation in three dimensions, and thus form a complete orthonormal set. The orthogonality relation is

$$\int_0^{2\pi}\int_0^{\pi} \overline{Y_{n_1}^{m_1}}(\theta,\varphi) Y_{n_2}^{m_2}(\theta,\varphi) \sin\theta \, \mathrm{d}\theta \, \mathrm{d}\varphi = \delta_{n_1,n_2}\delta_{m_1,m_2}. \tag{11.119}$$

The spherical harmonics are given explicitly by

$$Y_n^m(\theta,\varphi) = \mathcal{P}_n^m(\cos\theta)\mathrm{e}^{\mathrm{i}m\varphi} \tag{11.120}$$

with

$$\mathcal{P}_n^m(\cos\theta) = (-1)^m \sqrt{\frac{2n+1}{4\pi}\frac{(n-m)!}{(n+m)!}} P_n^m(\cos\theta) \tag{11.121}$$

for $n \geq 0$ and $-n \leq m \leq n$, where $P_n^m(\cos\theta)$ is an associated Legendre function. The lowest-order spherical harmonics are

$$Y_0^0(\theta,\varphi) = \frac{1}{\sqrt{4\pi}} \tag{11.122}$$

$$Y_1^0(\theta,\varphi) = \sqrt{\frac{3}{4\pi}}\cos\theta \qquad Y_1^{\pm}(\theta,\varphi) = \mp\sqrt{\frac{3}{8\pi}}\sin\theta\,\mathrm{e}^{\pm\mathrm{i}\varphi}. \tag{11.123}$$

In figure 11.21 we show a contour plot of the SO(3) invariant weight distribution for the particular case $W_0 = 1$, $W_1 = 3$ and $W_n = 0$ for $n \geq 2$. The contour plot represents the distribution joining neurons with the same spatial frequency (same latitude on the sphere). It can be seen that away from the pinwheels (poles of the sphere), cells with similar orientation excite each other whereas those with dissimilar orientation inhibit each other. This is the standard interaction assumption of the ring model; see section 11.3. However, around the pinwheels, all orientations uniformly excite, which is consistent with the experimental observation that local interactions depend on cortical separation [18]. That is, although the cells around a pinwheel can differ greatly in their orientation preference, they are physically close together within the hypercolumn.

How does the anisotropy in the lateral connections manifest itself when spatial frequency is taken into account, so that the internal symmetry is SO(3) rather than O(2)? The first point to make is that, unlike O(2), SO(3) does not have a faithful representation in $\mathbb{R}^2$. Nevertheless, it is possible to generalize equation (11.22) so that the weights are invariant with respect to the action of

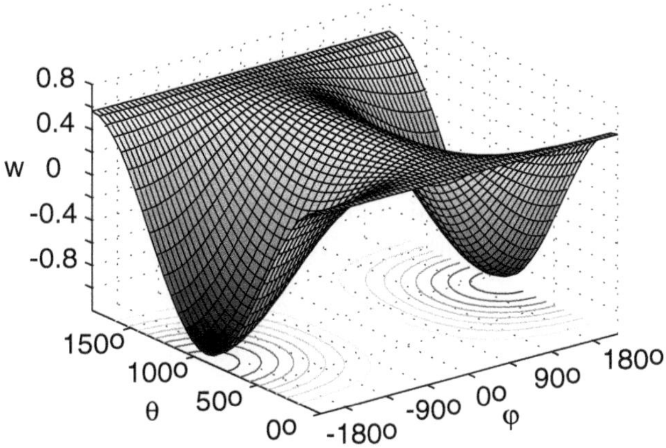

Figure 11.21. Plot of $w(\theta, \varphi|\theta', \varphi')$ given by the SO(3) invariant weight distribution (11.118) with $W_0 = 1$, $W_1 = 3$ and $W_n = 0$ for $n \geq 2$. Since w only depends on the difference $\varphi - \varphi'$, we set $\varphi' = 0$ and plot w as a function of θ and φ for $\theta' = \theta$.

$\mathbb{R}^2 \dot{+} \mathrm{O}(2) = \mathrm{E}(2)$ on $\{\mathbf{r}, \theta, \varphi\}$. That is, we consider a lateral weight distribution of the form

$$w_{\mathrm{lat}}(\mathbf{r}, P|\mathbf{r}', P') = \varepsilon J(T_{-\varphi/2}(\mathbf{r} - \mathbf{r}')) \frac{1}{\sin\theta'} \delta(\theta - \theta')\delta(\varphi - \varphi') \tag{11.124}$$

with $J(\mathbf{r})$ defined by equation (11.23) and $\varepsilon < 0$ (inhibitory lateral interactions). Such a distribution links neurons with the same orientation and spatial frequency label, with the function $p(\eta)$ again determining the degree of spatial spread (anisotropy) in the pattern of connections relative to the direction of their common orientation preference. An elegant feature of the spherical model is that it naturally incorporates the fact that, at the population level, there is less selectivity for orientation near pinwheels. In other words, any solution $a(\mathbf{r}, \theta, \varphi)$ expanded in terms of spherical harmonics is independent of φ at $\theta = 0, \pi$. This constrains the allowed structure of the spread function $p(\eta)$, in the sense that the horizontal weight distribution (11.124) has to be isotropic at the pinwheels. This follows from equations (11.124) and (11.23), which show that

$$\int w_{\mathrm{lat}}(\mathbf{r}, P|\mathbf{r}', P')a(\mathbf{r}', P')\,\mathrm{d}P' = \int_{-\infty}^{\infty} g(s) \int_{-\pi/2}^{\pi/2} p(\eta)a(\mathbf{r}+s\mathbf{r}_{\varphi/2+\eta}, P)\,\mathrm{d}\eta\,\mathrm{d}s. \tag{11.125}$$

It is clear that the right-hand side of this equation will be independent of φ at $\theta = 0, \pi$ if and only if $p(\eta) = 1/\pi$ at the pinwheels. In order to incorporate the known anisotropy away from the pinwheels, we conclude that the spread function has to be θ-dependent, $p = p(\eta|\theta)$. An example of a spread function that smoothly interpolates between isotropic connections at the pinwheels (high and low spatial frequencies) and strong anisotropy in the linear zones (intermediate spatial frequencies) with degree of spread η_0 is

$$p(\eta|\theta) = \frac{\cos^2\theta}{\pi} + \frac{1}{2\eta_0}\Theta(\eta_0 - |\eta|)\sin^2\theta. \tag{11.126}$$

The anisotropic weight distribution (11.124) is invariant under the following action of $\mathbb{R}^2 \dot{+} \mathrm{O}(2) = \mathrm{E}(2)$ on $\{\mathbf{r}, \theta, \varphi\}$:

$$\begin{aligned} \mathbf{s}\cdot(\mathbf{r},\theta,\varphi) &= (\mathbf{r}+\mathbf{s},\theta,\varphi) && \mathbf{s}\in\mathbb{R}^2 \\ \xi\cdot(\mathbf{r},\theta,\varphi) &= (T_\xi\mathbf{r},\theta,\varphi+\xi) && \xi\in S^1 \\ \kappa\cdot(\mathbf{r},\theta,\varphi) &= (\kappa\mathbf{r},\theta,-\varphi) \end{aligned} \tag{11.127}$$

where κ is the reflection $(x_1, x_2) \mapsto (x_1, -x_2)$.

11.6.2 Cortical patterns and spatial frequency tuning

A theory of spontaneous pattern formation in the coupled spherical model (11.117) can be developed along similar lines to that of the coupled ring model (11.24). Now each isolated hypercolumn generates a *tuning surface* for orientation and spatial frequency through spontaneous SO(3) symmetry breaking. The peak of this surface can be locked to a weakly biased stimulus from the LGN. Moreover, lateral interactions induce correlations between the tuning surfaces across the cortex leading to spatially periodic patterns of activity, which are consistent with the hallucinatory images presented in section 11.5.4. Rather than presenting the full nonlinear analysis here, we restrict ourselves to studying the linear eigenmodes that are excited when a homogeneous fixed point of the network becomes unstable.

First, we linearize equation (11.117) about the fixed-point solution $a(\mathbf{r}, \theta, \varphi, t) = a_0$. Setting $a(\mathbf{r}, \theta, \varphi, t) = a_0 + \mathrm{e}^{\lambda t} u(\theta, \varphi)\mathrm{e}^{\mathrm{i}\mathbf{k}\cdot\mathbf{r}}$ generates the eigenvalue equation

$$\lambda u(P) = -u(P) + \mu \int w(P|P')u(P')\,\mathrm{d}P' + \varepsilon \widetilde{J}(T_{\varphi/2}\mathbf{k})u(P) \tag{11.128}$$

where $\widetilde{J}(\mathbf{k})$ is the Fourier transform of $J(\mathbf{r})$ and $\mathbf{k} = q(\cos\zeta, \sin\zeta)$. Expanding $u(\theta, \varphi)$ in terms of spherical harmonics

$$u(\theta,\varphi) = \sum_{n\in\mathbb{Z}} \sum_{m=-n}^{n} \mathrm{e}^{-\mathrm{i}m\zeta} A_{nm} Y_n^m(\theta,\varphi) \tag{11.129}$$

leads to the matrix eigenvalue equation

$$[\lambda + 1 - \mu W_n]A_{nm} = \varepsilon \sum_{n'\in\mathbb{Z}} \sum_{m'=-n'}^{n'} J_{nm,n'm'}(q) A_{n'm'} \tag{11.130}$$

with

$$J_{nm,n'm'}(q) = \int_0^{2\pi}\int_0^{\pi} \overline{Y_n^m}(\theta,\varphi)Y_{n'}^{m'}(\theta,\varphi) \\ \times \left[\int_{\pi/2}^{\pi/2} p(\eta|\theta)\int_0^{\infty} g(s)\cos(sq\cos[\varphi/2+\eta])\,\mathrm{d}\eta\,\mathrm{d}s\right] \\ \times \sin\theta\,\mathrm{d}\theta\,\mathrm{d}\varphi. \tag{11.131}$$

We have included an additional phase-factor $\mathrm{e}^{-\mathrm{i}m\zeta}$ in equation (11.129) that simplifies the matrix equation (11.130) by eliminating any dependence on the direction ζ of the wavevector $\mathbf{k}$. Such a phase factor is equivalent to shifting φ, since $\mathrm{e}^{-\mathrm{i}m\zeta}Y_n^m(\theta,\varphi) = Y_n^m(\theta,\varphi-\zeta)$.

The matrix elements $J_{nm,n'm'}(q)$ are evaluated by substituting the explicit expression (11.120) for the spherical harmonics into equation (11.131):

$$J_{nm,n'm'}(q) = 2\pi G_{m-m'}(q)\Gamma_{nm,n'm'} \tag{11.132}$$

where $G_m(q)$ satisfies equation (11.73) and

$$\Gamma_{nm,n'm'} = \int_0^{\pi} \mathcal{P}_n^m(\cos\theta)\mathcal{P}_{n'}^{m'}(\cos\theta)\left[\int_{-\pi/2}^{\pi/2} p(\eta|\theta)\mathrm{e}^{2\mathrm{i}(m-m')\eta}\,\mathrm{d}\eta\right]\sin\theta\,\mathrm{d}\theta. \tag{11.133}$$

For the spread function (11.126) we have

$$\Gamma_{nm,n'm'} = \int_0^{\pi} \mathcal{P}_n^m(\cos\theta)\mathcal{P}_{n'}^{m'}(\cos\theta) \\ \times \left[\delta_{m-m'}\cos^2\theta + \frac{\sin[2(m-m')\eta_0]}{2(m-m')\eta_0}\sin^2\theta\right]\sin\theta\,\mathrm{d}\theta. \tag{11.134}$$

In the absence of lateral interactions ($\varepsilon = 0$), each hypercolumn can exhibit orientation and spatial frequency tuning through spontaneous symmetry breaking of SO(3). This is a generalization of the O(2) symmetry-breaking mechanism underlying orientation tuning in the ring model; see section 11.3. The orthogonality relation (11.119) shows that the eigenmodes are spherical harmonics with $\lambda = \lambda_n \equiv -1 + \mu W_n$ for $u(\theta,\varphi) = Y_n^m(\theta,\varphi)$, $-n \le m \le n$. Since λ is $\mathbf{k}$-independent, the full solution for the eigenmodes can be written in the form

$$a(\mathbf{r},\theta,\varphi) - a_0 = c_0(\mathbf{r})Y_n^0(\theta,\varphi) + \sum_{m=1}^{n}[c_m(\mathbf{r})Y_n^m(\theta,\varphi) + \bar{c}_m(\mathbf{r})\overline{Y_n^m}(\theta,\varphi)] \tag{11.135}$$

with the coefficients $c_m(\mathbf{r})$ arbitrary modulo the normalization condition

$$\sum_{m=0}^{n}|c_m(\mathbf{r})|^2 = B \tag{11.136}$$

where B is fixed. This normalization condition reflects the fact that the associated amplitude equations for the coefficients $c_m(\mathbf{r})$ are SO(3) equivariant [15].

Suppose that $W_1 > W_n$ for all $n \neq 1$. The fixed point $a = a_0$ then destabilizes at a critical value of the coupling $\mu_c = 1/W_1$ due to excitation of the first-order spherical harmonics. Sufficiently close to the bifurcation point, the resulting activity profile can be written as

$$a(\mathbf{r}, \theta, \varphi) = a_0 + c_0(\mathbf{r}) + \sum_{m=0,\pm} c_m(\mathbf{r}) f_m(\theta, \varphi) \tag{11.137}$$

for real coefficients $c_0(\mathbf{r}), c_\pm(\mathbf{r})$ with $\sum_{m=0,\pm} c_m(\mathbf{r})^2 = B$ and

$$f_0(\theta, \varphi) = \cos\theta \qquad f_+(\theta, \varphi) = \sin\theta\cos\varphi \qquad f_-(\theta, \varphi) = \sin\theta\sin\varphi. \tag{11.138}$$

Equation (11.137) represents a *tuning surface* for orientation and spatial frequency preferences for the hypercolumn at cortical position $\mathbf{r}$, which consists of a solitary peak whose location is determined by the values of the coefficients $(c_0(\mathbf{r}), c_+(\mathbf{r}), c_-(\mathbf{r}))$. Such a solution spontaneously breaks the underlying SO(3) symmetry. However, full spherical symmetry is recovered by noting that rotation of the solution corresponds to an orthogonal transformation of the coefficients $c_0, c_\pm$. Thus the action of SO(3) is to shift the location of the peak of the activity profile on the sphere, that is, to change the particular orientation and spatial frequency selected by the tuning surface. (This hidden SO(3) symmetry is explicitly broken by external stimuli, along similar lines to section 11.4.2.)

The tuning surface generated by our spherical model has the important property that it is not separable with respect to orientation and spatial frequency – the activity profile cannot be written in the form $u(\theta, \varphi) = \Theta(\theta)\Phi(\varphi)$. Consequently, selectivity for orientation varies with spatial frequency. If $c_\pm = 0$ then the activity profile is peaked at the pinwheel associated with high ($c_0 > 0$) or low ($c_0 < 0$) spatial frequencies, and there is no selection for orientation. However, if $c_0 = 0$ then the activity profile is peaked at intermediate spatial frequencies and there is strong selection for orientation. It is important to emphasize that the tuning surface represents the activity profile of a population of cells within a hypercolumn, rather than the tuning properties of an individual cell. Thus the absence of orientation selectivity at pinwheels is an aggregate property of the population. Indeed, it has been found experimentally that individual cells around pinwheels are orientation selective, but there is a broad distribution of orientation preferences within the pinwheel region so that the average response of the population is not orientation selective.

In order to solve the eigenvalue equation (11.130) for non-zero ε, we exploit the experimental observation that the lateral connections are weak relative to the local connections, and carry out a perturbation expansion in the small parameter ε. We show that there is a q-dependent *splitting* of the degenerate eigenvalue λ_1 that also separates out the first-order spherical harmonics. Denoting the characteristic size of such a splitting by $\delta\lambda = \mathcal{O}(\varepsilon)$, we impose the condition that $\delta\lambda \ll \mu\Delta W$, where $\Delta W = \min\{W_1 - W_m, m \neq 1\}$. This ensures that the perturbation does not excite states associated with other eigenvalues of the unperturbed problem, and we can then restrict ourselves to calculating perturbative corrections to the degenerate eigenvalue λ_1 and its associated eigenfunctions. Thus, introduce the perturbation expansions

$$\lambda = -1 + \mu W_1 + \varepsilon\lambda^{(1)} + \varepsilon^2\lambda^{(2)} + \cdots \tag{11.139}$$

$$A_{nm} = c_m\delta_{n,1} + \varepsilon A_{nm}^{(1)} + \varepsilon^2 A_{nm}^{(2)} + \cdots \tag{11.140}$$

and substitute these into the eigenvalue equation (11.130). We then systematically solve the resulting hierarchy of equations to successive orders in ε, taking into account the fact that the unperturbed problem is degenerate. Here we will only describe the lowest-order corrections.

Setting $n = 1$ in equation (11.130) yields the $O(\varepsilon)$ equations

$$\sum_{m,m'=-1}^{1} \widehat{J}_{mm'}(q)c_{m'} = \lambda^{(1)}c_m$$

where $\widehat{J}_{mm'}(q) = J_{1m,1m'}(q)$. Using equations (11.126), (11.132) and (11.134), we find that

$$\begin{pmatrix} G_0(q) & 0 & 0 \\ 0 & G_0(q) & 4\chi G_2(q)/5 \\ 0 & 4\chi G_2(q)/5 & G_0(q) \end{pmatrix} \begin{pmatrix} c_0 \\ c_1 \\ c_{-1} \end{pmatrix} = \lambda^{(1)} \begin{pmatrix} c_0 \\ c_1 \\ c_{-1} \end{pmatrix} \tag{11.141}$$

where χ is given by equation (11.69). Equation (11.141) has solutions of the form

$$\lambda^{(1)} = \lambda_0^{(1)}(q) \equiv G_0(q) \tag{11.142}$$

with eigenmode $c_0 = 1$, $c_\pm = 0$ and

$$\lambda^{(1)} = \lambda_\pm^{(1)}(q) \equiv G_0(q) \pm \tfrac{4}{5}\chi G_2(q) \tag{11.143}$$

with eigenmodes $c_0 = 0$, $c_{-1} = \pm c_1$.

As a concrete example, suppose that $g(s)$ is given by equation (11.74) and $\chi = \sin 4\eta_0/4\eta_0$. Plotting $G_m(q)$ for $m = 0, 2$ establishes that when there is sufficiently strong anisotropy in the linear zones ($\eta_0 < \pi/4$)

$$\min_q\{\lambda_-^{(1)}(q)\} < \min_q\{\lambda_0^{(1)}(q)\} < \min_q\{\lambda_+^{(1)}(q)\} \tag{11.144}$$

and the marginally stables modes are of the form

$$a(\mathbf{r}, \theta, \varphi) = \sum_i b_i \mathrm{e}^{\mathrm{i}\mathbf{k}_i\cdot\mathbf{r}} \sin\theta \sin(\varphi - \zeta_i) \tag{11.145}$$

where $\mathbf{k}_i = q_c(\cos\zeta_i, \sin\zeta_i)$ and q_c is the wavenumber that minimizes $\lambda_-^{(1)}(q)$. However, when there is weaker anisotropy within the linear zones ($\eta_0 > \pi/4$)

$$\min_q\{\lambda_+^{(1)}(q)\} < \min_q\{\lambda_0^{(1)}(q)\} < \min_q\{\lambda_-^{(1)}(q)\} \tag{11.146}$$

and the marginally stables modes are now of the form

$$a(\mathbf{r}, \theta, \varphi) = \sum_i b_i \mathrm{e}^{\mathrm{i}\mathbf{k}_i\cdot\mathbf{r}} \sin\theta\cos(\varphi - \zeta_i). \tag{11.147}$$

In both cases, these are states in which each hypercolumn has a tuning surface that peaks at intermediate frequencies. This regime is the one relevant to contour formation, and thus we recover the basic contoured patterns presented in section 11.5.4.

11.7 Future directions

There are many directions in which this work can be expanded. For example, it is now clear that many different features are mapped onto the visual cortex and beyond, in addition to retinal position, orientation and spatial frequency. Thus ocularity, directional motion, binocular disparity, and colour seem to be mapped [6, 34, 35, 38, 47]. It therefore remains to work out the symmetry groups associated with these features and thence to apply the mathematical machinery we have introduced here to compute the patterns which arise via spontaneous symmetry breaking.

There is also the fact that the neuron model used to formulate equations (11.4) is extremely simplified. It is nothing more than a low-pass RC filter followed by a sigmoidal current–voltage characteristic. One can easily make this model more complex and realistic by adding recovery variables in the style of the FitzHugh–Nagumo equations [63], or one can simplify it still further by using the well-known *integrate-and-fire* model [10]. Interestingly, in either case the dynamics is richer, and a variety of oscillatory phenomena are immediately apparent.

However, the most interesting new direction is to study, not spontaneous but *driven* pattern formation in neural networks of the type we have introduced. This amounts to studying the effects of external stimuli on such networks. In our case this means studying the effects of visual stimuli. Depending on the order of magnitude of such stimuli, various effects are possible. Thus external stimuli of order O(1) couple to the fixed points of equation (11.4), stimuli of order $O(\epsilon^{1/2})$ couple to the linearized equations (11.6), and $O(\epsilon^{3/2})$ stimuli couple directly to amplitude equations such as those of equation (11.52) and effectively unfold any bifurcation [10, 13]. The various effects produced by these and related stimuli remain to be studied in detail.

References

[1] Angelucci A, Levitt J B, Hupe J M, Walton E J S, Bullier J and Lund J S 2000 Anatomical circuits for local and global integration of visual information: Intrinsic and feedback connections *Eur. J. Neurosci.* **12** 285

[2] Angelucci A, Levitt J B and Lund J S 2002 Anatomical origins of the classical receptive field and modulatory surround field of single neurons in macaque visual cortical area V1 *Prog. Brain Res.* **136** 373–88

[3] Ben-Yishai R, Lev Bar-Or R and Sompolinsky H 1995 Theory of orientation tuning in visual cortex *Proc. Natl Acad. Sci.* **92** 3844–8

[4] Ben-Yishai R, Hansel D and Sompolinsky H 1997 Traveling waves and the processing of weakly tuned inputs in a cortical network module *J. Comput. Neurosci.* **4** 57–77

[5] Blasdel G G and Salama G 1986 Voltage-sensitive dyes reveal a modular organization in monkey striate cortex *Nature* **321** 579–85

[6] Blasdel G G 1992 Orientation selectivity, preference, and continuity in monkey striate cortex *J. Neurosci.* **12** 3139–61

[7] Bonhoeffer T, Kim D S, Malonek D, Shoham D and Grinvald A 1995 Optical imaging of the layout of functional domains in area 17/18 border in cat visual cortex *Eur. J. Neurosci.* **7** 1973–88

[8] Bosch Vivancos I, Chossat P and Melbourne I 1995 New planforms in systems of partial differential equations with Euclidean symmetry *Arch. Rat. Mech.* **131** 199–224

[9] Bosking W H, Zhang Y, Schofield B and Fitzpatrick D 1997 Orientation selectivity and the arrangement of horizontal connections in tree shrew striate cortex *J. Neurosci.* **17** 2112–27

[10] Bressloff P C, Bressloff N W and Cowan J D 2000 Dynamical mechanism for sharp orientation tuning in an integrate-and-fire model of a cortical hypercolumn *Neural Comput.* **12** 2473–511

[11] Bressloff P C, Cowan J D, Golubitsky M, Thomas P J and Wiener M 2001 Geometric visual hallucinations, Euclidean symmetry and the functional architecture of striate cortex *Phil. Trans. R. Soc.* B **356** 299–330

[12] Bressloff P C, Cowan J D, Golubitsky M and Thomas P J 2001 Scalar and pseudoscalar bifurcations: pattern formation on the visual cortex *Nonlinearity* **14** 739–75

[13] Bressloff P C and Cowan J D 2002 An amplitude equation approach to contextual effects in visual cortex *Neural Comput.* **14** 493–525

[14] Bressloff P C, Cowan J D, Golubitsky M, Thomas P J and Wiener M 2002 What geometric visual hallucinations tell us about the visual cortex *Neural Comput.* **14** 473–91

[15] Bressloff P C and Cowan J D 2002 SO(3) symmetry breaking mechanism for orientation and spatial frequency tuning in visual cortex *Phys. Rev. Lett.* **88** 078102

[16] Bullier J, Hupe′ J-M, James A J and Girard P 2001 The role of feedback connections in shaping the response of visual cortical neurons. *Prog. Brain Res.* **134** 193–204

[17] Cowan J D 1997 Neurodynamics and brain mechanisms *Cognition, Computation, and Consciousness* ed M Ito, Y Miyashita and E T Rolls (Oxford: Oxford University Press) pp 205–33

[18] Das A and Gilbert C 1999 Topography of contextual modulations mediated by short-range interactions in primary visual cortex *Nature* **399** 655–61

[19] Dionne B, Golubitsky M, Silber M and Stewart I 1995 Time-periodic spatially periodic planforms in Euclidean equivariant partial differential eqnarrays *Phil. Trans. R. Soc.* A **325** 125–68

[20] Douglas R, Koch C, Mahowald M, Martin K and Suarez H 1995 Recurrent excitation in neocortical circuits *Science* **269** 981–5

[21] Ermentrout G B and Cowan J D 1979 A mathematical theory of visual hallucination patterns *Biol. Cybernet.* **34** 137–50

[22] Ermentrout G B 1998 Neural networks as spatial pattern forming systems *Rep. Prog. Phys.* **61** 353–430

[23] Fitzpatrick D 2000 Seeing beyond the receptive field in primary visual cortex *Curr. Op. Neurobiol.* **10** 438–43

[24] Ferster D, Chung S and Wheat H 1997 Orientation selectivity of thalamic input to simple cells of cat visual cortex *Nature* **380** 249–81

[25] Gilbert C D and Wiesel T N 1983 Clustered intrinsic connections in cat visual cortex *J. Neurosci.* **3** 1116–33

[26] Gilbert C D 1992 Horizontal integration and cortical dynamics *Neuron* **9** 1–13

[27] Gilbert C D, Das A, Ito M, Kapadia M and Westheimer G 1996 Spatial integration and cortical dynamics *Proc. Natl Acad. Sci.* **93** 615–22

[28] Golubitsky M, Stewart I and Schaeffer D G 1988 *Singularities and Groups in Bifurcation Theory II* (Berlin: Springer)

[29] Häussler A and von der Malsburg C 1983 Development of retinotopic projections: an analytical treatment *J. Theoret. Neurobiol.* **2** 47–73

[30] Hirsch J D and Gilbert C D 1991 Synaptic physiology of horizontal connections in the cat's visual cortex *J. Physiol. Lond.* **160** 106–54

[31] Hubel D H and Wiesel T N 1962 Receptive fields, binocular interaction and functional architecture in the cat's visual cortex *J. Neurosci.* **3** 1116–33

[32] Hubel D H and Wiesel T N 1974 Sequence regularity and geometry of orientation columns in the monkey striate cortex *J. Comput. Neurol.* **158** 267–94

[33] Hubel D H and Wiesel T N 1974 Uniformity of monkey striate cortex: A parallel relationship between field size, scatter, and magnification factor *J. Comput. Neurol.* **158** 295–306

[34] Hubel D H and Livingstone M S 1987 Segregation of form, color, and stereopsis in primate area 18 *J. Neurosci.* **7** 3378–415

[35] Hubener M, Shoham D, Grinvald A and Bonhoeffer T 1997 Spatial relationship among thre columnar systems *J. Neurosci.* **15** 9270–84

[36] Issa N P, Trepel C and Stryker M P 2000 Spatial frequency maps in cat visual cortex *J. Neurosci.* **20** 8504–14

[37] LeVay S and Nelson S B 1991. Columnar organization of the visual cortex *The Neural Basis of Visual Function* ed A G Leventhal (Boca Raton, FL: Chemical Rubber Company) pp 266–315

[38] Livingstone M S and Hubel D H 1984 Anatomy and physiology of a color system in the primate visual cortex *J. Neurosci.* **4** 309–56

[39] Malach R, Amir Y, Harel M and Grinvald A 1993 Relationship between intrinsic connections and functional architecture revealed by optical imaging and in vivo targeted biocytin injections in primate striate cortex *Proc. Natl Acad. Sci.* **90** 10 469–73

[40] Marr D and Poggio T 1976 Cooperative computation of stereo disparity *Science* **194** 283–7

[41] Mundel T, Dimitrov A and Cowan J D 1997 Visual cortex circuitry and orientation tuning *Advances in Neural Information Processing Systems* vol 9, ed M C Mozer, M I Jordan and T Petsche (Cambridge, MA: MIT Press) pp 886–93

[42] Obermayer K and Blasdel G G 1993 Geometry of orientation and ocular dominance columns in monkey striate cortex *J. Neurosci.* **13** 4114–29

[43] Patterson A 1992 *Rock Art Symbols of the Greater Southwest* (Boulder, CO: Johnson Books)

[44] Pfleger B and Bonds A B 1995 Dynamic differentiation of $gaba_A$—sensitive influences on orientation selectivity of complex cells in the cat striate cortex *Exp. Brain Res.* **104** 81–8

[45] Roberts M, Swift J and Wagner D H 1986 The Hopf bifurcation on a hexagonal lattice *Contemp. Math.* **56** 283–318

[46] Rockland K S and Lund J 1983 Intrinsic laminar lattice connections in primate visual cortex *J. Comput. Neurol.* **216** 303–18

[47] Shmuel A and Grinvald A 1996 Functional organization for direction of motion and its relationship to orientation maps in cat area 18 *J. Neurosci.* **16** 6945–64

[48] Shmuel A, Korman M, Harel M, Grinvald A and Malach R 1998 Relationship of feedback connections from area V2 to orientation domains in area V1 of the primate *Soc. Neurosci. Abstr.* **24** 767

[49] Siegel R K and Jarvik M E 1975 Drug-induced hallucinations in animals and man *Hallucinations: Behaviour, Experience and Theory* ed R K Siegel and L J West (New York: Wiley) pp 81–161

[50] Silber M and Knobloch E 1991 Hopf bifurcation on a square lattice *Nonlinearity* **4** 1063–107

[51] Somers D C, Nelson S and Sur M 1995 An emergent model of orientation selectivity in cat visual cortical simple cells *J. Neurosci.* **15** 5448–65

[52] Swindale N V 1980 A model for the formation of ocular dominance stripes *Proc. R. Soc.* B **208** 243–64

[53] Swindale N V 1982 A model for the formation of orientation columns *Proc. R. Soc.* B **215** 211–30

[54] Swindale N V 1996 The development of topography in visual cortex: a review of models *Network* **7** 161–247

[55] Tass P 1997 Oscillatory cortical activity during visual hallucinations. *J. Biol. Phys.* **23** 21–66

[56] Toth L J, Rao S C, Kim D-S, Somers D C and Sur M 1996 Subthreshold facilitation and suppression in primary visual cortex revealed by intrinsic signal imaging *Proc. Natl Acad. Sci.* **93** 9869–74

[57] Turing A M 1952 The Chemical Basis of Morphogenesis *Phil. Trans. R. Soc.* B **237** 32–72

[58] Vidyasagar T R, Pei X and Volgushev M 1996 Multiple mechanisms underlying the orientation selectivity of visual cortical neurons *Trends Neurosci.* **19** 272–7

[59] von der Malsburg C 1973 Self-organization of orientation selective cells in the striate cortex *Kybernetik* **14** 85–100

[60] von der Malsburg C and Willshaw D J 1977 How to label nerve cells so that they can interconnect in an ordered fashion *Proc. Natl Acad. Sci., USA* **74** 5176–8

[61] von der Malsburg C and Cowan J D 1982 Outline of a theory for the ontogenesis of iso-orientation domains in visual cortex *Biol. Cybernet.* **45** 49–56

[62] Willshaw D J and von der Malsburg C 1979 A marker induction mechanism for the establishment of ordered neural mappings: its application to the retinotectal problem *Phil. Trans. R. Soc.* B **287** 203–43

[63] Wilson H R and Cowan J D 1972 Excitatory and inhibitory interactions in localized populations of model neurons *Biophys. J.* **12** 1–24

[64] Wilson H R and Cowan J D 1973 A mathematical theory of the functional dynamics of cortical and thalamic nervous tissue *Kybernetik* **13** 55–80

[65] Yoshioka T, Blasdel G G, Levitt J B and Lund J S 1996 Relation between patterns of intrinsic lateral connectivity, ocular dominance, and cytochrome oxidase–reactive regions in macaque monkey striate cortex *Cereb. Cortex* **6** 297–310

[66] Zweck J W and Williams L R 2001 Euclidean group invariant computation of stochastic completion fields using shiftable-twistable functions *Proc. 6th Eur. Conf. Computer Vision, Dublin 2000*

Chapter 12

Models for pattern formation in development

Bard Ermentrout
University of Pittsburgh

Remus Osan
University of Pittsburgh

The visual system is one of the most thoroughly studied sensory regions in the brain due to its importance to humans. (The standard reference for neuroscience is [10] and in particular, chapter 27, but see also chapter 11 by Bressloff and Cowan in this volume.) Indeed, unlike many other mammals, our understanding of the outside world is through our eyes. The transformation of this limited range in the electromagnetic spectrum into a conscious image is complex and involves many different regions of the brain. The inputs to the visual system start at the eye. Light enters the eye where it excites photoreceptors. These receptors connect to several layers of neurons within the retina ultimately exciting the retinal ganglion cells. Retinal ganglion cells respond best to spots of light. They are organized topographically so that nearby ganglion cells respond to nearby points of light in the visual field. The ganglion cells send axons into a region of the brain called the lateral geniculate nucleus (LGN). Images presented on the left-hand side end up projecting to the right LGN (see figure 12.1(A)) and vice versa. Topography is maintained in the LGN as well; that is, nearby ganglion cells project to nearby neurons in the LGN. The inputs remain segregated in different layers of the LGN according to whether they come from the left or right eye. Like ganglion cells, LGN neurons respond best to spots of light. LGN neurons then project to layer 4 of the visual cortex. Topography is maintained in the cortex but, additionally, there are new forms of organization in the projections from the cortex to the LGN. Inputs from the left and right eyes project to layer 4C of the cortex in a regular periodic pattern; see figures 12.1(B) and 12.3. Unlike LGN neurons which respond best to spots of light, cortical neurons are excited by oriented bars. Figure 12.1(C) shows the generally accepted mechanism. A bar of light excites an array of LGN neurons which are connected to a cortical neuron. If the light bar is of the appropriate orientation (in this case horizontal), then the cortical neuron will fire at a maximal rate. Thus, the simple diagrams in figures 12.1(B) and (C) suggest that if one records the activity in a cortical neuron, then whether it fires or not depends on the properties of the stimulus: e.g. the position in the visual field of the stimulus, the occularity (left or right eye), or the orientation. These tuned cortical neurons are not randomly dispersed across the visual area. Rather, their properties form well-ordered maps. There are at least three well-studied maps in the mammalian visual system:

(i) the topographic map,
(ii) the orientation map, and
(iii) the occular dominance map.

In this chapter, we will take the topographic map for granted and assume that it already exists. Each neuron in the cortex responds maximally to a particular orientation and thus, one can project these preferred orientations on the two-dimensional surface of the primary visual cortex. It is experimentally possible to see this map; for example, figure 12.2 shows the orientation map from the tree shrew. We point out several aspects of the map: (i) there are 'singularities' at spatially periodic intervals and (ii) there are linear regions where the orientation varies smoothly. The precise structure of the map is different in every animal so that it is not 'hardwired' genetically but rather arises during development of the visual system.

In addition to the topographic map and the orientation map, there is also the occular dominance map. Any particular cortical neuron will respond more strongly to stimuli presented to one eye than to the other. The mapping of preferences across the cortex results in a striped pattern. The occular dominance map can be seen experimentally by injecting a labelling agent into one eye and then reading the label on the surface of the cortex; figure 12.3 shows an example of such a map. The occular dominance map is very sensitive to environmental influences; animals who have one eyelid sutured develop distorted maps in which the uncovered eye occupies a much larger area. We emphasize that these maps are not independent. Indeed, there are striking relationships between the orientation and occular dominance maps. Figure 12.4 illustrates the relationship between the iso-orientation lines and the borders between the occular dominance stripes. The singularities tend to lie in the middle of the occular dominance stripes and the iso-orientation lines cross the occular dominance borders at right angles.

The main theoretical question we wish to address is what sorts of mechanisms underlie the formation of these maps. Mathematical models range from abstract pattern formation (e.g. [14]) to detailed models involving biophysics of single neurons [16]. Erwin and Miller [6] provide a nice review of the state of the art on these models and a biological review of the evidence for activity

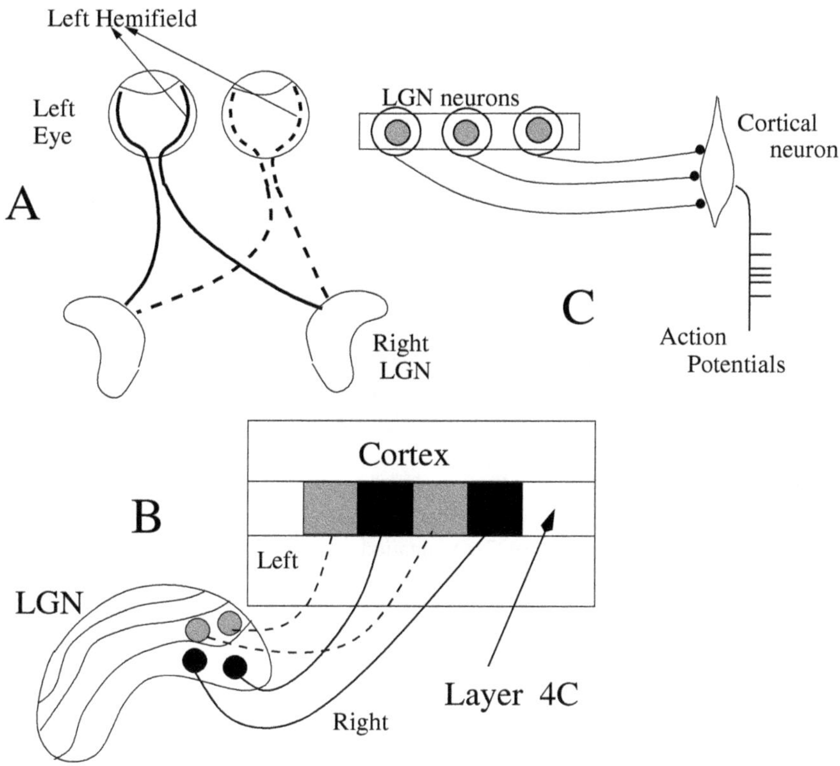

Figure 12.1. A summary of the organization of the visual system up to the cortex. (A) Mapping of the visual space into the eyes and onto the lateral geniculate nucleus (LGN). (B) Projection of the LGN neurons to layer 4C of the visual cortex showing organization of left–right inputs. (C) Orientation tuning begins in the cortex. A light bar (in this case horizontal) excites LGN neurons which project in an organized fashion to give an orientation-tuned cortical neuron.

dependence of orientation selectivity is given in [13].

We will briefly describe some of these models in the context of general pattern formation principles. In all cases, the ultimate mechanism for the formation of the periodicities is the Turing mechanism [18], whereby a band of spatially periodic modes becomes unstable. Thus, the main differences between models are how this instability is manifested and what determines the dominant spatial frequency.

One can view a cortical map as a pattern of connections from the geniculate to the cortex. Thus, when we look at a map we are actually looking at a distribution of connectivities. Hebb [9] first suggested that connections between neurons were made when the neurons fired together. This idea is called *Hebbian plasticity*. The obvious problem with this rule is that eventually every neuron

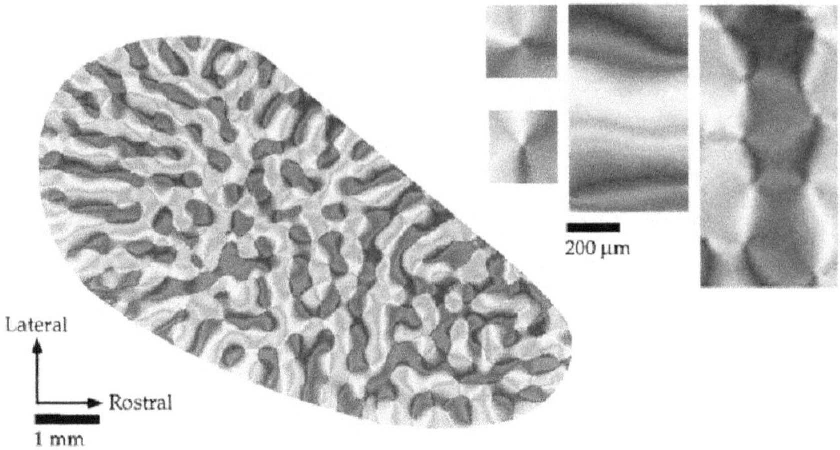

Figure 12.2. Orientation map for tree shrew. Reproduced from W H Bosking, Y Zhang, B Schofield and D Fitzpatrick 1997 Orientation selectivity and the arrangement of horizontal connections in tree shrew striate cortex *J. Neurosci.* **17** 2112–27 ©1997 by the Society for Neuroscience.

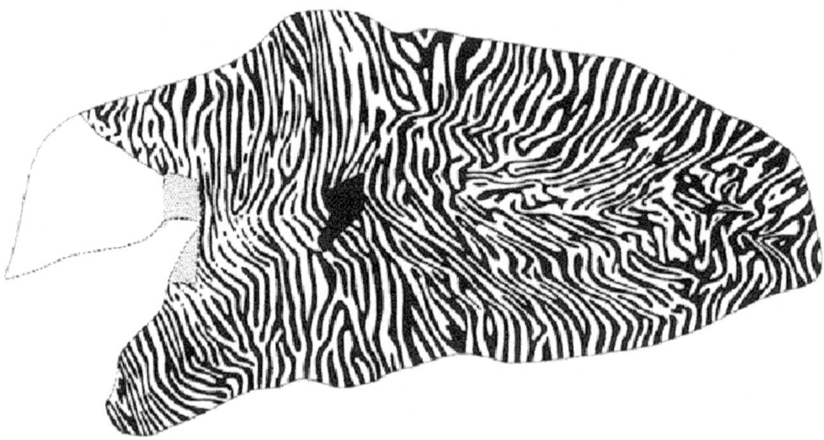

Figure 12.3. Occular dominance map. Reproduced from S LeVay, M Connolly, J Houde and D C van Essen 1985 The complete pattern of occular dominance stripes in the striate cortex and visual field of the macaque monkey *J. Neurosci.* **5** 486–501 ©1985 by the Society for Neuroscience.

will be connected to every other neuron and the growth of connections will be unbounded. Thus, in addition to a growth rule, models of development also incorporate some sort of decay of the connections. A typical method is

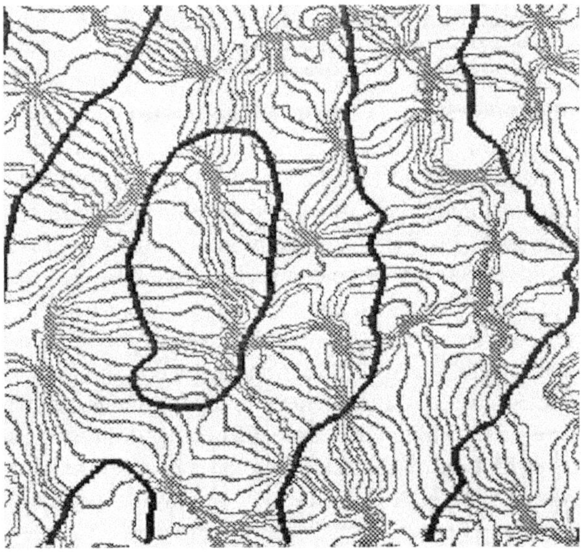

Figure 12.4. Interaction between occular dominance and orientation maps. Reproduced from K Obermayer and G G Blasdel 1993 Geometry of orientation and occular dominance columns in monkey striate cortex *J. Neurosci.* **13** 4114–29 ©1993 by the Society for Neuroscience.

to use *normalization of the weights* between neurons. For example, the total weights coming into a neuron is constrained to some fixed value. This sets up a competition and it is the interaction of this negative feedback with the positive Hebbian growth which determines the map.

The classes of models which have been studied can be reduced to two distinct types: (i) weight-based models and (ii) feature-based models. In weight-based models, one attempts to analyse the dynamics of functions, $w(x, y, \theta, z, t)$. Here w is the strength of a connection between a cortical cell at x, with favoured orientation, $\theta \in [0, \pi)$, occularity, $z \in \{L, R\}$ and responding to a point y in the visual field. Miller's, Swindale's and many other models are of this form. They have the advantage of being more closely connected to biology. In feature-based models, one instead studies the evolution of a vector of features, $[Y(x,t), \Theta(x,t), Z(x,t)]$. Here, for example, $\Theta(x,t)$ is the preferred orientation of a cortical neuron at point x in the cortex. Kohonen's self-organizing map [11] and the elastic net by Durbin and Mitchison [4] are examples of such models. These are more abstract but seem to capture the features of real maps better than the more 'realistic' models [7].

These mechanisms share certain generic properties. In particular, we address the periodicity of these maps and discuss how periodicity is imposed on the models. Models of Swindale and older models of Miller imposed the periodicity through cortical–cortical interactions. However, the elastic map, the Kohonen model and Dayan's recent model (see section 12.4) all obtain the periodicity through an interaction of the cortical interactions and the input correlations. We want to show how and why these models differ and suggest some new models that operate on the same principles but do not require any imposed normalization. We begin in section 12.1 with a description of the mechanisms that produce competition. This is illustrated by both simple models with global constraints and models involving competition for some factor. Section 12.2 describes mechanisms that set the spatial scales. We show that a combination of postsynaptic-dependent depression and Hebbian growth are sufficient to produce a clustered pattern whose spatial scale is a function of the input correlations and the intrinsic cortical connectivity. In section 12.3 we look at occular dominance and orientation selectivity in the context of competition and spatial interactions. We also consider joint development of both the occular dominance and orientation patterns. In section 12.4 we discuss the so-called self-organized feature maps. We reduce the question of orientation preference to a model which is similar to such a self-organizing map. Our new models are put into context in section 12.5.

12.1 Competition

Consider a single neuron which receives inputs from two cells, say L and R. We consider a linear neuron so that the output of the neuron is the weighted sum of its inputs:

$$V = w_L I_L + w_R I_R.$$

Hebb's rule states that the weights should grow in a manner proportional to the product of the inputs and the outputs:

$$\begin{aligned} w'_L &= K(w_L) I_L V \\ w'_R &= K(w_R) I_R V. \end{aligned}$$

If K is independent of the weights, then there is nothing to keep the weights bounded (or even positive). Swindale solves this problem by letting K depend on the weights, e.g. $K(u) = ku(1-u)$. This automatically keeps the weights bounded between 0 and 1. If we assume that the inputs are correlated white noise,

$$I_L = C_S \xi_1 + C_D \xi_2 \qquad I_R = C_S \xi_2 + C_D \xi_1$$

then depending on whether or not C_D is positive or negative, the weights will either both grow to 1 or one will grow to 1 and the other to 0. If we assume that the growth rate is slow compared to the rapidly changing random inputs, then we can average the growth:

$$\begin{aligned} \langle I_L V \rangle &= a w_L + b w_R \\ \langle I_R V \rangle &= a w_R + b w_L \end{aligned}$$

where

$$a = C_S^2 + C_D^2 \qquad \text{and} \qquad b = 2C_S C_D.$$

Thus, on average, the weights satisfy

$$w_L' = K(w_L)(aw_L + bw_R) \qquad w_R' = K(w_R)(aw_R + bw_L).$$

Supposing for the moment that K is a constant, we see that the weights grow along the eigenvectors, $(1, 1)$ corresponding to the eigenvalue, $a+b = (C_S + C_D)^2$ and $(1, -1)$ corresponding to the eigenvalue, $a - b = (C_S - C_D)^2$. If C_S, C_D are the same sign, then the fastest growth is along the *same* eigenvector and there is no competition; both weights go to their maximum. If C_S, C_D have opposite sign, then the fastest growth is along the *different* eigenvector and one weight 'wins' the competition. Thus, for this simple model, in order to get competition, there must be negative correlations between L and R. This is biologically unlikely, and one question is how to get competition with positive correlations.

One way to solve this is to assume an active form of decay. Furthermore, in order to keep everything bounded, we will assume a kind of Markovian or mass action model for weight growth. The 'kinetic' approach to weight growth prevents negative weights and separates the decay processes from the growth processes. It has been successfully applied to synaptic and ion channel processes [3]. The model is simple:

$$f \underset{K^-}{\overset{K^+}{\rightleftharpoons}} w.$$

Here, f is the available pool of substance for the formation of the weight and w is a fully formed weight. Thus, $f + \sum w = M$ where M is a constant which we can set to 1 without loss of generality and the sum is over all weights which compete for that particular pool. The weight satisfies:

$$w' = K^+\left[1 - \sum w\right] - K^- w.$$

In most of the models described in this paper, each synapse will compete for a unique pool. However, there will be instances where several synapses compete. Unless otherwise indicated, all synapses have their own pool and it is finite. Since the rates, $K^\pm$ must be non-negative, this guarantees that the weights will always lie between 0 and $M > 0$. For example, suppose that K^+ is a function of the Hebbian term $I\ V$ while K^- is a function of the postsynaptic activity only. Then, after averaging, the model becomes:

$$\begin{aligned} w_L' &= K^+(aw_L + bw_R)[1 - w_L] - K^-(w_L + w_R)w_L \qquad (12.1) \\ w_R' &= K^+(aw_R + bw_L)[1 - w_R] - K^-(w_L + w_R)w_R. \end{aligned}$$

Here, we first averaged and then applied the function. This is not strictly correct but is a useful approximation. Note that $\langle V \rangle \approx \bar{I}(w_L + w_R)$, where $\bar{I}$ is the mean input strength. Note also that if the two synapses compete for the same pool, then the terms, $[1 - w_j]$ will be replaced by $[1 - w_L - w_R]$. This change will not lead to substantial differences in the model's behaviour. We assume that $K^{\pm}$ are monotonic functions of their arguments. The parameters a, b characterize the correlations of the left and right inputs. Clearly, one solution to (12.1) is $w_L = w_R = w$ where w is such that

$$0 = K^{+}(cw)(1 - w) - K^{-}(2w)w \equiv q(w).$$

and $c = a + b$. Since $q(0) = K^{+}(0) > 0$ and $q(1) = -K^{-}(0) < 0$ there is at least one root in the interval (0, 1). Note that all weights remain bounded in this interval and there is no need to normalize the weights. The easiest way to check for competition is to show that there is a symmetry-breaking instability. The linearization of (12.1) about the fixed point is

$$y_L' = Ay_L + By_R \qquad y_R' = Ay_R + By_L$$

where

$$A = aK^{+'}(cw)[1 - w] - K^{+}(cw) - K^{-}(2w) - K^{-'}(2w)w$$
$$B = bK^{+'}(cw)[1 - w] - K^{-'}(2w)w.$$

There are two eigenvalues corresponding to the eigenvectors $(1, 1)^T$ and $(1, -1)^T$, which are the symmetric and antisymmetric states. The symmetric eigenvalue is

$$\lambda_s = A + B = (a + b)K^{+'}(cw)[1 - w] - K^{+}(cw) - K^{-}(2w) - 2K^{-'}(2w)w$$

and the antisymmetric eigenvalue is

$$\lambda_a = A - B = (a - b)K^{+'}(cw)[1 - w] - K^{+}(cw) - K^{-}(2w).$$

We want the antisymmetric eigenvalue to become positive while the symmetric one remains negative. There are two ways that this can happen since $K^{\pm'} \geq 0$. Either $b < 0$, which means that there are negative correlations between the eyes or $K^{-'}(2w)$ is large. We have already pointed out that negative correlations between eyes is not biologically reasonable. Thus, we need to have a strong activity-dependent synaptic depression as manifested by the dependence of the decay rate K^{-} on the total weights $w_L + w_R$. For example, we take $c = 1$, $K^{+}(u) = 1/(1 + \exp(-r(u - 0.5)))$ and $K^{-}(u) = 1/(1 + \exp(-r(u/2 - 0.5)))$ so that $w = \frac{1}{2}$. We fix $a > b = 1 - a > 0$ so that there are positive correlations between the eyes and choose, r as the bifurcation parameter. This is the sharpness of the dependence of the rates on the activity. Figure 12.5(a) shows the bifurcation diagram. As r increases, the symmetric state loses stability at a subcritical pitchfork. This turns around, leading to a large amplitude branch of asymmetric solutions in which one synapse dominates. There is a small region

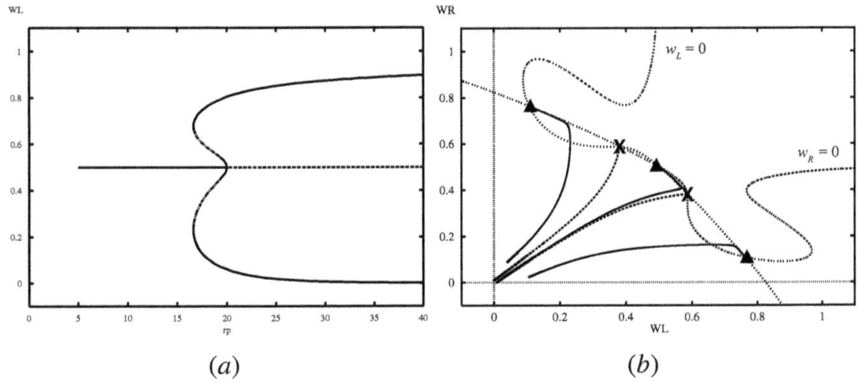

Figure 12.5. Simple competition between synapses: the bifurcation diagram shows the bistability (*a*) and the phase-plane for a fixed parameter value in the bistable regime (*b*). The points ▲ are stable and × are saddle equilibria.

of bistability between the symmetric and antisymmetric states. Figure 12.5(*b*) shows the nullclines and representative trajectories in the phase-plane for a value of r in this bistable regime. We also show one side of the stable manifolds of the two unstable asymmetric states. These accumulate onto the origin. Thus, there is a very narrow range of initial conditions that converge to the *binocular* solution in which both weights are equal.

Harris *et al* [8] introduced a model in which the competition between synapses was implemented assuming that the formation of synapses required neurotrophic factors. Adopting this approach, let n_L, n_R denote the amount of neurotrophin available to form the synapse. Equation (12.1) is then replaced by

$$w_L' = (aw_L + bw_R)n_j(1 - w_L) - \beta_1(w_L + w_R)w_L$$
$$\tau n_L' = [N - (n_L + n_R)]w_L - \beta_2 n_L$$

and a pair of similar equations exists for (w_R, n_R). Here, the rate functions are all linear so that β_1 and β_2 are both constants. The bifurcation parameter is N, the total amount of neurotrophic factor available. If N is large, it is easy to show that the weights converge to equal values. If N is very small, then no growth occurs and both weights converge to zero. Finally, for intermediate values of N, there is competition and one or the other weight dominates. If we let τ be small, then the four-dimensional system can be reduced to a planar one and analysed easily. Note that we still need the dependence of the decay on the total synaptic weight in this model. Without this all solutions tend to the symmetric state.

12.2 Spatial scales

In order to look at spatial pattern formation, we must move beyond a single neuron and look at a *population* of neurons in the cortex. In this section, we derive spatial models from the same general principles that we used earlier. Let $w(x, y, t)$ denote the synaptic weight from a visual field point y to a cortical location x at time t. Let $I(x, y, t)$ be the inputs from y to x at time t. Suppose that there are weak cortical-cortical interactions $\epsilon J(x, x')$. Then the linear model for cortical activity takes the form:

$$\tau_c \frac{\mathrm{d}V(x,t)}{\mathrm{d}t} = -V(x,t) + \epsilon \int J(x,x')V(x',t)\,\mathrm{d}x' + \int w(x,y',t)I(x,y',t)\,\mathrm{d}y.$$

We could solve for the steady states, but we would have to invert the linear operator $1-\epsilon J$. If ϵ is small, we can approximate this inversion. Letting $M(x, x')$ be the inverse, we obtain

$$V(x) = \int M(x,x')\,\mathrm{d}x' \int w(x',y',t)I(x',y',t)\,\mathrm{d}y' \tag{12.2}$$

where $M(x, x') \approx \delta(x - x') + \epsilon J(x, x')$. Consider first the mean value of the output over the input space. This becomes

$$\langle V(x)\rangle = \bar{I} \int M(x,x')\,\mathrm{d}x' \int w(x',y',t)\,\mathrm{d}y'.$$

The Hebbian term is more complicated:

$$\begin{aligned}\langle V(x)I(x,y)\rangle &= \int M(x,x')\,\mathrm{d}x' \int w(x',y',t)\langle I(x',y',t)I(x,y,t)\rangle\,\mathrm{d}y' \\ &= \iint M(x,x')C(x-x',y-y')w(x',y')\,\mathrm{d}x'\,\mathrm{d}y'\end{aligned}$$

where $C(\cdot,\cdot)$ is the correlation function. Assuming that the growth model has the same form as the previous section, namely

$$w' = K^{+}(\text{Hebbian})(1-w) - K^{-}(\text{Activity})w$$

the model we look at has the following form:

$$\begin{aligned}\frac{\partial w(x,y,t)}{\partial t} = {}& K^{+}\left[\iint M(x,x')C(x-x',y-y')w(x',y')\,\mathrm{d}x'\,\mathrm{d}y'\right] \\ &\times [1-w(x,y)] - K^{-}\left[\bar{I}\iint M(x,x')w(x',y')\,\mathrm{d}y'\,\mathrm{d}x'\right].\end{aligned} \tag{12.3}$$

We note that if there is a fixed resource pool at each spatial location x rather than a fixed pool at each synapse, then the term $[1 - w(x, y)]$ is replaced by

$[1 - \int_y w(x, y)\,\mathrm{d}y]$. Finally, we assume homogeneity in the cortical–cortical interactions so that $M(x, x') = M(x - x')$. There are many possible types of solutions that we could look for in this model. In fact, (12.3) is a very general model in that the variable y appearing in $I(x, y)$ is really nothing more than a label which need not be space. For example, it could take on the two discrete values, $\{L, R\}$ representing the left and right cortices. Similarly, it could also just represent orientations in which case it lies on the circle.

Let us first look at the case of spatial clustering. We will assume that the topographic map is already set and it is near an identity map. Then we expect that $I(x, y)$ is a function of x alone and the correlations are only a function of $x - x'$. This implies that the weights are functions of their cortical locations only and that they satisfy

$$\frac{\partial w(x,t)}{\partial t} = K^{+}\left[\int M(x-x')C(x-x')w(x',t)\,\mathrm{d}x'\right](1 - w(x,t)) - K^{-}\left[\overline{I}\int M(x-x')w(x',t)\,\mathrm{d}x'\right]w(x,t). \tag{12.4}$$

We look for pattern formation in this system; in particular, steady-state patterns. We assume that the system is homogeneous, and thus translation invariant, in order to do the requisite linear analysis. This occurs, for example, if the domain is infinite or periodic. Thus, the analysis should be regarded as an approximation since a real cortical sheet is finite with non-periodic boundaries. Let $Q(x) = C(x)M(x)$ be the product of the input correlation and the cortical connectivity. Suppose that

$$\overline{M} = \int M(x')\,\mathrm{d}x' \qquad \text{and} \qquad \overline{Q} = \int Q(x')\,\mathrm{d}x'$$

are both positive and $\overline{I} = 1$. Since $K^{\pm}$ are non-negative functions, there is at least one constant solution to (12.4), say, $w(x, t) = \overline{w}$. The stability is found by linearizing about this constant solution. The linearized equation is:

$$\frac{\partial w(x,t)}{\partial t} = K^{+'}(\overline{Qw})(1-\overline{w})Q \star w - K^{-'}(\overline{Iw})\overline{w}M \star w - [K^{+}(\overline{Qw}) + K^{-}(\overline{Iw})]w$$

where $\star$ is the convolution symbol. The assumption of translation invariance implies that solutions to the linearized equations are:

$$w(x,t) = \mathrm{e}^{\lambda t}\mathrm{e}^{\mathrm{i}k\cdot x}$$

and that

$$\lambda(k) = A^{+}\widehat{Q}(|k|) - A^{-}\widehat{M}(|k|) - A_0.$$

Here,

$$A^+ \equiv K^{+'}(\overline{Qw})(1-\overline{w})$$
$$A^- \equiv K^{-'}(\overline{Iw})\overline{w}$$
$$A_0 \equiv K^+(\overline{Qw}) + K^-(\overline{Iw})$$

and $\widehat{Q}$, $\widehat{M}$ are the Fourier transforms of the interaction functions.

First, consider the case in which there is no activity-dependent suppression of the weights, that is, K^- is constant, so $A^- = 0$. Then the maximal eigenvalue occurs at the maximum value of $\widehat{Q}$. If both the cortical interaction function and the input correlation matrix are Gaussian, then so is their product and the transform has a maximum at $k = 0$. This situation does not lead to pattern formation. Suppose, instead, that the cortical interaction function M is of lateral-inhibitory type so that the product with C is similar. Then the maximal wavenumber will be non-zero in magnitude. Hence, if A^+ is sufficiently large, we can force the uniform state to lose stability at a non-zero wavenumber and pattern formation will occur.

Now suppose that the suppression of weights is also activity dependent, i.e. $K^{-'} \neq 0$. Then, *even* with monotone cortical interactions $M(x)$, it is still possible to get pattern formation. To see why this is expected, recall that $Q(x)$ is the *product* of the stimulus correlations $C(x)$ with the intracortical interactions $M(x)$. If both of these are Gaussians, then their product is also a Gaussian but with a smaller variance than $M(x)$. Therefore, $\widehat{Q}(|k|)$ is broader than $\widehat{M}(|k|)$ so that, if $K^{\pm'}$ are sufficiently large, $\lambda(k)$ will be maximal and positive for a non-zero value of k.

Therefore, in order to get pattern formation, we need either a 'Mexican-hat'-type interaction within the cortical network or activity-dependent suppression of synaptic weights. As an example, consider the following:

$$Q(x) = \frac{1}{5.6}\mathrm{e}^{-0.1x^2}$$
$$M(x) = \frac{1}{12.3}\mathrm{e}^{-0.02x^2}$$
$$K^\pm(u) = \frac{1}{1+\exp(-r^\pm[u-0.5])}.$$

With these choices, $w(x) = 0.5$ is a constant solution, which is the solution where all weights are equal. The coefficients are:

$$A^\pm = \tfrac{1}{8}r^\pm \qquad A_0 = 1.$$

The effective spatial kernel, $J_{\text{eff}}(x) = A^+Q(x) - A^-M(x)$ is shown in figure 12.6 on the left. If we set $r^- = \frac{1}{2}r^+$, for example, then there is only one parameter, r^-, which is the sharpness of the dependence of the rates on the interactions. It is not hard to see that the uniform state loses stability as r^+ increases. The right panel in figure 12.6 shows the steady state values of the weights for a finite domain with 'reflecting' conditions at the end-points.

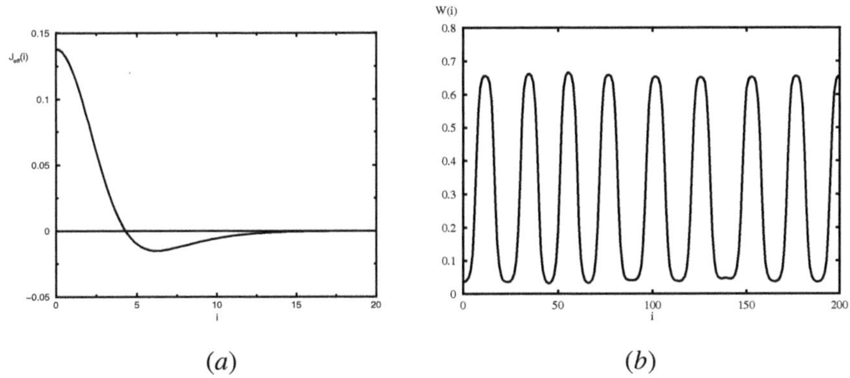

Figure 12.6. Clustering model: the effective interaction function is a 'Mexican hat' (*a*); and the final values of the weights for $r^+ = 12.5$ and $r^- = \frac{1}{2}r^+$ (*b*).

12.2.1 Occular dominance

We now turn to the more complicated issue of occular dominance (OD). Recall (12.3) and that the second variable y codes the specific features such as retinal location or occularity. Let us suppose that this is a discrete variable coding $\{L, R\}$ for the left and right eyes. Furthermore, we assume that $C(x, y) = C_1(x)\, C_2(y)$, where C_2 is just a 2×2 matrix representing the correlations between the two eyes. We assume that C_2 is symmetric such that the diagonal terms are the same eye correlations c_s and the off-diagonal terms are the different eye correlations c_d. In the present formulation, there is no dependence on correlations for the decay of the weights; this depends only on activity. The resulting model takes the following form:

$$\begin{aligned}
\frac{\partial w_L}{\partial t} &= K^+[Q(x) \star (\beta w_L(x) + (1-\beta) w_R(x))][1 - w_L(x) - \nu w_R(x)] \\
&\quad - K^-[M(x) \star ((1-\gamma) w_L(x) + \gamma w_R(x))] w_L(x) \qquad (12.5) \\
\frac{\partial w_R}{\partial t} &= K^+[Q(x) \star (\beta w_R(x) + (1-\beta) w_L(x))][1 - w_R(x) - \nu w_L(x)] \\
&\quad - K^-[M(x) \star (\gamma w_L(x) + (1-\gamma) w_R(x))] w_R(x).
\end{aligned}$$

To simplify notation, we absorbed the mean input $\overline{I}$ into the function $M(x)$. We also introduced three additional parameters: c_s and c_d are replaced by a single parameter β, using $c_s + c_d = 1$ without loss of generality, because they are both positive. This means that $\beta = \frac{1}{2}$ corresponds to no difference in the same eye and different eye correlations. The parameter ν is zero when the synapses compete for a local pool. If all synapses at the same spatial location must *share* some substance, then $\nu = 1$. The parameter γ is $\frac{1}{2}$ if the decay terms depends only on the total activity and there are no correlations in the decay. It can never be 1. Our derivation of the equations is predicated on the idea that $\gamma = \frac{1}{2}$. When $\gamma \neq \frac{1}{2}$, we are dealing with a more abstract form of weight decay which is still Hebbian. (Actually, if $\gamma = \frac{1}{2}$, the term in the decay is off from the derived term by a factor of 2. This can be absorbed into the weight function $M(x)$ as well.)

We now show that we need either Mexican hat interactions in the cortical interactions in order to get OD patterns or correlations in the decay of the weights. As in the cluster mode earlier, we note that there is a homogeneous steady-state solution, $w_L(x) = w_R(x) = \overline{w}$. The linearized equations have the form:

$$\begin{aligned} w_L' &= A^+ Q(x) \star [\beta w_L + (1-\beta) w_R] - A^- M(x) \star [\gamma w_L + (1-\gamma) w_R] \\ &\quad - [A_0 w_L + A_1(w_L + \nu w_R)] \\ w_R' &= A^+ Q(x) \star [\beta w_R + (1-\beta) w_L] - A^- M(x) \star [\gamma w_R + (1-\gamma) w_L] \\ &\quad - [A_0 w_R + A_1(w_R + \nu w_L)]. \end{aligned}$$

The constants $A^\pm$, A_0, A_1 all have the obvious values. In particular, both A_0, A_1 are positive while $A^\pm \geq 0$ could vanish if the rates $K^\pm$ are independent of activity. As before, we look for the solutions to the linearized equations. One major difference is that there are two equations rather than a single scalar equation. However, because of symmetry, we can decompose this into a pair of uncoupled scalar equations corresponding to symmetric $(1, 1)$ and anti-symmetric $(1, -1)$ perturbations of the weights (w_L, w_R). This means that there are two eigenvalues, $\lambda_s(k)$ and $\lambda_a(k)$, for each wavevector k. In order to get occular dominance patterns, we require that the maximum of these occur at $|k| > 0$ and that this maximum be the antisymmetric case. The eigenvalues are:

$$\begin{aligned} \lambda_s(k) &= A^+ \widehat{Q}(k) - A^- \widehat{M}(k) - A_0 - A_1(1+\nu) \\ \lambda_a(k) &= A^+ \widehat{Q}(k)(2\beta - 1) - A^- \widehat{M}(k)(2\gamma - 1) - A_0 - A_1(1-\nu). \end{aligned}$$

Several points are immediately clear. Suppose that there is no correlated decay of the synaptic weights, i.e. $\gamma = \frac{1}{2}$. Then OD pattern formation can only occur if $\widehat{Q}$ has a maximum at a non-zero k and $\beta > 1$. The former condition implies that the cortical interactions are of Mexican-hat type and the latter implies that the correlations between different eyes, $c_s = 1 - \beta$ are negative. These conditions are found in some of the models of Swindale and Miller. The problem with these assumptions is that they state that the activity in the left eye is *negatively* correlated with the activity in the right eye. This seems to be somewhat unrealistic. The problem with *lateral inhibition* in the cortical interactions is that this implies that the spacing between stripes depends on the range of the lateral inhibition rather than the correlations of the inputs.

Hence, we conclude that a strictly activity-dependent decay is not sufficient and we require some correlated decay of the weights. Suppose that $\gamma \neq \frac{1}{2}$. If $\gamma > \frac{1}{2}$, then the decay is more sensitive to the same eye weights, while $\gamma < \frac{1}{2}$ means it is more sensitive to different eye weights. Suppose also that the coupling functions Q, M are Gaussians. If $\gamma < \frac{1}{2}$, then the asymmetric eigenvalue is larger for all k since both k-dependent terms are positive. Furthermore, it is clear that

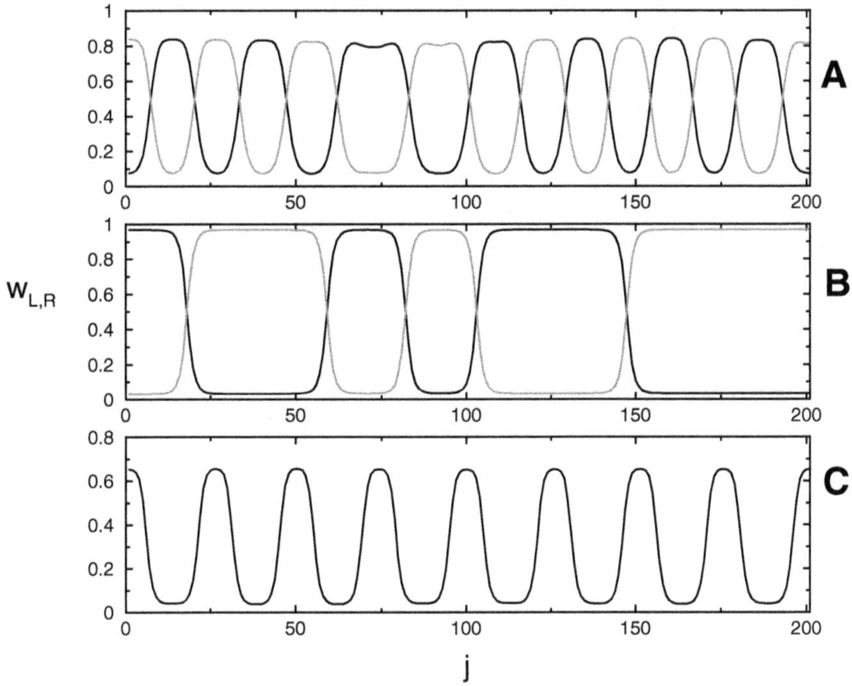

Figure 12.7. Simulation of (12.5) with different parameters: (A) $\gamma = 0.7$, $\nu = 1$ showing OD stripes with a well-defined periodicity; (B) $\gamma = 0.5$ leading to OD columns with long wave and ill-defined periodicity; (C) $\nu = 0$ leading to symmetric periodic weights.

$k = 0$ will be maximal. Thus, if $\gamma < \frac{1}{2}$, the zero wavenumber will be the first to bifurcate and it will be in the asymmetric state. The cortical network will be completely right- or left-eye dominant.

Suppose, instead, that $\gamma > \frac{1}{2}$ so that the weight decay depends more on the same eye weight. Then, it is possible (although not guaranteed) to choose parameters so that the maximal eigenvalue occurs at a non-zero k and in the asymmetric state. If $\nu = 1$, as would be the case if local left and right weights compete for the same pool of weight-forming substance, then it is much easier to find parameters that destabilize the asymmetric state. This is because the symmetric eigenvalue has a stabilizing term $-A_1(1 + \nu)$ when $\nu = 1$.

Figure 12.7 shows the result of a one-dimensional simulation of (12.5) with

$$K^{\pm}(u) = 1/(1 + \exp(\rho^{\pm}(u - 0.5)))$$

and Gaussian weights. Panel A shows the model with $\nu = 1$, $\gamma = 0.7$ in the parameter regime required for asymmetric pattern formation. Normal alternating stripes of left and right dominance form. However, if we eliminate the side-dependent differences in the weight decay ($\gamma = \frac{1}{2}$) then patterns form but they

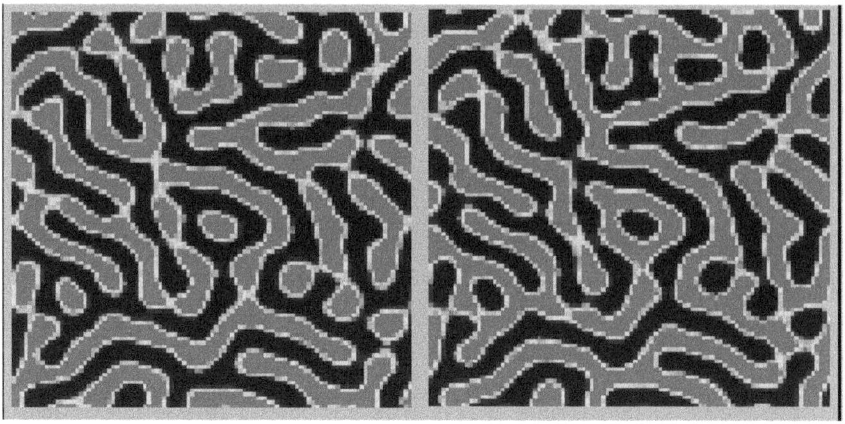

Figure 12.8. Simulation of (12.5) in a two-dimensional domain with the same parameters as in figure 12.7(A). Grey scales indicate the strengths of the weights: dark grey is large and black is small. Left represents w_L and right w_R.

are irregular (panel B). Bifurcation theory predicts that homogeneous asymmetric patterns will arise—that is, one of the two eyes is dominant. This behaves like a discretely coupled symmetric bistable medium so that local domains form but there is no spatial scale. Finally, if we keep $\gamma > \frac{1}{2}$, but eliminate ν, then the dominant instability has spatial structure. However, there is insufficient competition to separate the left and right weights and a symmetric clustered state appears.

The one-dimensional model sheds insight into the general behaviour of these models. However, in two dimensions there are still many questions about what types of behaviour can be expected. In figure 12.8 we show the results of a simulation of a 100×100 network with parameters identical to those in figure 12.7(A) except that the space constants for the Gaussians are half as large to compensate for the fact that there are 200 units in the one-dimensional simulations. The pattern is complicated and is not dominated as much by stripes as in the biological figure 12.3. However, there is a range of patterns found across different species that vary from stripes to a more irregular pattern. Thus, the simple model with positive correlations and non-negative cortical interactions is able to capture the qualitative behaviour of these patterns.

12.3 Orientation maps and feature maps

The activity dependence of orientation maps is more controversial, because experimental evidence is somewhat ambiguous on the post-natal activity dependence of orientation maps. However, we can use essentially the same formulation to derive equations for orientation preferences. As we did with the occular dominance equations, we will generalize (12.3) so that is includes both local competition for weights and correlated decay of the weights. We will also assume translation invariance. This leads to a generalized feature model:

$$\begin{aligned}\frac{\partial w(x,y)}{\partial t} &= K^{+}\left[\iint Q^{+}(x-x',y-y')w(x',y')\,\mathrm{d}x'\,\mathrm{d}y'\right]\\ &\quad\times\left(1-\int w(x,y')\,\mathrm{d}y'\right)\\ &\quad- K^{-}\left[\iint Q^{-}(x-x',y-y')w(x',y')\,\mathrm{d}x'\,\mathrm{d}y'\right]w(x,y).\end{aligned} \tag{12.6}$$

Here, $w(x, y)$ is the strength of the connection of feature y to the point x in the cortex. One can think of $w(x, y)$ as the probability of encountering feature y at position x. If Q^- is independent of y, then we recover the model in which weight decay is depending only on the total activity at location x. The term $1 - \int w(x, y')\,\mathrm{d}y'$ could be replaced by $1 - w(x, y)$ if the resource pool for the formation of synapses is completely localized. The integral form specified in (12.6) presumes that there is a shared pool for each location x. By making the obvious assumptions about homogeneity, there will be a constant steady state $w(x, t) = \overline{w}$ independent of both the feature and the spatial location, just as with the OD models. Hence, we can again study the stability of this state by linearizing about the constant fixed point and then performing bifurcation analysis. This is very similar to the analysis of the OD model. Indeed, one would like to get patterns which depend non-trivially on both the spatial and the feature variables. To facilitate the analysis of the model, we assume that the functions $Q^{\pm}(x, y)$ are products of two independent functions, that is,

$$Q^{\pm}(x,y) = S^{\pm}(x)\Gamma^{\pm}(y).$$

We assume, without loss of generality, that all of these functions integrate to 1 over their respective domains. Let $\widehat{Q}^{\pm}(k)$, $\widehat{\Gamma}^{\pm}(\ell)$ denote the Fourier transforms of the coupling and correlation functions, respectively. Then, as before, we determine the stability of the constant state by linearizing and looking for solutions to the linearized problem of the form

$$w(x,y,t) = \mathrm{e}^{\lambda t}\mathrm{e}^{\mathrm{i}k\cdot x}\mathrm{e}^{\ell y}$$

and finding λ as a function of (k, ℓ). The result of this calculation is

$$\lambda(k,0) = A^{+}\widehat{Q}^{+}(k)\widehat{\Gamma}^{+}(0) - A^{-}\widehat{Q}^{-}(k)\widehat{\Gamma}^{-}(0) - A_0^{-} - A_0^{+}N \tag{12.7}$$

$$\lambda(k,\ell) = A^{+}\widehat{Q}^{+}(k)\widehat{\Gamma}^{+}(\ell) - A^{-}\widehat{Q}^{-}(k)\widehat{\Gamma}^{-}(\ell) - A_0^{-} \tag{12.8}$$

where

$$N = \int 1\,dy$$
$$A^+ = K^{+'}(\bar{w})(1 - N\bar{w})$$
$$A^- = K^{-'}(\bar{w})\bar{w}$$
$$A_0^\pm = K^\pm(\bar{w})$$

are all positive constants. Note that N is the measure of the total feature space (e.g. π in the case of orientation). From (12.7) we can see that the effect of the competition for a resource at the feature level is the additional stabilizing term $A_0^+ N$. If this is large, then the most unstable mode will be $\ell > 0$ corresponding to a pattern along the feature direction. As in the case of OD models, if there is no feature-dependent decay of the weights, then

$$\Gamma^-(y) = 1$$

and there will be no A^- term in equation (12.8) since $\widehat{\Gamma}(\ell)$ is 1 for $\ell = 0$ and zero otherwise. Thus, as with the OD model, the only way to get spatial pattern formation in absence of feature-dependent weight decay is to have Mexican-hat cortical connectivity. The maximal eigenvalue will correspond to the most unstable mode of the spatial interaction function $\widehat{Q}^+(k)$ and the most unstable *non-zero* mode of the function $\widehat{\Gamma^+}(\ell)$. The key here is that typically all feature–feature correlations are non-negative so that the Perron–Frobenius theorem implies that the maximal eigenvalue is zero. Thus, for a monotonically decreasing function $\Gamma(y)$ the most unstable non-zero mode is the first one. For example, in the case of occular dominance, this is just the eigenvector $(1, -1)$ and in the case of orientation, it is $\cos(2y)$, a singly-peaked function on the interval $[0, \pi)$.

This implies that in any local spatial domain, each orientation is represented exactly once. Indeed, it is interesting to note that this type of pattern occurring in the *activity* models recently proposed by Cowan and Bressloff (see chapter 11) for visual hypercolumns as well as related to the orientation-spatial models they suggest for fine-structure hallucinations. If there is feature-dependent decay of the weights, then it is possible to get spatial patterns in conjunction with feature patterns. For example, suppose that $\Gamma^+(y) = \Gamma^-(y) \equiv \Gamma(y)$. Then the eigenvalues are

$$\lambda(k, 0) = \widehat{\Gamma}(0)[A^+\widehat{Q}^+(k) - A^-\widehat{Q}^-(k)] - A_0^- - A_0^+ N$$
$$\lambda(k, \ell) = \widehat{\Gamma}(\ell)[A^+\widehat{Q}^+(k) - A^-\widehat{Q}^-(k)] - A_0^-.$$

Thus, the maximum will occur at the first mode of Γ and at the maximum of the Mexican-hat-like function

$$R(k) \equiv A^+\widehat{Q}^+(k) - A^-\widehat{Q}^-(k).$$

However, if the competition for synapses is completely local, then the term $1 - \int w(x, y')\,\mathrm{d}y'$ in (12.6) will be replaced by $1 - w(x, y)$. The eigenvalues now become

$$\lambda(k, \ell) = \widehat{\Gamma}(\ell)[A^+\widehat{Q}^+(k) - A^-\widehat{Q}^-(k)] - (A_0^- + A_0^+).$$

This has a maximum at $\ell = 0$ so that the only type of pattern observed will be spatial clustering independent of features.

If we have feature competition *and* feature-dependent weight decay, then we can get patterns that depend non-trivially on both features and spatial position. As with all linearized analysis, all we can get is an overall picture of the periodicity. One can apply a full bifurcation analysis to obtain the requisite normal form. If the spatial system is on a periodic domain and the feature is orientation, then the general form for the solutions is:

$$w(x, y, t) = \bar{w} + v_1^-(t)\mathrm{e}^{\mathrm{i}(kx_1-2y)} + v_1^+(t)\mathrm{e}^{\mathrm{i}(kx_1+2y)}$$
$$+ v_2^-(t)\mathrm{e}^{\mathrm{i}(kx_2y)} + v_2^+(t)\mathrm{e}^{\mathrm{i}(kx_1+2y)} + \text{c.c.}$$

where k is the critical wavenumber. The normal form for this bifurcation consists of four complex equations with real coefficients that have the form:

$$z_j' = z_j(\nu - a_j|v_1^+|^2 - b_j|v_1^-|^2 - c_j|v_2^+|^2 - d_j|v_2^-|^2)$$

where z_j is each of the four $v_{1,2}^{\pm}$ and ν is the bifurcation parameter.

Because of symmetry, the fixed points to this equation are quite easy to find. Among the possible solutions are those in which all the vs are non-zero. This leads to weights that have a modulation proportional to

$$\cos(2y)[\cos(kx_1) + \cos(kx_2)].$$

This solution is unsatisfactory since the orientation $y = \pi/4$ never has a maximum. That is, there is no neuron whose tuning curve has $\pi/4$ as its maximum. Another possibility is a solution with a single non-zero v leading to a weight modulation:

$$\cos(kx_1 - 2y).$$

Each orientation is represented here but the orientation preference is independent of the second spatial coordinate x_2, which does not agree with the data. Finally, there is a solution in which $v_1^+ = v_2^+ = v$ and all others vanish. This leads to a solution of the form:

$$\cos(kx_1 - 2y) + \cos(kx_2 - 2y).$$

This solution is again unsuitable, because it leads to orientations that are constant along the lines $x_1 \pm x_2 = K$.

Hence, there appears to be no way to obtain the kinds of patchy solutions seen in the data, at least near the bifurcation point. However, full simulations indicate that it *is* possible to obtain patchy orientations, as is shown in figure 12.9.

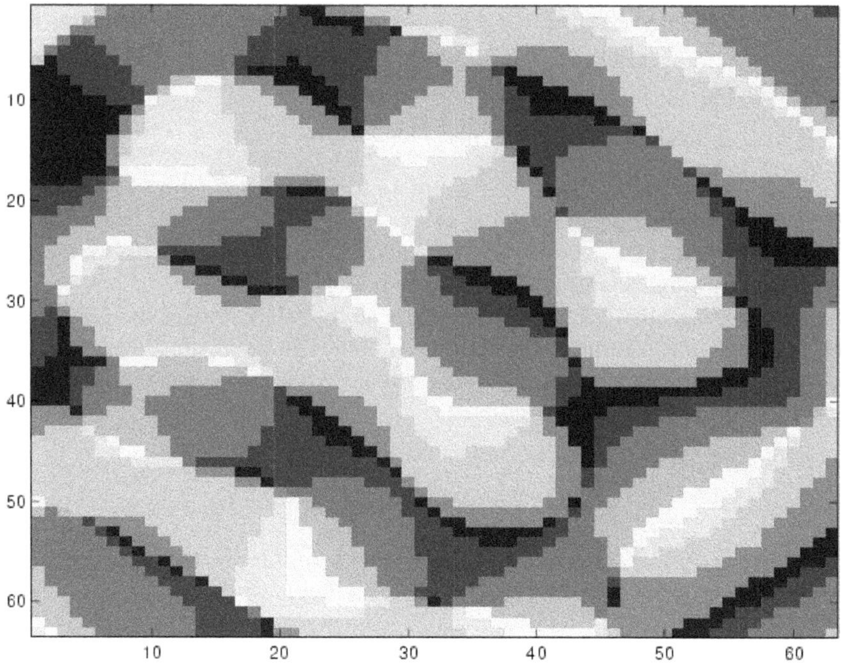

Figure 12.9. Development of orientation in a 64 × 64 cell network with eight orientations.

12.3.1 Combined orientation and occular dominance models

While there have been numerous models for OD and also a number for orientation maps, there are far fewer attempts at creating a combined map that has both orientation and OD. Swindale developed a model for both maps and successfully simulated it; see the review in [17]. However, his model was rather abstract and it is hard to connect it with a mechanistic interpretation. Erwin and Miller [6] have one of the most complete models. If the orientation and OD maps develop sequentially, then they can explain most of the relevant experimental phenomena. As with most models, they use a Mexican-hat connectivity function so that the periodicity is built in through the cortical interactions. Our formulation, notably (12.6) makes no assumptions about the nature or dimensionality of the feature space y (although we did treat it as either discrete for OD or one-dimensional for orientation). Thus, one can readily generalize the model to include both OD and orientation development. Figure 12.10 shows the results of a simulation of the joint development of the maps. Note that the 'singularities' lie in the centres of the OD bands and the borders of the OD bands are orthogonal to the contours of the orientation regions. A mathematical explanation for why this should be expected remains to be discovered.

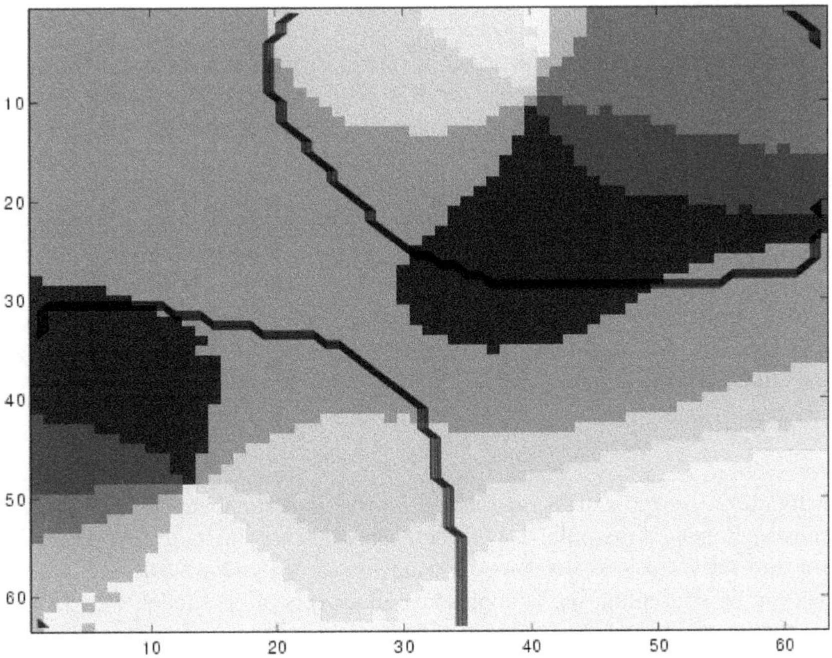

Figure 12.10. Development of both occular dominance and orientation.

12.4 Kohonen maps and abstract feature models

The model described in the previous section as well as the models of Miller and Swindale and their collaborators describe the evolution of weights $w(x, y, t)$ that code the fraction or strength of connections with feature y at cortical position x. Experimental cortical maps for a particular feature generally produce a picture showing the preferred value of that feature at each location x. That is, they depict the value of y which maximizes $w(x, y, t)$ at position x or the mean over the feature space of the weights. For this reason, some modellers suggested more abstract models in which one looks at the evolution of the preferred feature in the cortical space. The so-called self-organized feature maps (SOFMs) that were first proposed by Kohonen are best known; see [11] for a review. These algorithms have been shown to reproduce the experimental results better than the more biologically motivated models [7]. Another closely related class of models is the elastic net [4]. In this section, we briefly describe the SOFM. Based on it, we derive a continuum model and provide a stability analysis. We then show how to take an equation such as (12.6) and reduce it to a model which is similar to the SOFM. We apply this to orientation maps.

12.4.1 SOFMs

Consider a vector of features denoted by $\vec{f}(x)$. Each component is the preferred value of that particular feature at spatial position x. In the model analysed by Obermayer and Blasdel [15] there are five components (y_1, y_2, z, r, θ) to the feature map, representing the retinal coordinates $(y_1, y_2) \in \mathbb{R} \times \mathbb{R}$, the occularity $z \in [-1, 1]$, and the orientation preference as well as the degree of such preference $(\theta, r) \in [0, \pi) \times [0, 1]$, respectively. Note, that the occular dominance takes values between -1 and 1, with -1 associated with left dominance and $+1$ with right. An occularity $z = 0$ corresponds to no preference.

The Kohonen algorithm is as follows. Randomly pick a vector from the feature space $\vec{F}$. Find the cortical position x^* for which $|\vec{F} - \vec{f}(x)|$ is minimized. Then, update all the features as:

$$\vec{f}(x) = \vec{f}(x) + \delta H(|x - x^*|)(\vec{F} - \vec{f}(x)).$$

The function H is typically a Gaussian and δ is a small number. The idea is that all feature vectors near the point x^* will be pushed toward the input feature $\vec{F}$. This algorithm does not lend itself to the usual types of dynamic stability analysis. However, by 'softening' the winner-takes-all aspects of the model, we can apply the usual types of stability analysis. To simplify things, we will assume that the cortex and retina are both one-dimensional structures. We will also assume that the retino-cortical map (the topographic map) is already set up and is the identity. That is $y(x) = x$. Finally, we will only study the emergence of occular dominance.

Suppose that we present a feature (Y, Z) corresponding to a retinal position Y and occularity Z. The output of the cortical neuron at point x is

$$U(x) = H_T(y(x) - Y)H_O(z(x) - Z)$$

where H_j is, for example, a Gaussian:

$$H(a) = \mathrm{e}^{-a^2/\sigma_j^2}.$$

(Here T stands for topography and D stands for oscularity.) Thus, the output of the neuron is largest for inputs that are closest to the input feature. The parameters $\sigma_{T,O}$ describe the sharpness of the dependence on the feature. We normalize these functions so that their integrals are 1 over their respective feature spaces.

To set up the competition, we divide the output by the output of every neuron in the cortical domain:

$$\widehat{U}(x) = \frac{U(x)}{\int U(x')\,\mathrm{d}x'}.$$

Thus, for small values of $\sigma_{T,O}$, the function $\widehat{U}(x)$ is sharply peaked at a point x^* for which the feature at x is closest to the input feature. This is a softer version of the Kohonen competition. Finally, due to the cortical interactions, the actual activity is

$$V(x; Z, Y) = \int J(x - x')\widehat{U}(x')\,\mathrm{d}x'$$

where J is the cortical interaction function. We emphasize that this activity depends on the input features. The equations for the occularity feature are now:

$$z(x) \mapsto z(x) + \delta V(x; Z, Y)(Z - z(x)).$$

We average this over all possible features and proceed to continuous time to obtain the following model:

$$\frac{\mathrm{d}z}{\mathrm{d}t} = \int_{-1}^{1}\int \left[\frac{\int J(x - x')H_T(x' - Y)H_O(z(x')\,\mathrm{d}x' - Z)}{\int H_T(x' - Y)H_O(z(x')\,\mathrm{d}x' - Z)}\right](Z - z(x))\,\mathrm{d}Y\,\mathrm{d}Z. \tag{12.9}$$

For simplicity, all integrals with respect to x', Y are taken over the real line.

A model like this was analysed by Dayan [2]. We only sketch the stability analysis. The parameter of interest is σ_O, the degree or sharpness of the occular dominance tuning. Since the function H is symmetric in each of its arguments, it follows that $z(x) = 0$ is a solution to (12.9). We linearize about this solution and obtain the following linear problem:

$$\frac{\mathrm{d}z}{\mathrm{d}t} = -2z + \nu \int [J(x - x') - Q(x - x')]z(x')\,\mathrm{d}x'$$

where

$$\nu = \int_{-1}^{1} Z\frac{H_O'(Z)}{H_O(Z)}\,\mathrm{d}Z$$

$$Q(x) = \int J(x - x')H_T(x')\,\mathrm{d}x'.$$

If H_O is a normalized Gaussian then $\nu = \frac{4}{3}\sigma^{-3}$, and if J is also a Gaussian then Q is a Gaussian which is broader than J and normalized. Thus, $J - Q$ is a Mexican hat and for σ_O small enough, the $z = 0$ state loses stability such that spontaneous patterns arise. The wavelength of these patterns depends both on the cortical distance, say, σ_C and on the spatial input correlations σ_O. In fact, the two relevant space scales are σ_C and $\sigma_C + \sigma_O$, the result of convolving the two Gaussians. This contrasts with the models we described in section 12.3.1. There, the relevant space scales are the cortical scale σ_C and the effective scale

$$\sigma_E \equiv \frac{\sigma_C}{\sqrt{1 + (\sigma_T/\sigma_C)^2}}$$

due to the product of J with the input correlation. Note that σ_E is narrower than σ_C so that SOFM predicts wider stripes than the model in (12.5).

12.4.2 Reduction of kinetic-based models to SOFM-like models

We conclude this section with a reduction of a kinetic-based model to an abstract feature map. To simplify the derivation, consider only a single feature instead of a vector of features. Furthermore, we assume that $Q^{\pm}(x, y) = J^{\pm}(x)C(y)$ in (12.6), so that the correlations for weight growth and decay are the same. We can view the weight $w(x, y, t)$ as a tuning curve for the feature y. That is, at a fixed point in time and at location x, the weight $w(x, y, t)$ describes the output of a neuron when presented with the feature y. We assume that this tuning has a single peak at $y = \hat{y}(x, t)$ and that it is stereotypical in shape. That is, we suppose that

$$w(x, y, t) = W(\hat{y} - y(x, t))$$

where W is a Gaussian-like function. We define $\hat{y}$ by the condition that $w_y(x, y, t) = 0$ at $y = \hat{y}$ and, furthermore, that $w_{yy}(x, \hat{y}, t) < 0$ so that $\hat{y}$ is a local maximum. Differentiate $w_y(x, \hat{y}(x, t), t) = 0$ with respect to time to obtain

$$w_{yy}(x, \hat{y}(x, t), t)\frac{\partial \hat{y}}{\partial t} + w_{yt}(x, \hat{y}, t) = 0. \qquad (12.10)$$

If we now differentiate (12.6) with respect to y and evaluate at $\hat{y}$, this leads to:

$$\begin{aligned} w_{yt} = {} & (1 - W_{\max})K^{+'}(\cdot)\iint J^{+}(x - x')C'(\hat{y}(x, t) - y') \\ & \times W(\hat{y}(x', t) - y')\,\mathrm{d}x'\,\mathrm{d}y' \\ & - W_{\max}K^{-'}(\cdot)\iint J^{-}(x - x')C'(\hat{y}(x, t) - y')W(\hat{y}(x', t) - y')\,\mathrm{d}x'\,\mathrm{d}y'. \end{aligned}$$

Here $W_{\max}$ is the maximum weight and we assume that it is constant across the spatial domain so that all tuning cures look the same; only the maximum feature differs. We can evaluate the integrals over y':

$$\begin{aligned} \int C'(\hat{y}(x, t) - y')W(\hat{y}(x', t) - y')\,\mathrm{d}y' &= \int C'(\hat{y}(x, t) - \hat{y}(x', t) - s)W(s)\,\mathrm{d}s \\ &=: H(\hat{y}(x', t) - \hat{y}(x, t)). \end{aligned}$$

If the correlations and the tuning curve are even and peaked at the origin then H is an odd function and $H'(0) > 0$. Finally, since $K^{\pm'}$ are positive, we approximate them by constants and we also assume that w_{yy} is a negative constant $-\alpha$ when evaluated at the peak of the tuning curve. Combining these calculations and using our approximations, we obtain:

$$\alpha\frac{\partial \hat{y}}{\partial t} = \int J(x - x')H(\hat{y}(x) - \hat{y}(x'))\,\mathrm{d}x' \qquad (12.11)$$

where

$$J(x) = (1 - W_{\max})K^{+'}(\cdot)J^{+}(x) - W_{\max}K^{-'}(\cdot)J^{-}(x).$$

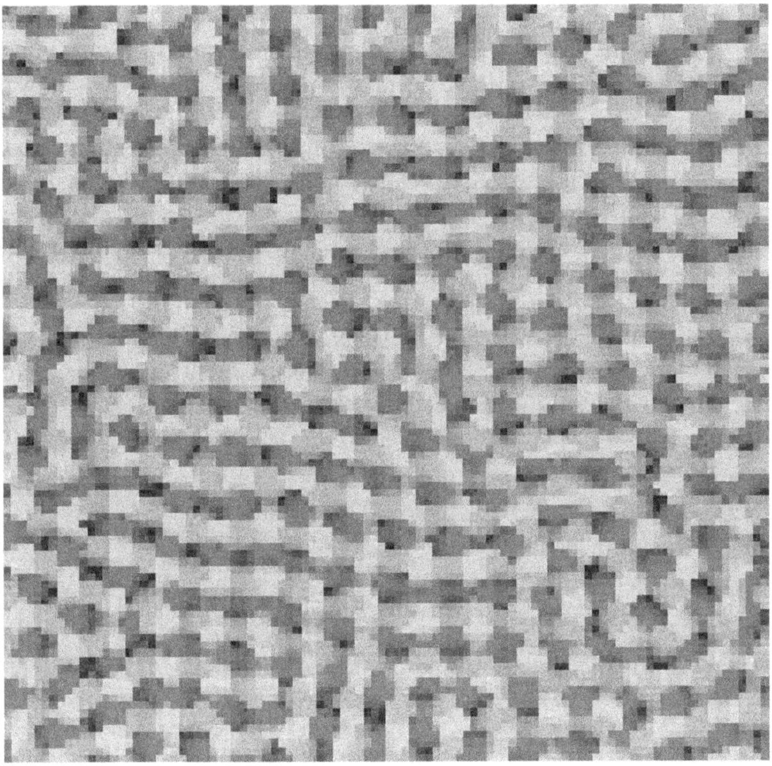

Figure 12.11. Solution to (12.11) on a 100×100 grid; see also colour plate 5.

Equation (12.11) is identical in form to models of continuum phase oscillators studied by numerous authors; for example, see [5] and references therein. Suppose that the feature of interest in orientation. Then H is a π-periodic function of its argument. As an example of the kinds of patterns found, we take

$$H(u) = \sin(2u) + \tfrac{1}{2}\sin(4u)$$

and $J(x)$ as a Mexican hat. An example output is shown in figure 12.11. This compares favourably with figure 12.2. In contrast to SOFMs, the reduction to (12.11) provides biological meaning to each of the terms.

12.5 Conclusions

We have looked at a number of models for the development of maps in visual cortex. Our approach is equivalent to the law of mass action and constraints are built in, so that there is never any problem with unbounded or negative solutions. By assuming correlated decay of weights as well as growth, we are able to obtain pattern formation even if the cortical weights are positive and the correlations between features are non-negative. Thus, our models share this aspect with the SOFM which also assumes positive interactions. However, the growth of weights and the interpretation of the weights is akin to the more biologically motivated models by Miller and Swindale. As with many pattern formation problems, notably animal coat markings (e.g. [14]), it is difficult to use the final patterns obtained from a given model to distinguish between mechanisms. This is obvious near instabilities, since all spatial models have the same normal form. Behaviour may provide a means of distinguishing mechanisms far from equilibria, but based on the results described here and in other work, this is unlikely. As certain aspects of the biology remain controversial (such as whether or not activity—read Hebbian models—is required at all for orientation), we will have to wait before a definitive answer can be given for map formation in cortex.

References

[1] Bosking W H, Zhang Y, Schofield B and Fitzpatrick D 1997 Orientation selectivity and the arrangement of horizontal connections in tree shrew striate cortex *J. Neurosci.* **17** 2112–27

[2] Dayan P 2000 Competition and arbors in ocular dominance *NIPS 2000* (Cambridge, MA: MIT Press) pp 203–9

[3] Destexhe A, Mainen Z F and Sejnowski TJ 1994 Synthesis of models for excitable membranes, synaptic transmission and neuromodulation using a common kinetic formalism *J. Comput. Neurosci.* **1** 195–230

[4] Durbin R and Mitchison G 1990 A dimension reduction framework for cortical maps *Nature* **343** 644–7

[5] Ermentrout G B 1992 Stable periodic solutions to discrete and continuum arrays of weakly coupled nonlinear oscillators *SIAM J. Appl. Math.* **52** 1665–87

[6] Erwin E and Miller K D 1998 Correlation-based development of ocularly-matched orientation and ocular dominance maps: determination of required input activities *J. Neurosci.* **18** 9870–95

[7] Erwin E, Obermayer K and Schulten K 1995 Models of orientation and occular dominance columns in visual cortex: A critical comparison *Neural Comput.* **7** 425–68

[8] Harris A E, Ermentrout G B, and Small S L 1997 A model of ocular dominance column development by competition for trophic factor *Proc. Natl Acad. Sci., USA* **94** 9944–9

[9] Hebb D O 1949 *The Organization of Behaviour: A Neuropsychological Theory* (New York: Wiley)

[10] Kandel E R, Schwartz J H, and Jessel T M 2000 *Principles of Neural Science* (New York: McGraw-Hill)

[11] Kohonen T 1995 *Self-Organizing Maps* (New York: Springer)

[12] LeVay S, Connolly M, Houde J and Van Essen D C 1985 The complete pattern of occular dominance stripes in the striate cortex and visual field of the macaque monkey *J. Neurosci.* **5** 486–501

[13] Miller K D, Erwin E and Kayser A 1999 Is the development of orientation selectivity instructed by activity? *J. Neurobiol.* **41** 44–57

[14] Murray J D 1989 *Mathematical Biology* (New York: Springer)

[15] Obermayer K and Blasdel G G 1993 Geometry of orientation and occular dominance columns in monkey striate cortex *J. Neurosci.* **13** 4114–29

[16] Song S, Miller K D and Abbott L F 2000 Competitive Hebbian learning through spike-timing-dependent synaptic plasticity *Nature Neurosci.* **3** 919–26

[17] Swindale N V 2000 How many maps are there in visual cortex? *Cereb. Cortex* **10** 633–43

[18] Turing A M 1952 The chemical basis for morphogenesis *Proc. R. Soc.* B **237** 37–72

Chapter 13

Spatio-temporal nonlinear dynamics: a new beginning

William L Ditto
University of Florida

Progress in nonlinear dynamics to date has been in the area of low-dimensional systems with no spatial extent. The first column of table 13.1 shows a partial list of commonly used analytical tools developed for this setting, that is, for the study of temporal nonlinear dynamics. They have been applied with success and led to novel applications of nonlinear dynamics. Representative applications are listed in the second column of table 13.1. In particular, I would like to mention communications (which is also the topic of chapter 5 by Van Wiggeren, Garcia-Ojalvo and Roy), control and computation.

While the study of low-dimensional dynamical systems has been successful, there are serious theoretical and practical problems in extending those studies towards understanding dynamical systems that are noisy, non-stationary, inhomogeneous and spatio-temporal. In my view, the key issue is the *interplay between temporal and spatial dynamics.* As shown in the third column of table 13.1, there are very few tools for understanding spatio-temporal nonlinear dynamics. When we veer away from isolated complex systems into coupled and continuous systems, or systems that demonstrate adaptation and evolution, we encounter a world that is analytically intractable yet remarkably rich with undiscovered concepts. In our ignorance, however, we remain unsuccessful in developing new techniques to explore these concepts. Regardless of the difficulty, spatio-temporal nonlinear dynamics is a new beginning; a new frontier that requires a truly interdisciplinary approach. To navigate this frontier we need to answer a central question.

Question 13.1. *How do spatially extended nonlinear systems self organize and why?*

Table 13.1. Tools for temporal dynamics and some areas of their application, contrasted with the limited list of tools for spatio-temporal dynamics.

Temporal tools	Temporal applications	Spatio-temporal tools
Time series	Communications	Statistics
Phase space	Controls	Patterns
Dimensions	Encryptions	Information measures
Lyapunov exponents	Encoding	
Information measures	Computing with chaos	
Symbolic dynamics	Synchronization	
Statistics	Signal detection	
Unstable periodic orbits		
Templates		
Basins		
Fractals		
Intermittencies		
Bifurcations		
Surrogates		

Let us explore this with two very different examples.

13.1 Fibrillation of the heart

Sudden cardiac death accounts for more than 300 000 deaths annually in North America [15]. It is the leading cause of death in people aged 20 to 64 years, and frequently attacks at the prime of life at a median age of only 59 years [7]. In one sense, sudden cardiac death can be considered an electrical accident, representing a tragic interplay between anatomic and functional substrates modulated by the transient events that perturb the balance. Most such incidents involve ventricular fibrillation (VF), which is a frenzied nonlinear dynamical state where the electrical activity of the heart is very fast and very irregular (but not random!). The events surrounding the onset of ventricular fibrillation in humans, even after 50 years of study, remain opaque. To understand ventricular fibrillation we need to visualize and characterize the nonlinear behaviour of the cardiac electrical system.

Applications of nonlinear dynamical techniques to fibrillation have until recently yielded contradictory results, primarily because of the inadequacies of current techniques to resolve determinism in short and noisy temporal datasets. Previous quantitative temporal measures for determinism, such as Fourier spectra, fractal dimension and Lyapunov exponents, along with other statistical techniques, have proven uniformly inadequate for characterizing VF. This is compounded here by the fact that the heart is a spatio-temporal system. Clearly, the following question arises.

Figure 13.1. Single frame of a 60 000-frame movie of merging spiral waves of electrical activity on the surface of a perfused in vitro porcine heart; see also colour plate 3.

Question 13.2. *How much of the underlying spatio-temporal nature of the heart does one need to know in order to properly interpret time series measured at different locations of the heart?*

It is now possible to measure electrical waves on the entire surface of the heart, and we have recently performed such measurements. This is work we started with the late Frank Witkowski. Frank was a genius who was an absolute master of these techniques; see also chapter 10 by Winfree. Different dynamical regimes can be identified as different types of electrical patterns travelling over the surface of the heart. To measure these patterns experimentally, we perfused a voltage-sensitive fluorescent dye through the ventricles of dog and pig hearts. We measured the ongoing electrical activity by picking up the delta fluorescence with a high-speed high-sensitivity video camera [14]. The particular camera used in the most recent of these experiments has a resolution of 80×80 pixels at 2000 frames per second. An example of an electrical wave on a fibrillating porcine ventricle is shown in figure 13.1.

The key question is to make sense of the images we recorded and to develop methods to bring unwanted and deadly dynamics under control. Much of this imaging work was motivated by our previous experiments on real-time feedback control of fibrillation in humans. While in humans it is too dangerous to perform VF experiments, we could study the related condition atrial fibrillation (AF). We

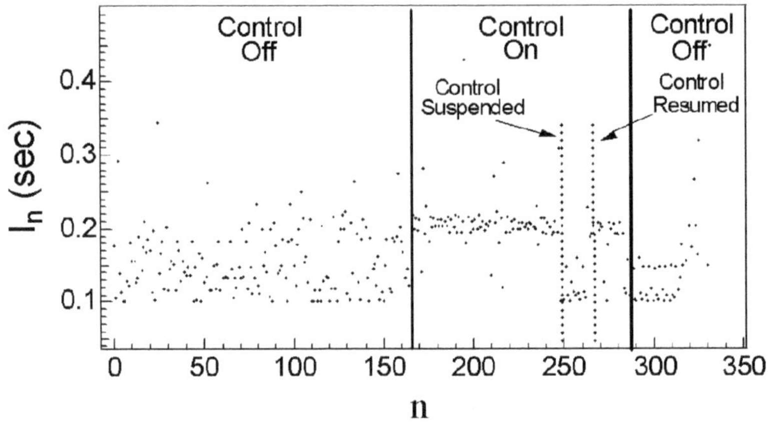

Figure 13.2. Interbeat interval versus interval number of human atrial fibrillation before, during and after the applications of chaos control.

put the atria, the upper chambers of the human heart, into fibrillation by means of an electrical stimulus in 27 test patients. As opposed to ventricular fibrillation, atrial fibrillation is not immediately life-threatening in healthy humans, allowing for our experiments. However, it is interesting in its own right, because long-term atrial fibrillation is the leading cause of strokes in the USA according to the American Heart Association.

Through the implementation of chaos control algorithms guiding small electrical stimuli applied to the human atria, we were able to initiate control to regularize and slow down the beating of the atria; this is shown in figure 13.2. To demonstrate that we had achieved chaos control, we turned the controls off to confirm a resumption of chaotic fast beating. We then turned the control back on to demonstrate the reacquisition and efficacy of the control. Finally, the control is discontinued, which (in this patient) led to a resumption of a fast but more regular pattern that dynamically transitioned into a more normal, slow beating pattern. The results from the full sequence of patients can be found in [3]. Another demonstration of dynamical control in humans can be found in [2].

It is of great medical interest to be able to bring spatio-temporal phenomena, such as the spiral wave of a fibrillating porcine heart in figure 13.1, under control and suppress them. For practical reasons this control must be achieved by placing probes on a small number of sites on the heart. This raises a profound general question.

Question 13.3. *Is it possible to develop a coherent control theory for spatio-temporal systems that allows one to construct effective control strategies based on a limited (hopefully tiny) number of measurement and control sites?*

To date, this question remains open! Until it is answered, thousands of people a day will continue to succumb to sudden cardiac death.

13.2 Neuro computing

Existing computational paradigms do not approach the capabilities of even simple organisms in terms of adaptability and real-time control. There are computational mechanisms and network architectures in living neural systems that are missing from even the most sophisticated artificial computing systems. This leads to another central question.

Question 13.4. *Are there alternative and better ways of building computers inspired or extracted from biology, using spatially extended nonlinear dynamical systems?*

If only we could achieve the goal of reverse engineering the most efficient and flexible computer known: the biological brain. It is an adaptable collection of cells and tissue that learns, stores and processes information. While an enormous amount of work has been performed to see how the brain behaves, from the intracellular level to the whole brain, (see again chapters 11 and 12) it seems that it performs its amazing computational feats through the use of coupled biological elements and the dynamics of their interaction. Again spatio-temporal nonlinear dynamics is the key concept.

A more moderate goal is to engineer biological systems to work as computers. At present we are considering ideas from dynamical systems that may provide a way for us to *program neural tissue*. We have been successful in getting dynamical systems to self-organize and solve problems in a variety of situations and simulations [9–12]. Ironically, these simulations were performed on conventional computers utilizing conventional computational paradigms. To avoid such irony we are now looking at hybrid systems that combine traditional computing with living neurons.

Recent technological developments regarding the patterned growth of neuronal cultures [1] and the long-term electrophysiological recording of cultured neuronal signals [4, 6, 13] have inspired us to investigate methods to exploit living neuronal networks specifically designed to encode and process information. Recently we were able to show as an experimental proof-of-concept that analogue operations and fundamental digital logic operations can be attained through the use of living neurons (in this case, leech *[Hirudo medicinalis]* neurons) coupled via an artificial excitatory synapse [8]. Details of our experiments can be found in [5]. Figure 13.3(*a*) shows the dynamic clamp in the electrophysiology rig and figure 13.3(*b*) is a blow-up of leech nerve cord and ganglia. We incrementally increased the leak conductance via the dynamic clamp. This incremental increase in depolarizing input resulted in an essentially linear relationship between increasing leak conductance and firing frequency, which we exploited to perform

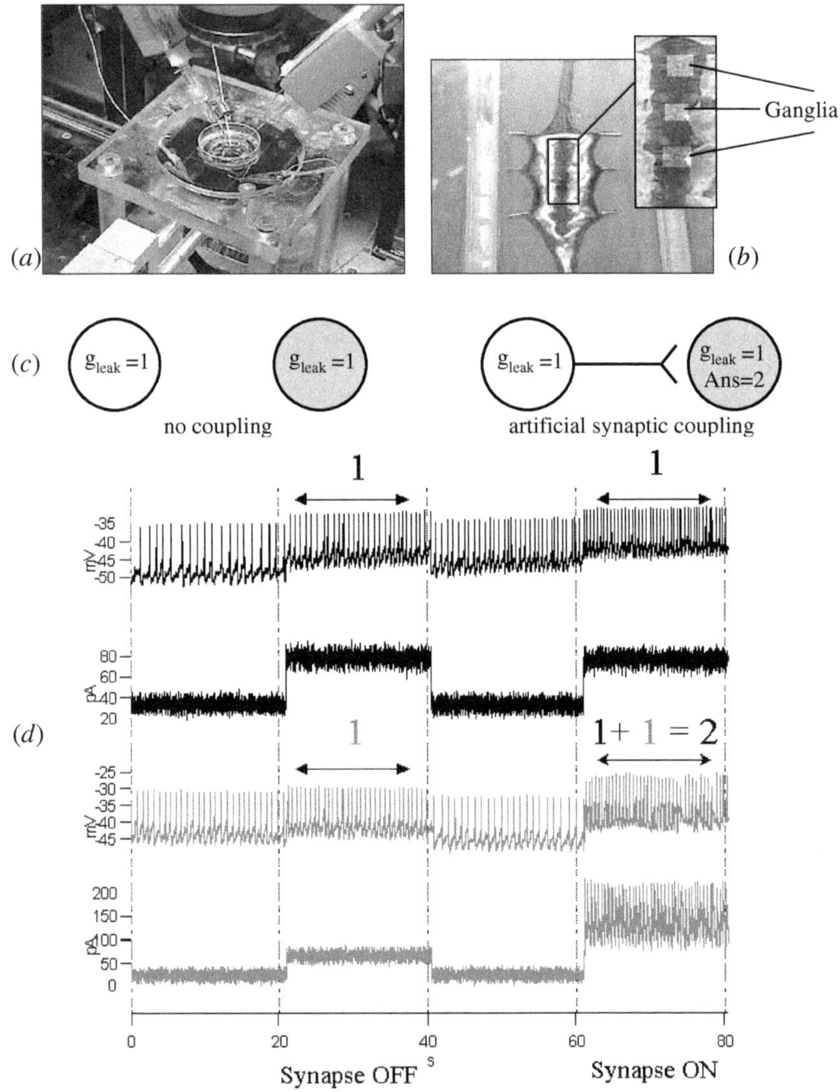

Figure 13.3. Experimental set-up of dynamic clamp in leech neurons in the electrophysiology rig (*a*), a blow-up of leech nerve cord and ganglia in our experimental preparation (*b*), experimental schematic diagram (*c*) and the results of simple addition (*d*). Each cell receives numeric input (as a change in leak conductance). The postsynaptic cell firing frequency represents the sum of these two numeric integer inputs when the synapse is active; for details see [5].

the analogue and digital logic operations. Figure 13.3(*c*) shows the experimental schematic diagram for the example of addition, the results of which are shown in figure 13.3(*d*). The firing frequency of the first and second cells both represent 1. After artificial synaptic coupling between the two cells is switched on, the firing of the second cell now represents the answer: superimposed onto the number initially represented in the firing of the second cell is now the influence from the first cell. This experiment, while rudimentary, served as a warm-up for our future experiments on living neuronal networks.

We have now embarked upon projects that will result in the development of computational systems that incorporate both living neuronal networks and circuitry. These hybrid neuronal systems will provide a platform for discovering, exploring and using the computational dynamics of living neuronal networks to perform real-time tasks in the physical world. Unlike brains in animals, *in vitro* networks are amenable to detailed observation and manipulation of every cell in the network. By embodying such hybrid networks with actuators and sensors, we hope to map the spatio-temporal dynamical landscape of neuronal networks under the conditions for which they evolved: continuous real-time feedback for adaptive behavioural control.

13.3 Where do we go from here?

In the spirit of this book, I would like to speculate a little bit on the future of the field. In my view the fibrillating heart in figure 13.1 is pointing the way, the merging spiral waves spelling out *S* for spatio-temporal. Until we come to grips with that *S*, we will be chained to problems we can solve rather than pressing on with problems that need to be solved. We should move on from the world of linear systems and those nonlinear systems that we can only get a glimpse of through the narrow slits of statistical mechanics and temporal nonlinear dynamics. So where do we go from here? I don't know about you, but I will be staring at figure 13.1 for a very long time and dare to dream…

Acknowledgments

The results on the leech neuron reported in section 13.2 are the PhD work of Paul Garcia and I would like to thank him for figure 13.3 and his help.

References

[1] Branch D W, Wheeler B C, Brewer G J and Leckband D E 2000 Long-term maintenance of patterns of hippocampal pyramidal cells on substrates of polyethylene glycol and microstamped polylysine *IEEE Trans. Biomed. Eng.* **47** 290–300

[2] Christini D J, Stein K M, Markowitz S M, Mittal S, Slotwiner D J, Scheiner M A, Iwai S and Lerman B B 2001 Nonlinear-dynamical arrhythmia control in humans *Proc. Natl Acad. Sci., USA* **98** 5827–32

[3] Ditto W L, Spano M L, In V, Neff J, Meadows B, Langberg J J, Bolmann A and McTeague K 2000 Control of human atrial fibrillation *Int. J. Bif. Chaos* **10** 593–601

[4] Egert U, Schlosshauer B, Fennrich S, Nisch W, Fejtl M, Knott T, Muller T and Hammerle H 1998 A novel organotypic long-term culture of the rat hippocampus on substrate-integrated multielectrode arrays *Brain Res. Protocols* **2** 229–242

[5] Garcia P S, Ellis, R T, Calabrese R L, DeWeerth S P and Ditto W L 2002 Computing with Leech Neurons *Preprint*

[6] Jimbo Y, Kawana A, Parodi P and Torre V 1998 The dynamics of a neuronal culture from dissociated cortical neurons of neonatal rats *Eur. J. Neurosci.* **10** 403

[7] Lown B 1979 Sudden cardiac death: the major challenge confronting contemporary cardiology *Am. J. Cardiol.* **43** 313–28

[8] Sharp A A, O'Neil M B, Abbott L F and Marder E 1993 The dynamic clamp: artificial conductances in biological neurons *Trends Neurosci.* **16** 389–94

[9] Sinha S and Ditto W L 1998 Dynamics based computation *Phys. Rev. Lett.* **81** 2156–9

[10] Sinha S and Ditto W L 1999 Computing with distributed chaos *Phys. Rev.* E **60** 363–77

[11] Sinha S, Munakata T and Ditto W L 2002 Parallel computing with extended dynamical systems *Phys. Rev.* E **65** 036214

[12] Sinha S, Munakata T and Ditto W L 2002 Flexible parallel implementation of logic gates using chaotic elements *Phys. Rev.* E **65** 036216

[13] Thiebaud P, deRooij N F, Robert F, Correges P, Duport S and Stoppini L 1997 Simultaneous multi-electrode recording of field potential and extracellular medium monitoring in brain slice cultures *Eur. J. Cell Biol.* **74** 162

[14] Witkowski F X, Leon L J, Penkoske P A, Giles W R, Spano M L, Ditto W L and Winfree A T 1998 Spatiotemporal evolution of ventricular fibrillation *Nature* **392** 78–82

[15] Zipes D P and Wellens H J 1998 Sudden cardiac death *Circulation* **98** 2334–51

Author index

- Uwe an der Heiden
  Institut für Mathematik
  Universität Witten/Herdecke
  Stockumer Str. 10
  58448 Witten
  Germany
  adheiden@uni-wh.de
- Paul Bressloff
  Department of Mathematics
  155 South 1400 East, 233 JWB
  University of Utah
  Salt Lake City, UT 84112
  USA
  bressloff@math.utah.edu
- Jack D Cowan
  Department of Mathematics
  University of Chicago
  5734 S. University Avenue
  Chicago, IL 60637
  USA
  cowan@math.uchicago.edu
- William L Ditto
  Department of Biomedical
  Engineering
  College of Engineering
  University of Florida
  363 NEB, PO Box 116131
  Gainesville, FL 32611-6131
  USA
  william.ditto@bme.ufl.edu
- G Bard Ermentrout
  Department of Mathematics
  University of Pittsburgh
  Pittsburgh, PA 15260
  USA
  bard+@pitt.edu
- Mathias Fink
  Laboratoire Ondes et Acoustique
  ESPCI—10 rue Vauquelin
  75231 PARIS Cedex 05
  France
  mathias.fink@espci.fr
- Jordi Garcia-Ojalvo
  Departament de Fisica
  i Enginyeria Nuclear
  Universitat Politecnica
  de Catalunya
  Colom 11
  E-08222 Terrassa
  Spain
  jordi.g.ojalvo@upc.es
- John Guckenheimer
  Department of Mathematics
  Malott Hall
  Cornell University
  Ithaca, NY 14853-4201
  USA
  gucken@cam.cornell.edu
- Christopher K R T Jones
  Lefschetz Center for Dynamical
  Systems
  Division of Applied Mathematics
  Box F
  Brown University
  Providence, RI 02912
  USA
  ckrtj@cfm.brown.edu
- Edgar Knobloch
  School of Mathematics
  University of Leeds
  Leeds LS2 9JT
  United Kingdom
  knobloch@maths.leeds.ac.uk

- Yoshiki Kuramoto
  Nonlinear Dynamics Group
  Department of Physics
  Kyoto University
  Kyoto 606-8502
  Japan
  kuramoto@ton.scphys.
  kyoto-u.ac.jp
- Robert S MacKay
  Mathematics Institute
  University of Warwick
  Coventry CV4 7AL
  UK
  r.s.mackay@warwick.ac.uk
- Tom Mullin
  Manchester Centre for Nonlinear
  Dynamics
  University of Manchester
  Oxford Road
  Manchester M13 9PL
  UK
  tom.mullin@man.ac.uk
- Remus Osan
  Physics & Astronomy Department
  University of Pittsburgh
  Pittsburgh, PA 15260
  USA
  remus@phyast.pitt.edu
- Rajarshi Roy
  Department of Physics *and*
  Institute for Physical Science and
  Technology
  University of Maryland
  College Park, MD 20742
  USA
  rroy@glue.umd.edu
- Gregory D Van Wiggeren
  School of Physics
  Georgia Institute of Technology
  Atlanta, GA 30332
  USA

  Current address
  Agilent Laboratories
  Palo Alto, CA 94304
  USA
  greg_vanwiggeren@agilent.com
- Arthur T Winfree
  Ecology and Evolutionary Biology
  University of Arizona
  BioSciences West
  Tucson, AZ 85721
  USA
  winfree@email.arizona.edu

For Product Safety Concerns and Information please contact our EU
representative GPSR@taylorandfrancis.com
Taylor & Francis Verlag GmbH, Kaufingerstraße 24, 80331 München, Germany

www.ingramcontent.com/pod-product-compliance
Ingram Content Group UK Ltd.
Pitfield, Milton Keynes, MK11 3LW, UK
UKHW020317270926
475702UK00050B/940

* 9 7 8 0 7 5 0 3 0 8 6 2 5 *